CHARLES N BROWN

LOST
CONTINENTS

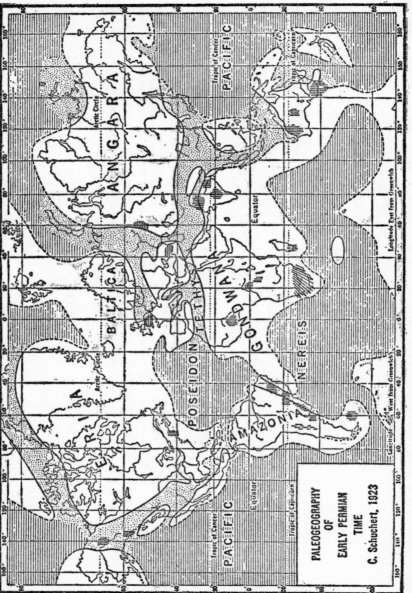

Fig. 1. MAP OF THE WORLD IN THE PERMIAN PERIOD, according to the American geologist Schuchert. Dotted areas are epeiric seas. (Reproduced by permission from *Outlines of Historical Geology*, 2d ed., by C. Schuchert, publ. by John Wiley & Sons, Inc., 1931.)

For Charlie Brown —
from L. Sprague de Camp
10/30/81

LOST

CONTINENTS

The Atlantis Theme
in History, Science,
and Literature

L. SPRAGUE DE CAMP

The Gnome Press Inc., New York

to Catherine Crook de Camp,
best of wives.

CONTENTS

ILLUSTRATIONS

PREFACE

WHEN something like two thousand books and articles have been written about Atlantis and other supposed lost continents, why write another? For several reasons: Because of these previous books most, including all the best ones, are either in foreign languages or out of print. Because a lot of people have told me they were interested in the subject. Because I have been interested in it myself for many years, and when a writer has a book on anything in him he has to get it out of his system or it will give him no peace. And finally because I hope to better the work of my predecessors and to make a few original contributions to the subject.

It would be nice, of course, to write the definitive work, the last word, on any subject, but I am not rash enough to hope I have done that. In fact I shall be lucky if some of my more extreme fellow-members of the Atlantist cult do not quote me in support of opinions with which I strongly disagree.

A word about the spelling of Greek names: Many Greek names can be spelled in two or three different ways —by direct transliteration from the Greek (e.g. Ptolemaios), or with the Latinized form (Ptolemaeus), or sometimes in addition with an Anglicized form (Ptolemy). Now, I prefer the "Greek" forms, which seem to be getting commoner among Classical scholars and which may eventually supplant the others. However, one can't use them consistently because some names (e.g. Homer, Thucydides) have become so widely known in Latin or English dress that to write "Homeros" and "Thoukydides" would bewilder most readers. Moreover in the Bibliography and in references to it in the Notes it is necessary to use the form actually employed in the edition of each book listed, which is usually the Latin form. That is how the same name is sometimes spelled two different ways on the same page. If the inconsistency bothers you, remember that we get along quite happily speaking in the same sentence of Ivan the Terrible (not John) but Peter the Great (not Piotr).

I acknowledge with thanks the permission of Mr. Conrad Aiken to quote from his poems *Priapus and the Pool* and *Senlin*; that of the Macmillan Company to quote from John Masefield's *Fragments*; that of the J. B. Lippincott Company to quote from Alfred Noyes's *Forty Singing Seamen;* that of Otis Kline Associates, literary agents for the estate of Robert E. Howard, to quote from one of Howard's poems; and that of John Wiley & Sons to reproduce one of the paleogeographical maps from Dr. Charles Schuchert's *Outlines of Historical Geology.* Parts of this book have appeared as articles in *Astounding Science Fiction, Galaxy Science Fiction, Natural History Magazine, Other Worlds Science Stories,* and the *Toronto Star* (permission of whose publishers to reprint is gratefully acknowledged) and have been used in lectures. For criticism and suggestions I am grateful to Everett F. Bleiler, Bernard De Voto, Evelyn Payne Hatcher, L. Don Leet, Willy Ley, Garrett Mattingly, P. Schuyler Miller, and Victor Wolfgang von Hagen; and for the loan of much source material from his library to Oswald Train.

L. SPRAGUE DE CAMP
Wallingford, Penn.

LOST
CONTINENTS

CHAPTER I

THE STORY OF ATLANTIS

There was an island in the sea
That out of immortal chaos reared
Towers of topaz, trees of pearl
For maidens adored and warriors feared.

Long ago it sunk in the sea;
And now, a thousand fathoms deep,
Sea-worms above it whirl their lamps,
Crabs on the pale mosaic creep.*

Aiken

MEN have always longed for a land of beauty and plenty, where peace and justice reigned. Failing to make one in the real world, they have often sought consolation by creating imaginary Edens, Utopias, and Golden Ages. Formerly they located these ideal commonwealths in the distant past or in undiscovered parts of the world. Now, however, that the unexplored places left on earth are few and uninviting, and the history of the remote past is fairly well known, they prefer to place their utopias in the distant future or even on other planets.

Many such dreams have been written up, and to lend extra interest to their stories the writers have sometimes pretended that their tales were literally true. This practice has had the unfortunate effect of convincing some readers (who have enough trouble distinguishing fact from fiction anyway) that such indeed was the case. For instance, when the noted sixteenth-century idealist Sir Thomas More published his famous *Utopia*—a story about an imaginary island where people led lives of simple virtue—the conscientious

---

* From *Priapus and the Pool*, by Conrad Aiken, in *Selected Poems* (Scribner's, 1929); copr. 1918-1929 by Conrad Aiken; by permission of the author.

1

More was much disconcerted when one of his pious contemporaries, the learned Budé, wrote him urging that missionaries be sent to convert the Utopians to Christianity! And when G. B. McCutcheon wrote his Graustark novels, he was deluged with fan mail asking how to get to Graustark, or taking exception to the author's statements about his imaginary Balkan kingdom. None of his correspondents, evidently, thought to look at a map.

Of all these creators of imaginary worlds, the one with the widest and most lasting influence was the Greek philosopher Aristokles the son of Ariston, better known by his nickname of Plato, the inventor or historian of Atlantis. Although his Atlantis story made but little stir at the time he wrote it, it became so popular in later centuries that to this day the name "Atlantis" evokes a picture of a beautiful world with a high and colorful culture (now, alas, gone forever) in the minds of thousands of people who never heard of Plato.

Nearly two thousand books and articles have been written about Atlantis and other hypothetical continents, ranging in tone from the soberest science to the wildest fantasy. Explorers have travelled thousands of miles looking for traces of the Atlantean culture described by Plato, and geologists have devoted thousands of hours to study of the earth's crust to find out whether continents do rise and sink, and if so when and why. And many plain men to whom it is no great matter whether

> . . . a strange city lying alone
> Far down within the dim West

ever sank beneath "the dragon-green, the luminous, the dark, the serpent-haunted sea" have turned their attention to Atlantis.

Recently a group of English newspapermen voted the reëmergence of Atlantis as the fourth most important news story they could imagine—five places ahead of the Second Coming of Christ. Astronomers have bestowed the name "Atlantis" (along with many others from Classical mythology) upon an area on the planet Mars. The Woods Hole Oceanographic Institution calls the little ship it explores with the *Atlantis,* and the name has served as a title for several periodicals, a theatrical company, a hotel in Miami, and several establishments (a book shop, an engineering firm, and a restaurant) in London. Finally, in 1951 the Minsky's Burlesque show at the Rialto Theatre in Chicago included an aquatic strip-act called "Atlantis, The Sea-Nymph." Evidently the Atlantis theme has a grip upon the fancy all out of proportion to its practical importance—though I wouldn't say it were entirely trivial, since the question of Atlantis does enter into the general problem of the origins of man and of civilization.

Perhaps the very impracticality of Atlantism constitutes part of its charm. It is a form of escapism that lets people play with eras and continents as a child plays with blocks.

As many people have heard vaguely of the lost-continent theories, and as many seem interested in them without ever having looked into the historical and scientific sides of the question, I will try to tell the story of the Atlantis concept and its progeny such as Mu and Lemuria. Where did the Atlantis story originate? Is the tale fact, fiction, or fiction founded on fact? What is there to the various lost-continent theories? Did Atlantis or Lemuria ever exist? If not, what then is this curious hold that the idea of lost continents has upon the minds of men? I can't guarantee final answers to all these questions, but that is the chance you take.

First, to consider the story of Atlantis in its earliest known form: About the year 355 B.C. Plato wrote two Socratic dialogues, *Timaios* and *Kritias,* wherein he set forth the basic story of Atlantis. At that time Plato was in his seventies and had been through a lot, including enslavement and liberation and an unsuccessful attempt to apply his theories of government at the court of the Tyrant of Syracuse. For many decades he had lectured at Athens, during which time he wrote a number of dialogues: little plays representing his old teacher Socrates and his friends sitting around and discussing such problems as politics, morals, and semantics.

Although Socrates is the chief talker in many of these

dialogues, we cannot be sure which of the ideas set forth
are really those of Socrates and which those of Plato. While
we find many of Plato's ideas unsympathetic today—he
sneered at experimental science, glorified male homo-
sexuality, and advocated a type of government that we
should call "fascistic" or "technocratic"—he pioneered in
some departments of human thought. Furthermore he
wrote with such poetic charm and vivacity that he seduced
many later thinkers into exaggerating his solid contributions
to man's intellectual growth.

Some years previously he had written his best-known
dialogue, *The Republic,* in which he gave his prescrip-
tion for an ideal state. Now he undertook *Timaios* as a
sequel to *The Republic,* since the same cast of characters
is assembled in the house of Plato's great-uncle or distant
cousin Kritias on the day following the conversation of
The Republic.

The time is about 421 B.C., when in real life Socrates
was not yet fifty and Plato a small child; and early in June,
during the festival of the Lesser Panathenaia, just after
that of the Bendideia. The characters are Socrates, Kritias,
Timaios, and Hermokrates. Kritias, Plato's relative, was a
talented historian and poet on one hand, and on the other
a scoundrelly politician, a leader of the Thirty Tyrants who
inflicted a reign of terror on Athens after that state had
been defeated by Sparta in the Peloponnesian War. Timaios
(not to be confused with the later historian of the same
name) was an astronomer from Locri in Italy, while Hermokrates was an exiled Syracusan general.

Timaios was intended as the first book of a trilogy,
in which, first, Timaios lectured on the creation of the
world and the nature of man according to the Pythagorean
philosophy; second, Kritias told the tale of the war between Atlantis and Athens; and third, Hermokrates was
to have spoken on some similar subject—perhaps the military side of the Atheno-Atlantean war after Kritias had
finished with the theological and political aspects. It is
quite likely that Plato, a thorough militarist, should have
had some such discourse in mind. However, the trilogy was
never completed. Some time (perhaps years) after finishing
Timaios, Plato began a rough draft of *Kritias,* but dropped

the whole project before he had finished this piece and instead wrote his last dialogue, *The Laws*.

Timaios starts with Socrates and Timaios recalling the discourse of Socrates on the previous day; that is, the dialogue of *The Republic*. Hermokrates then says that Timaios, Kritias, and he himself are all ready to give speeches on man and the universe, especially Kritias, who has already "mentioned to us a story derived from ancient tradition."

Pressed for details, Kritias tells how, a century and a half before, the half-legendary Athenian statesman Solon heard the story in Egypt, whither he had withdrawn because of the unpopularity he had incurred by his reform of the Athenian constitution. On his return to Athens he repeated it to his brother Dropides, Kritias's great-grandfather, who in turn passed it on down to his descendants. Solon had intended to make an epic poem of the narrative, but had never found enough time off from politics to complete the work.

During his Egyptian tour Solon stopped off at Saïs, the capital of the friendly King Aahmes. (Here is a discrepancy; Solon is supposed to have made his trip between 593 and 583 B.C., whereas Aahmes II reigned from 570 to 526. But never mind that now.) Here Solon got into a discussion of ancient history with a group of priests of the goddess Neïth or Isis, whom the Greeks identified with their own wise and warlike Athena. When Solon tried to impress them by telling them some of the Greek traditions, like that of Deukalion and Pyrrha and their Flood, the oldest priest (named Sonchis, according to Plutarch) laughed at him. The Greeks were children, he said; they had no ancient history because their records had all been destroyed by the periodical catastrophes of fire and flood that overwhelm the world—all but Egypt, which, being proof against such misfortunes, had kept records from the Creation on down.

The priest went on to tell Solon that Athena had founded a great Athenian empire 9000 years previously (that is, about 9600 B.C.) divinely organized along the lines that Plato had sketched in his *Republic*. A communistic military caste had ruled the state, and everybody was brave, handsome, and virtuous.

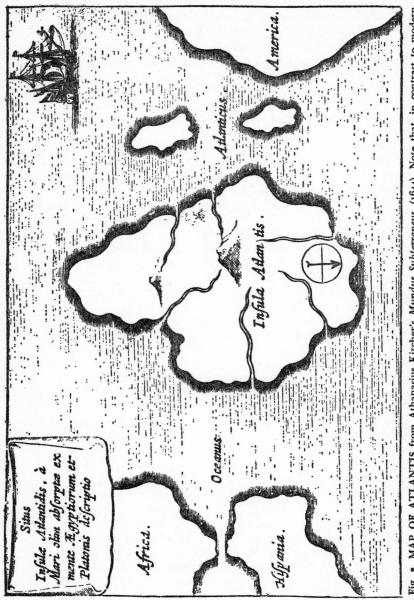

Fig. 2. MAP OF ATLANTIS from Athanasius Kircher's *Mundus Subterraneus* (1644). Note that, in contrast to modern maps, north is down and south up.

There had also been a mighty empire of Atlantis, centering upon an island west of the Pillars of Herakles (the Strait of Gibraltar) larger than North Africa and Asia Minor combined, and surrounded by smaller islands. In those days you could, by traversing this great archipelago island by island, reach the super-continent, beyond Atlantis, which surrounded the ocean that encircled the inhabited world.

The Atlanteans, not satisfied with ruling their own islands and parts of the outer "true" continent, had tried to conquer the whole Mediterranean region. They had extended their rule as far as Egypt and Tuscany when they were defeated by the brave Athenians, who led the fight against them and persisted in it even after their allies had fallen away. Then a great earthquake and flood devastated Athens, swallowed the Athenian army, and caused Atlantis to sink between the waters of the Atlantic Ocean. Forever after the waters west of Gibraltar were unnavigable because of the shoals left by the sinking of Atlantis.

Kritias then says he lay awake all night trying to remember the details of the story, since it would illustrate the theories that Socrates had propounded the previous day. In fact, he goes on: "the city with its citizens which you described to us yesterday, as it were in a fable, we will now transport hither into the realm of fact; for we will assume that the city is that ancient city of yours, and declare that the citizens you conceived are in truth those actual progenitors of ours, of whom the priest told."

Socrates is enthusiastic, especially since the account "is no invented fable but genuine history." However, Kritias prefers that Timaios deliver his discourse first. So Timaios takes up the conversation and devotes the rest of this long dialogue to Pythagorean scientific theories: the movements of the Solar System, the shape of the atoms comprising the Four Elements, the creation of mankind, and the working of the human body and soul.

In the next dialogue, *Kritias,* Kritias resumes his narration: When the gods divided up the world, Athena and Hephaistos received Athens and set up the Athenian state on suitably Platonic lines. Not only were its workers and farmers incredibly industrious, and its "Guardians" inhumanly noble, but Greece itself, also, was larger and more fertile in those days. Another of Plato's inconsistencies is that in *Timaios* Athens (and, by implication, Atlantis) was founded 9000 years before Solon's time, while in *Kritias* it is said that "many generations" after this founding Atlantis was sunk—also 9000 years before Solon's time.

Meanwhile Poseidon, the leading sea-god and also the god of earthquakes and horsemanship, had received Atlantis as his share. The population of Atlantis then consisted of a couple who had sprung from the earth, Euenor and Leukippë, and their daughter Kleito. When the old couple died, Poseidon (undeterred by the fact that he was already married to Amphitritë, one of the daughters of the other sea-god Okeanos) set up housekeeping with Kleito on a hill in Atlantis. To keep his sweetheart safe he surrounded the hill with concentric rings of land and water. He also supplied the hill with hot and cold fountains from underground streams. (The Greeks had exaggerated ideas about subterranean watercourses, believing for instance that the Alpheios River in western Greece, the original of Coleridge's sacred River Alph, ducked under the Adriatic to reappear as the spring Arethusa in Sicily.)

Poseidon, a fertile fellow like all the gods, begat ten sons—five pairs of twins—upon Kleito, and when they grew up divided the land and its adjacent islands among them to rule as a confederacy of kings. The eldest of the first pair, Atlas (after whom Atlantis was named) he made the chief king over all. Atlas's brother Gadeiros (or to translate it into Greek, Eumelos, "rich in sheep" or "in fruit") received as his portion the region of Gadeira (later Gades or Cadiz) in Spain. Plato says that Solon translated the original Atlantean names into Greek to make them easier for his hearers. Somebody was evidently a poor linguist, since "Gadeira" is actually from a Phoenician word for "hedge."

Since the kings were prolific and the land rich in vegetation, minerals, and elephants, Atlantis in time became a mighty power. The kings and their descendants built the city of Atlantis on the south coast of the continent. The city took the form of a circular metropolis about 15 miles in diameter, with Kleito's ancient hill in the center. Around this hill the rings of land and water still existed (two of land and three of water) forming a circular citadel three miles in diameter. The kings built bridges connecting the land rings and tunnels big enough for ships connecting the rings of water. The city's docks were located on the outer margin of the outermost ring. A canal of enormous size ran straight through the city, connecting the harbor works with sea at the south end, and with a great rectangular irrigated plain, 230 by 340 miles in dimensions and surrounded by lofty mountains, at the north end. This plain was divided into square lots which were assigned to the leading farmers, who in turn had to furnish men for the Atlantean army—light and heavy infantry, cavalry, and chariots (thousands of years before cavalry and chariots were invented).

On the central island the kings erected a royal palace and, over the sacred spot where Poseidon had dwelt with Kleito, a temple surrounded by a golden wall. The rings of land were covered with parks, temples, barracks, racecourses, and other public facilities. These structures were all lavishly decorated with gold, silver, brass, tin, ivory, and the mysterious metal *oreichalkōn* which "glowed like fire."

Note that Plato says nothing about the explosives, searchlights, or airplanes with which some imaginative modern Atlantists have credited the Atlanteans. The only ship he mentions is the trireme (or *triērē,* a type of ship said to have been invented by Ameinokles of Corinth about 700 B.C.) and, except for *oreichalkōn,* he described no technics not known to his own time. While orichalc (literally "mountain bronze") has not been identified beyond doubt, Classical writers like Hesiod casually mention it in a way that implies that the term was applied to some unusually good grade of bronze or brass.

The kings met at alternate intervals of five and six years to consult on matters of state. They assembled in the sacred precincts of Poseidon, captured a bull with a noose, sacrificed the animal with much ceremony, and held a banquet. Afterwards they wrapped themselves in dark-blue robes, put out the embers of the sacrificial fire and, sitting about in a circle, spent the rest of the night giving judgments. The following morning these decisions were recorded on tablets of gold for the benefit of posterity.

For many centuries the Atlanteans were virtuous like the Athenians of that elder day. But in time, as the divine blood they inherited from Poseidon became more and more diluted, they suffered a moral decline. Zeus, the king of the gods, observing their wicked ambition and greed, decided to chastise them that they might do better in the future. (Drowning the lot seems a curious method of reforming them, however.) Therefore he called the gods together in his palace at the center of the Universe to discuss the matter, ". . . and when he had assembled them, he spake thus: . . ."

Here the dialogue ends in mid-sentence. We never do learn the details of the Atheno-Atlantean war.

Now, what have we here? A myth that Plato made up to illustrate his philosophical ideas? A true tale of the foundering of an ancient continent, handed down via the Egyptian priesthood and Solon in the manner described? Or a real tradition subsequently embroidered by the Egyptians, or Solon, or Dropides, or Plato?

There are many possibilities. Perhaps the story was told to Solon as described, but contains no truth, having been made up by lying priests to entertain visitors. Again, some writers assert that Plato, as well as Solon, visited Egypt and talked with several priests, including one named Pateneit at Saïs—not impossible, for Plato was noted as a traveller though there is no proof of such a visit in Plato's writings, and at the time the statement was made it was customary to credit all Greek philosophers with an Egyptian tour whether they had one or not. If Plato did make such a trip, he might have heard the story there, and, when he came to write it up, have introduced it with the tale of Solon and the priest of Neïth to lend it a superficial glamor of antiquity.

It is worth noting, however, that there is no mention by any writer before Plato of any sunken island in the Atlantic, and no evidence outside of Plato's word that Solon's unfinished epic ever existed. There is nothing either in the rather scanty remains of pre-Platonic Greek literature; nothing in any of the surviving records of Egypt,

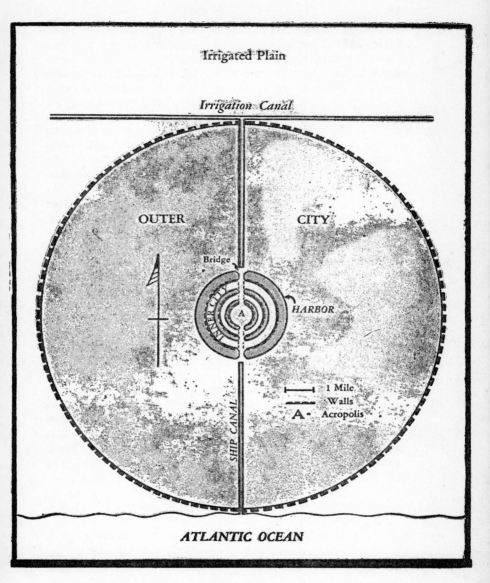

Fig. 3. THE CITY OF ATLANTIS, as described in Plato's *Kritias*.

Phoenicia, Babylonia, or Sumeria, which go back many centuries before the beginnings of Greek civilization.

Of course that does not prove that no such account
existed. Only a small fraction of the original Greek literature has come down to us, owing to the ravages of time
and neglect, and the bigotry of early Christians like Pope
Gregory I who destroyed pagan literature wholesale lest it
distract the faithful from the contemplation of heaven.
When books were hand-written upon awkward and easily-
torn scrolls of brittle papyrus, only exceptional merit or
luck enabled a work to survive more than a few centuries.
The loss of important Greek works is to some extent made
up for by the Greeks' habit of quoting each other with
credit; but still, hundreds of books that would tell us
things we should like to know are gone for good.

Certain *elements* of Plato's story, even if not the story
itself, do appear before Plato's time.

For instance, Atlas himself was an old figure of Greek
myth: a son of the Titan Iapetos (one of the snake-legged
giants who attacked the gods on Olympos) and a brother
of Prometheus and Epimetheus. Homer speaks of him as
"crafty Atlas, who knows the depths of the whole sea, and
keeps the tall pillars that hold heaven and earth asunder."
Literally the poet says that Atlas "has" (*echei*) the pillars,
which in Greek might mean "possesses," "has charge of," or
"holds up." Later writers, assuming the last meaning to be
intended, made the description more definite. Thus the poet
Hesiod had Atlas supporting the heavens with his head
and arms, and the Greek playwrights added the detail that
Atlas was condemned by Zeus for his part in the Titans'
rebellion to perform this wearisome work in the West:

> Where the Hesperides their song
> Attune: no mariner can thence prolong
> The voyage, for, his daring bark t'impede
>   Neptune those hallowed bounds maintains
> Where Atlas with unwearied toil sustains
>   The heavens' incumbent load . . .

Later Classical writers rationalized this myth by making
Atlas a scholar-king who founded the science of astronomy,

and some Christian theologians identified him with the Biblical Enoch, the son of Cain.

By Pleionë, another daughter of Okeanos, Atlas had the seven daughters called Atlantides. (The singular, "Atlantis," means "daughter of Atlas.") These, Alkyonë, Meropë, Kelaino, Elektra, Steropë, Taÿgetë, and Maia, eventually became the stars of the Pleiades. The other sea-god Poseidon begat a son Lykos on Kelaino, and settled this son in the Islands of the Blest, somewhere to the West. Thus in the Classical version of the Atlas myth Poseidon was Atlas's son-in-law, though being immortal he could no doubt be Atlas's father (as Plato makes him) at the same time.

In any case it is pointless to expect consistency from any body of primitive myths. They are always exhuberantly contradictory, even though with the rise of civilization pious scholars try to rationalize them. Furthermore, as a parallel to the Poseidon-Kleito romance, it was said that on the island of Rhodes, which was supposed to have arisen from the depths, Poseidon "became enamored of Halia, the sister of the Telchines, and lying with her he begat six male children and one daughter, called Rhodos, after whom the island was named."

Subsequently Herakles, in the course of his Labors, induced Atlas to fetch him the Golden Apples of the Hesperides (also in the West) in return for relieving him of his burden. Returning with the apples, Atlas did not want to resume his task, but Herakles tricked him into doing so by persuading him to support the sky long enough for Herakles to put a pad over his head, and then decamping with the fruit. Finally Perseus turned the poor giant to stone by showing him the Gorgon's head, and thus Mount Atlas originated.

It does not follow that all these elements in the Atlas myth are equally old. In fact, there is reason to think that the idea of Atlas's having charge of the heaven-supporting pillars—a sort of divine building-and-grounds superintendent—existed in Greece centuries before the Greeks had any clear notion of the geography of the Western Mediterranean. Locating him in the West and identifying him with a real mountain were probably afterthoughts.

The rivalry of Poseidon and Athena, which in Plato

takes the form of the Atlanto-Athenian war, also goes back a long way. In the standard body of Greek myth, both Athena and Poseidon claimed Athens. The sea-god asserted his claim by striking the Akropolis with his trident, causing (like Moses) a spring to flow from the spot, while Athena caused an olive-tree to spring up and was adjudged the winner. This divine contention was a common theme of Greek art, and we might say that Plato presented Poseidon with Atlantis as a consolation-prize.

In addition to the mythological antecedents of Plato's story, various Classical writers from Herodotos on down describe primitive tribes in northwest Africa, which they call Atlantes, Atarantes, or Atlantioi, all names like that of Atlas.

However, since they tell us that these Atlantes have no names and never dream, and that their neighbors include men with snakes for feet (like the Titans) and headless men with faces in their chests, it would seem that the authors were not writing from any first-hand knowledge. Diodoros of Sicily tells a long tale of the North African Amazons (whom he carefully distinguishes from Homer's Asiatic Amazons) who dwelt on the island Hespera in the marsh Tritonis, and their queen Myrina who conquered the neighboring Atlantioi and the latter's enemies the Gorgons. According to this voluminous but uncritical first-century historian, the Atlantioi asserted that Atlas, a son of the elder god Ouranos, named both Mount Atlas and themselves after his own name. In later times, they added, Herakles exterminated both the Gorgons and the African Amazons, and the Tritonian Marsh disappeared as a result of an earthquake.

Furthermore, Classical authors often mention islands in the Atlantic Ocean and continents beyond it. Homer began the practice by scattering islands like Aiaia and Ogygia about the seas west of Greece in more or less complete disregard of the actual geography of those parts. Later authors wrote of Atlantic islands, some highly fanciful, inhabited by people with horses' feet, or satyrs, or folk with ears so large they used them in lieu of clothes. Other such references are plainly to the real Canaries, Madeiras, and (possibly) Azores, generally called the Fortunate Isles. These were known to the Carthaginians, but were subsequently lost until rediscovered by Genoese explorers in the thirteenth and fourteenth centuries.

They also believed, as did Plato, that the ocean surrounding the known world of Europe, Asia, and Africa was in turn encircled by a great continent. The historian Theopompos, a younger contemporary of Plato, wrote of a conversation between King Midas of Phrygia (he of the Golden Touch) and the satyr Silenos. The latter described the Outer Continent as inhabited by a people twice as large and twice as long-lived as those of the known world. One section of this land, known as Anostos ("No-Return") was veiled in red mist, and drained by a River of Pleasure and a River of Grief. Once a warlike tribe of these giants crossed the ocean to invade the civilized world, but when they got to the land of the Hyperboreans and found that the latter had nothing worth stealing, they concluded that it was useless to go further and returned home in disgust.

A similar Iranian legend made Zoroaster the descendant of such invaders from the Outer Continent, and perhaps both stories are late versions, much modified, of a much older migration-legend now lost forever in its original form.

Finally, the Greeks were familiar with the general concept of the emergence of lands from the sea and their sinking back into it. Myths told of the Island of Rhodes rising out of the sea. Herodotos on his Egyptian tour noticed fossil sea-shells in the hills, and, in the earliest geological observation on record, correctly inferred that the country must once have been under water. Furthermore it was generally held that Sicily had been torn loose from Italy by an earthquake, and that the Strait of Gibraltar had been opened up by a similar convulsion of nature.

And the Greeks also cherished a whole battery of Deluge-legends, like that in which a single pair, Deukalion and Pyrrha, warned by Zeus, escape by making a box in which they floated away. This myth, like that of Noah, may be derived from an old Sumerian legend, which in turn may be based upon real floods that once overwhelmed the Euphrates Valley.

Obviously, if any intelligent man of Plato's time had

wanted to write a piece of fiction like the Atlantean story, he had ample material available for his plot. Devout Atlantists take it for granted that these hints and rumors of unknown countries and people and submergences refer to a real Atlantis, any differences being due to distortion of the true historical narrative that appears in Plato's dialogues. Skeptics, on the other hand, retort that the distortion is the other way round—that Plato borrowed ideas from a number of these geographical, historical, and mythical sources and combined them to compose his fictional masterpiece.

We shall come back to this question later. Meanwhile I will merely point out that, although many of these references to Atlantes, Atlantic islands, and submergences suggest details of Plato's narrative, and might have something to do with it, not one of them says in plain language that an island named Atlantis once supported a civilized state but later sank beneath the Atlantic waves. Moreover the Atlas myths and the tales of the African Atlantes and the Western Islands are *known* to have existed long before Plato's mention of Atlantis.

After Plato's time, however, a number of writers commented upon the Atlantis story. Although no surviving original writings mention Atlantis for over three centuries after Plato's time, thereafter we begin to learn from such authors as Strabo what Plato's immediate successors thought of the tale. At first most of them either were cautiously noncommittal, or took it for granted that Atlantis was a fiction, an allegory by which Plato meant to expound his social ideals. The latter opinion is quite in keeping with what we know of Plato's character.

First, Plato's prize pupil, Aristoteles of Stagyra, grew up into a lisping, dandyfied encyclopedist who quarreled with his teacher, set up his own school (the Academy) and, with none of Plato's charm but a vastly greater grasp of fact, wrote a series of huge dry tracts on man and the universe. In these he covered nearly all the scholarly knowledge of his time and in some fields, notably biology and logic, made important additions to human knowledge. It was not Aristotle's fault that for a millenium and a half after his time most philosophers preferred quoting him as an infallible authority to using his pioneer work as a springboard to new discoveries.

Aristotle's only known remark upon his former master's Atlantis tale was in a lost work, quoted by Strabo, in which he ironically said that as Homer, for reasons of plot, had been compelled first to erect the wall of the Achaeans about their ships on the beach at Troy and then to wash it away, so in the case of Atlantis, "he who invented it also destroyed it."

A little over two centuries later the Stoic philosopher Poseidonios, friend and tutor of Cicero, was nettled by this dig. He accordingly wrote that in view of the known effects of earthquakes and erosion, it seemed to him more reasonable to say that: "it is possible that the story about the island of Atlantis it not a fiction," which for him was cautious. Strabo, though he considered Poseidonios a credulous enthusiast in other respects, approved this remark about Atlantis.

Later still, in the first century of the Christian Era, Gaius Plinius Secundus, "Pliny the Elder," said that Atlantis sank "if we are to believe Plato." His contemporary Plutarch told of Solon's attempt to make an epic poem of "the history or fable of the Atlantic island," which he never lived to complete, and how Plato later tried to improve on Solon's attempt with little better success. If skeptical about the truth of the story, Plutarch at least showed a sound appreciation of its literary merit: ". . . and the reader's regret for the unfinished part is the greater, as the satisfaction he takes in that which is complete is extraordinary."

Up to this point most of the commentators had viewed the story with a coldly critical eye. Under the later Roman Empire critical standards, which had never been high according to modern ideas, declined still further, with the result that people like Proklos the Neoplatonist began to take the story seriously. The Neoplatonists, followers of Plotinos and Porphyrios, constituted one of the many semi-magical and semi-philosophical cults that arose in brilliant Hellenistic Alexandria, flourished under the Roman Empire, became more and more magical, and finally disappeared (partly by absorption) with the triumph of Christianity. Careful discrimination was not among their virtues.

Proklos averred that Krantor, an early follower of Plato, took the story as straight history and claimed to have confirmed it by the testimony of Egyptian priests who showed tourists columns on which the account was inscribed, though since the visitors could not read hieroglyphics they had to take their guides' word for the content of the inscriptions. Moreover, said Proklos, the geographer Marcellus (first century B.C.) told in his *Ethiopic History* of islands in the Atlantic, three large and seven small, whose people preserved traditions of Atlantis and its empire. Others such as the Neoplatonist Porphyrios and the Church Father Origen considered Plato's story an allegory to which they attributed symbolic meanings, as that the Atlantean war symbolized conflicts among the spirits that animate the universe. The Neoplatonist Iamblichos and Proklos himself, by a Herculean effort, convinced themselves that the story was "true" in the literal and the figurative senses at the same time. Classical Alexandria was a hotbed of the vice of allegorization; the Alexandrine philosopher Philo the Jew (a believer in Atlantis) and the early Church Fathers rejoiced in ascribing symbolic meanings to their sacred writings, even, absurdly enough, asserting that every passage was both literally true and allegorically significant.

Proklos did the same with Plato in his *Commentary on the Timaios* — dreadful stuff; a vast mass of meaningless mystical "interpretation." Amidst a lot of maunderings about the significance of certain shawls embroidered with pictures of the gods trouncing the giants and the Athenians butchering the barbarians, used in Athenian religious festivals, Proklos dropped the remark that Kritias had woven a myth worthy of the festival of the Lesser Panathenaia, supposed in Plato to be in progress at the time of *Timaios*.

Later a scholiast on Plato's *Republic* misunderstood this passage to state that it was the custom at the Lesser Panathenaia to embroider a shawl with pictures of the war between Athens and Atlantis. The scholiast thereby gave the unfounded impression that the tale was known long before Plato's time, confusing an already dark subject further.

However, we cannot blame the scholiast, considering that Proklos was about the most obscure philosopher that ever put pen to papyrus.

For the most part the Church Fathers who commented upon the Atlantis story showed no more critical sense than the Neoplatonists, taking the tale at face value. Then the rise of Christianity and the decline of the Roman Empire shifted intellectual interest from the things of this world to those of the next. Interest in remote events of mundane history, including Atlantis, declined, despite the fact that for several centuries *Timaios,* having been translated into Latin by Chalcidius, was the only work of Plato with which the West was familiar.

The last comment on Atlantis, before the long night of the Age of Faith closed down upon the Western World, was that of a certain Kosmas, called Indikopleustes ("voyager-to-India") a sixth-century Egyptian monk. Having been a travelling merchant in his younger days, Kosmas, grown old and pious, undertook to refute pagan geographical ideas in a treatise called *Christian Topography.* Into this "monument of unconscious humor" he dragged the Atlantis tale as part of a strenuous effort to prove the earth not round, as the Greeks claimed, but flat.

Kosmas maintained that the Universe was shaped (as the ancient Egyptians thought) like the inside of a box, the Hebrew tabernacle built under the direction of Moses being a model thereof. Our "earth" was an island on the floor of this container, surrounded by the Ocean, which in turn was encompassed by a rectangular strip of land where the walls of the box came down to join the floor. Paradise (later a standard fixture on medieval European maps) lay in the eastern part of this outer land, wherein men dwelt before the Flood. As for Atlantis, Kosmas affirmed that it was merely a garbled version of the Biblical Flood story which Timaios had picked up from the Chaldeans and fictionized to suit himself.

After Kosmas, Atlantis would seem to have sunk for the second time, since, save for a brief mention in the medieval encyclopedia *De Imagine Mundi* by Honorius of Autun (about 1100) nothing more is heard of it for many centuries. Still, the cult was not dead, merely dormant.

When Europeans burst the intellectual bonds of the Church and the geographical bonds of their little peninsula, interest in far places and remote times would revive, and Atlantis would rise again into the consciousness of men.

CHAPTER II

THE RESURGENCE OF ATLANTIS

Push off, and sitting well in order smite
The sounding furrows; for my purpose holds
To sail beyond the sunset, and the baths
Of all the western stars, until I die.
It may be that the gulfs will wash us down:
It may be we shall touch the Happy Isles,
And see the great Achilles, whom we knew.
Tennyson

WHEN the aggressive and acquisitive young giant, Western Civilization, was a baby, his mother, the Roman Catholic Church, soothed him with stories of God and Heaven and saints and angels and miracles and Madonnas. As soon as he could toddle, however, he began reaching out in all directions to find out things for himself. And one of the things he reached for was geographical knowledge.

During the Dark and Middle Ages there was much talk of lands in or beyond the Atlantic Ocean, some based on fact and some on fancy rich with tales "of the Cannibals that each other eat, The Anthropophagi, and men whose heads Do grow beneath their shoulders." There was for instance the tale (which first appeared in the fifteenth century) of seven bishops who fled Spain with their flocks in 734 A.D., at the time of the Saracen conquest and, sailing westward, found an island on which they erected seven cities. This island was sometimes identified with the large rectangular island called Antilha, Antillia, or Antigla ("the counter-island") which appeared upon many pre-Columbian maps. Antillia corresponded so closely in size, shape, and direction with the real Cuba that when the latter and its neighbors were finally found they were promptly named the Antilles. And eighteen years before

Columbus set out on his first American voyage, the astronomer Toscanelli wrote him suggesting Antillia as an ideal
stopover on the way to Cathay.

Babcock the geographer thought that the pre-Columbian Antillia was evidence of a pre-Columbian voyage that
actually touched Cuba—not, perhaps, entirely impossible.
There is a long-standing controversy on theories of possible
voyages to the New World before Columbus, based upon
hints in pre-Columbian maps and travel tales, but nothing
you can definitely pin down. Hjalmar Holand, for instance,
has been writing for years arguing that the expedition under Paul Knudson, sent to Greenland by King Haakon VI
of Norway in mid-fourteenth century, continued on to
North America, where they built the mysterious Round
Tower of Newport and inscribed the dubious Kensington
Rune Stone. At least one medieval map-maker in 1455 identified Antillia with Plato's Atlantis, notwithstanding that
Atlantis was supposed to have sunk.

It was also said that about the same time as the flight
of the seven bishops, an Irish monk named Brendan set out
to find an ideal site for a monastery. At any rate later
geographers speckled the Atlantic with St. Brendan's Isles,
and romancers embroidered his tale with demons, dragons,
sea-serpents, and volcanic islands, many of his adventures
showing a suspicious resemblance to episodes in the *Odyssey*
and the stories of Sindbad the Sailor. If such a man ever
did go a-voyaging, which is more than doubtful, he may
have rediscovered the Fortunate Isles.

Meanwhile the authors of the Arthurian legend-cycle
described their king as convalescing from the wound he
got in the Battle of Camlan on the fairy island of Avalon
in the West, waiting, like Barbarossa in his Kyffhäuserberg
and Marko Kralyevich under his pine-trees, until the day
when he should return to lead his people. With him were
Olaf Tryggvasson, the Christianizing king of Norway, and
Ogier the Dane, one of Charlemagne's legendary paladins.

To come down from fiction to fact, the Norse discovered North America about 1000 A.D. despite Lord Raglan's attempt to prove the stories of Leif Eiriksson and
Thorfinn Karlsevni mere myths. Even if Leif was another
Irish sun-god in human guise, the authors of the sagas

could hardly have hit upon so accurate a description of the American Indians ("dark men and ugly, with unkempt hair . . . large eyes and broad cheeks . . ." wearing leather jackets, paddling skin boats, fighting with bows, slings, and clubs, and obviously ignorant of cloth, iron, and cattle) unless somebody had been to America and returned to tell the tale.

This discovery made enough talk for the Pope to appoint a Bishop of Vinland, and although it was not followed up by permanent colonization it was never quite completely forgotten. Some have thought Columbus might have heard of it on the trip to Iceland he made in his early years while following the family business of travelling drygoods salesman. He is also said to have been impressed by the hint of transoceanic lands in Seneca's *Medea,* quoted at the head of Chapter VII.

A century after the Norse voyages, the Arab geographer Edrisi told of another Atlantic voyage: A party of eight "deluded folk" sailed from Lisbon (then a Muslim city), found stinking shallow water covering dangerous reefs, then a region of darkness in which lay an island inhabited only by sheep. Twelve days' sail further south they came upon an island inhabited by tall tan people whose king, after questioning them sharply through an Arabic-speaking interpreter, sent them blindfolded to the African coast, whence they found their way home. Had these wanderers touched at Madeira and the Canaries, and in the latter place been interrogated by a Guanche chief? Possibly.

The scattering of hypothetical islands about the Atlantic really got started in the years following the voyages of Columbus. Then rumors of new lands ran riot, as the explorers were often far from accurate in their reports. As a result, the maps became covered with a swarm of geographical chimeras. Faulty navigation often caused a real land to be reported in several different places; and clouds, driftwood, and simple eyestrain led to many reports of rocks and islands where none existed. Thus Ortelius's world map of 1570 included the fictitious Isle of Brazil, St. Brendan's Isle, Isle of the Seven Cities, Green Island, Isle of the Demons, Vlaenderen, Drogio, Emperadada, Estotiland, Grocland, and Frisland—the last an imaginary twin

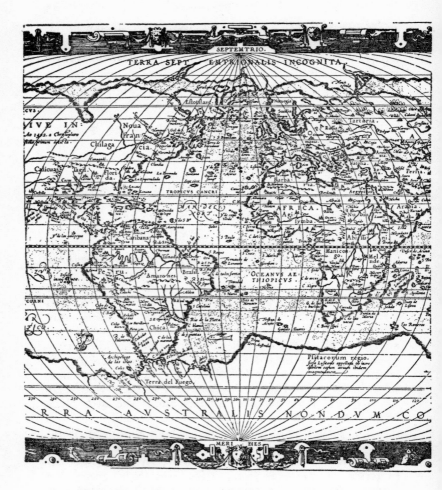

Fig. 4. ORTELIUS'S WORLD MAP OF 1570, center section, showing
imaginary Atlantic islands (Brazil etc.) and imaginary polar continents.
One of the captions on Terra Australis, the south polar continent, states
that it is inhabited by gigantic parrots.

of the real Iceland, begotten by confusion between Iceland and the Faeröes.

The ghostly Isle of Brazil, which haunted the maps for many years, was not finally exorcised until the nineteenth century. It was usually located a few hundred miles west of southern Ireland, and described as circular—in one case as a ring of islands. This phantom land was kept alive by reports like that of a Captain Nisbet who in 1674 arrived in Scotland with some "castaways" he claimed to have rescued from the Isle of Brazil. He said that the island was inhabited by huge black rabbits and by a magician who had been keeping the castaways captive in his castle until the gallant captain broke the spell that bound them. Alas for romance! there never was any such island.

This map of Ortelius also showed the great southern continent—*Terra Australis Incognita*—which geographers from the time of Claudius Ptolemy had been putting in the southern Indian and South Pacific Oceans. Belief in this circumpolar land—like Antarctica but ten times as big—originated in the mass of speculations of Classical times about possible continents in the Western and Southern Hemispheres.

To the early Greek geographers, Europe was *the* continent, with Asia and Africa mere peninsulas dangling from it. That is how it happens that we still speak, absurdly enough, of Europe as a continent instead of, more accurately, as a peninsula attached to *the* continent of Asia. Even after discovery had expanded Asia and Africa to continental size, the ancient map-makers continued to show them as having only a fraction of their true dimensions. In the days when people thought the earth was flat, men supposed that these three "continents" (Europe, Asia, and Africa) were surrounded by a river or a ring-shaped sea, called "the Ocean," and beyond "the Ocean" lay an unknown land stretching away to nobody knew where, as Silenos told Midas in Theopompos's fairytale.

When the round earth came in, the Alexandrine astronomers calculated that the three "continents" altogether occupied less than a quarter of the globe, which was thought to be divided into five zones: two polar zones too cold for life, an equatorial zone so hot the ocean boiled, and two intermediate temperate belts, one north and one south of the equator.

Now, philosophers of the Pythagorean school of thought (including Plato) supposed that the gods built the universe along lines of artistic symmetry. Therefore it seemed natural to them that there should be other land masses besides the group with which they were familiar, to balance the latter—three others according to Krates of Mallos (second century B.C.) occupying the three remaining quarters of the globe. The great Alexandrine scientist Hipparchos thought Ceylon might be the northern tip of such a land. Some geographer named this hypothetical continent *Antichthōn* ("counter-Earth") having borrowed the name from the Pythagoreans, who used it to designate the imaginary planet they had invented to bring the number of the movable heavenly bodies up to the mystic ten. Aristotle, although he rejected the idea of other inhabited planets, thought such continents likely. A few centuries later Strabo admitted the possibility, but deemed such speculations useless until somebody had been there to see.

Then, in the second century, the Egyptian astronomer and mathematician Klaudios Ptolemaios ("Claudius Ptolemy," no known relation to the royal Ptolemies) wrote a great *Geography* wherein he made a couple of major blunders. One was to reverse the mistake of his forerunners by making Asia and Africa much larger than they really were. The other was to infer, probably from a wrong report that the Indian Ocean had no tides, that this ocean must be an inland sea like the Mediterranean. Therefore he made Africa curve around to the east at its southern end, and Asia curve south at its eastern end where the Malay Peninsula exists in fact, until they met and inclosed the Indian Ocean.

Some later geographers (including Columbus) expanding on Ptolemy's ideas, assumed that the land surface of the globe was much larger than its water surface, so that all the oceans were merely oversized ponds completely surrounded by this land. It took the Age of Exploration to prove that these folk were wrong and Aristotle right: that the water surface was much the larger, and that the continents were islands entirely surrounded by water.

Of course the Phoenician sailors whom King Niku II of Egypt had sent on a perilous voyage around Africa had disproved the idea of the southern Asian-African connection seven centuries before, but for a long time the story of their voyage was looked upon with skepticism. Similar attempts by the Iranian noble Sataspes (who undertook the trip to escape execution for rape, failed, and was executed anyway) and the intrepid explorer Eudoxos of Kyzikos ended in defeat. Hence belief in the southern Afrasian connection was not finally killed until Vasco da Gama's voyage around Africa to India in 1497-98.

Early Christian thought was generally hostile to belief in transoceanic continents. Besides the anti-scientific extremists like Kosmas who tried to revive belief in a flat earth, others, willing to admit the roundness of the earth, could not bring themselves to swallow the thought of the Antipodes—the land where people walked with their feet pointing up. Many Christian theologians like St. Augustine and Isidore of Seville opposed belief in the Antipodes on the ground that the gospel had never been preached there, and neither Christ nor the Apostles had gone there, notwithstanding St. Paul's statement that "verily, their sound went unto all the earth, and their words unto the ends of the world." However, the anti-scientific wing of Christianity never really stamped out belief in lands on the other side of the globe, and the discoveries of the Age of Exploration quickly revived the idea of the Unknown Southern Land.

Even da Gama's journey did not dispose of the Southern Continent, for erroneous beliefs of that sort, when disproved, have a way of donning false whiskers and trying again and again to get back into intellectual respectability in disguise. Thus Ortelius's map showed Terra Australis as including the actual Antarctica, Australia, Java, and Tierra del Fuego; one of his captions even asserted that it was inhabited by gigantic parrots. Magellan had indeed thought while sailing through the strait named for him that the land on his left was a headland of Terra Australis. He named it Tierra del Fuego, "Land of Fire," because at night he saw so many of the campfires wherewith the primitive Fuegians were trying to keep warm in their lean-tos.

And even the discovery of the fact that Tierra del Fuego was an island failed to kill Terra Australis. For instance, in 1576 Juan Fernandez reported a continent about where Easter Island lies, with "the mouths of very large rivers . . . the people so white and so well clad" that he was amazed. Presumably he had sighted Easter Island, and an optimistic imagination worthy of a realtor had converted this little grassy island into a well-watered continent and its handful of Polynesian fishermen into a mighty nation of Caucasoids. Juan had achieved notoriety some years before by sailing wide of the South American coast in order to take advantage of the winds, and in consequence made such good time from Callao to Chile (thirty days) that on his arrival he was seized by the Inquisition on suspicion of witchcraft. Having convinced the Holy Office that he had not traded his soul to the Devil for seamanship, he wanted to go back for another look at his continent, but died before he got around to it.

Even after Australia was found, stories of a much larger continent in the South Pacific persisted until Captain James Cook settled the matter in the 1760's and 70's by sailing back and forth over the area until no large unexplored sections remained. It is a pity that Terra Australis does not exist, for it would be much more interesting than the vast wet waste of howling winds and mountainous waves that occupies most of the area fancy had assigned to this mysterious southland.

Interest in Atlantis revived with the discovery of the Americas, and since then has grown to the proportions of a neurosis. In 1553, sixty-one years after Columbus's discovery, the Spanish historian Francesco López de Gómara, in his *General History of the Indes,* suggested that Plato's Atlantis and the new continents were one and the same; or at least that Plato must have heard some rumor of real transatlantic continents and based his romance upon it. This theory received further impetus from the discovery that the Aztecs had a legend of having migrated from a place called Aztlan ("Plan of Reeds") and that *atl* was a common syllable in their language, meaning "water" by itself. The idea therefore caught hold. In 1561 Guillaume de Postel made the pleasing suggestion of naming one of the new continents "Atlantis," and in 1580 the English wizard John Dee applied this name to America on one of his maps. In 1689 the French cartographers Sanson, and in 1769 Robert de Vangoudy, went Dee one better by publishing maps of America showing how Poseidon divided the land up among his ten sons, for which Vangoudy was justly ridiculed by Voltaire.

The Atlantis-in-America theory had a fair run, being adopted by Sir Francis Bacon for his unfinished utopian romance *The New Atlantis,* and explained by John Swan in his *Speculum Mundi* (1644) as follows: ". . . this I may think may be supposed, that *America* was sometimes part of that great land which *Plato* calleth the Atlantick island, and that the Kings of that island had some intercourse between the people of *Europe* and *Africa.* . . . But when it happened that this island became a sea, time wore out the remembrance of remote countreys: and that upon this occasion, namely by reason of the mud and dirt, and other rubbish of this kind. For when it sunk, it became a sea, which at first was full of mud; and thereupon could not be sailed, untill a long time after: yea so long, that such as were the sea-men in those dayes, were either dead before the sea came to be clear again, or else sunk with the island: the residue, being little expert in the art of navigation, might, as necessitie taught them, sail in certain boats from island to island; but not venturing further, their memorie perished. . . . Yet that such an island was, and swallowed by an earthquake, I am verily perswaded: and if *America* joyned not to the West part of it, yet surely it could not be farre distant, because *Plato* describes it as a great island: neither do I think that there was much sea between *Africa* and the said island."

Subsequently the theory was accepted by Buffon in the eighteenth century and by Jacob Krüger and Alexander von Humboldt in the ninteenth. About 1855 the German poet Robert Prutz not only located Atlantis in America, but also worked out an elaborate scheme to show that the Phoenicians discovered America, using all the scraps of ancient literature that could be bent to his purpose.

However, since Prutz's time few have adhered to this

theory. From the scientific point of view it has too many facts against it: that the native American civilizations were barely rising from barbarism in Plato's time; that they never developed the material technics to mount military expeditions across an ocean; that before Plato's time Mediterranean ships could not go across the Atlantic and back. They could not row it because they could not carry enough food and water for the rowers, and they could not sail it because tacking against the wind had not yet been invented.

On the other hand, from the romantic-occult point of view Atlantis-in-America is too tame and prosaic a solution of the problem. During the last century the romantic and occult Atlantists have preferred to regard the native Amerindian civilizations as colonies of Atlantis.

A body of speculative thought like Atlantism, as it grows, branches out in all directions like a bush, and only rarely does a branch bear the flower of a genuine scientific or historical discovery. In studying this particular bush we shall have to follow each branch out to its tip, then backtrack to a fork and follow another branch. Although the Atlantis-in-America branch has long since ended in a dry twig, the related branch that conceived both the Amerindian and Old-World civilizations as offshoots of Atlantis is still a gaudy green, and perhaps the lustiest branch on the whole unearthly shrub; devotees of the theory include the late Charles Gates Dawes, onetime Vice-President of the United States. Let us trace this branch out.

This school of thought, like Atlantis-in-America, goes back to the years following the voyages of Columbus. The finding of the Americas loosed a regular flood of pseudoscientific speculation about the origin of the Amerinds. Theologians who had been denying that people could live in the Antipodes, embarrassed by the new discoveries, came forth with the opinion that the redskins were a separate species of mankind whom the Devil had made for his own sinful purposes. This theory was not unpopular with our own Anglo-Saxon forebears, since it gave them an excuse to slaughter the poor aborigines on sight.

Others again have supposed the Indians to be the decendants of Egyptians, Negroes, Phoenicians, Assyrians, East

Indians, Polynesians, and the natives of lost continents, or have even believed them to have evolved from a distinct line of apes or monkeys. When Europeans began to explore the New World, persons meeting the American natives for the first time jumped to the premature conclusions that they were speaking Welsh and were therefore the descendants of Prince Madoc and his band, who in Welsh legend were supposed to have crossed the Atlantic in 1170; or were practicing Hebrew religious rites and were the Lost Ten Tribes of Israel. These speculations have continued down to the present, despite the fact that science has pretty well established that (as Sir Paul Rycaut surmised in the seventeenth century) the American Indians belong to the Mongoloid or Yellow Race along with the Eskimos, Chinese, and Malays, and that they came from Siberia via Alaska.

Much of this pseudo-scientific activity stems from the work of one man, Diego de Landa, a Spanish monk who came in with the *conquistadores* and became, first Prior of the monastery of Izmal, then Provincial, and finally Bishop of Yucatán. The Mayan Indians over whom he presided had a considerable native literature, written on books made of long strips of native paper folded zig-zag and bound between a pair of wooden covers, which dealt with their history, astronomy, and other subjects. Landa, determined to wipe out "heathen" culture and substitute Christian European civilization, burnt all these books he could find from 1562 onwards. He explains that: "We found a large number of books in these characters and, as they contained nothing in which there were not to be seen superstition and lies of the devil, we burned them all, which they regretted to an amazing degree, and which caused them much affliction." For this vandalism Landa was criticized by some of his own Spanish colleagues, and has been consigned to his own Christian Hell by scholars ever since.

Subsequently Landa became interested in the Mayan culture and undertook to learn the Mayan writing. The Mayas used a complicated system of ideographic signs compounded together with some phonetic elements to make complex word-glyphs, something like the systems employed in early Egyptian and modern Japanese writing. Landa, however, assumed that Mayan, like the Spanish and Latin he knew, was written with a phonetic alphabet.

Apparently his method was to drag in some literate Maya, explain what he wanted, and bark: *"Qué es A?"*

The poor Amerind, doubtless shaking in his sandals for fear of being burnt as a heretic, thought the terrible old man wanted the sign for *aac,* "turtle." Therefore he drew the ideograph, a turtle's head.

"Qué es B?"

Now, *be* in Mayan means "road," so the Maya drew the ideograph for "road" — a pair of parallel lines representing a path, and between them the outline of a human footprint. And so on through the alphabet, until Landa had 27 signs and a few compounds, most of which however did not mean what he thought they did. He also took down a correct explanation of the Mayan numerals.

In the 1560's Landa was recalled to Spain on charges of exceeding his authority, and in preparing his defence (a successful one, I am sorry to say) he wrote a great treatise on the Maya civilization, *Relación de las Cosas de Yucatán,* or *Account of the Affairs of Yucatán,* wherein he set forth his "Mayan alphabet." This book is still one of the most valuable sources on Mayan culture and history. He wrote: "Some of the old people of Yucatán say that they have heard from their ancestors that this land was occupied by a race of people, who came from the East and whom God had delivered by opening twelve paths through the sea. If this were true, it necessarily follows that all the inhabitants of the Indes are descendants of the Jews. . . ."

Thus he gave the initial push to another pseudo-scientific theory: that the Amerinds were the Lost Ten Tribes of Israel. For many centuries men had been speculating about the eventual fate of the 27,290 Israelites whom Sargon of Assyria had deported from the northern Hebrew kingdom of Israel about 719 B.C. and settled "in Halah and in Habor by the river of Gozan, and in the cities of the Medes." Jewish and Christian speculators supposed these displaced tribesmen to be lurking in the wilds of Asia or Africa, and looked forward, the Jews with hope and the Christians with fear, to their reappearance on the stage of the world.

The Jewish-Indian theory was floated by Landa and

Fig. 5. THE "MAYAN ALPHABET" transcribed by Bishop Diego de Landa for his *Relación de las Cosas de Yucatán.*

some other Spanish missionaries like Durán, and later by
the adventurer Aaron Levi, who told the learned rabbi
Manasseh ben Israel of Amsterdam a romantic tale of his
visit to a brotherhood of Jewish Indians in Peru, descendants of the Tribe of Reuben. Manasseh published this tale
as a book, *The Hope of Israel,* which interested some Puri-
tan ministers in England. The latter wrote Manasseh and
brought the theory to Oliver Cromwell's attention. Cromwell invited Manasseh to England and formed such a firm
friendship with him that he tried to have the Jews read-
mitted to England, though he was able to effect this on a
small scale only.

It was even suggested that the Amerinds should return
to Palestine, but luckily for the peace of the world they
showed no interest in the idea.

The Jewish-Indian theory led an active life for over
two centuries, being adopted by William Penn among
others. The last prominent Amerind-Israelite, Lord Kingsborough, spent his entire fortune of £40,000 a century ago
publishing *The Antiquities of Mexico,* a monumental work
in nine immense volumes containing reproductions of Aztec
picture-writings and art-objects, and notes expounding the
Jewish-Indian theory. Kingsborough's obsession was literally the death of him, for it landed him in Dublin's debtor's
prison for non-payment of the bills incurred in publishing
this work, and there he expired.

Since then others have identified the Lost Tribes with
the Zulus, Burmese, Japanese, Papuans, and other peoples.
During the last century the noisiest branch of the cult has
been that which finds the Lost Ten in the present in-
habitant of the British Isles: the Anglo-Israelites or British-
Israelites. This sect was founded about 1795 by Richard
Brothers, an ornament of several lunatic-asylums who also
proclaimed himself the Nephew of God (a relationship to
puzzle the acutest theologian) and the divinely appointed
Prince of the Hebrews and Ruler of the World. Brothers
even tried to induce the mad George III to abdicate so that
he, Brothers, could take over, and was locked up as a dangerous madman.

All these identifications lack the slightest scientific
merit, being based upon wrong ideas of history, an-

thropology, linguistics, and Biblical interpretation. There is not the least doubt that Sargon's deportees did just what other tribes broken up by conquest and migration have done: They settled down in their new homes, intermarried with their neighbors, and gradually lost their original language, culture, and religion. As a result, the people of modern Iran and Iraq are all partly descended from them.

Such arguments have not stopped the hunt for the Lost Ten Tribes, however, just as no amount of argument will stop the search for Atlantis, since such activities are inspired by an emotional animus that mere facts cannot touch. While Ten Tribism impinges upon Atlantism at several points, the two cults are somewhat mutually exclusive, since Ten Tribism assumes a Fundamentalist-Christian view of the Bible, while Atlantism takes a cavalier attitude towards that discovered anthology of early Hebrew literature and, reversing the argument of Kosmas Indikopleustes, asserts that the Biblical Flood story is but a distorted account of the sinking of Atlantis.

To return to the Mayas and their oppressors: After Landa's time knowledge of Mayan writing was lost, as the Catholic priests continued their campaign against Mayan literature and the Mayas themselves dropped their native writing for the easier Latin alphabet, which the priests adapted to the Mayan language. (They used the letter *x* to represent the *sh* sound; hence all those *x*'s in Middle American names like Uxmal, pronounced "*oosh*-mahl," and Xiu, "she-*yoo*.") Only three Mayan books survived: the Dresden Codex, the Codex Perezianus in Paris, and the Tro-Cortesianus Codex in two sections (since reunited) in Madrid. Nobody could read any of them, nor could they read any of the many inscriptions which survived on Mayan monuments. Landa's treatise also disappeared, and the original has never been found though competently searched for.

Mayan writing remained completely mysterious until in 1864 a diligent but erratic French scholar, the Abbé Charles-Étienne Brasseur (called "de Bourbourg") discovered an abridged copy of Landa's *Relación* in the library of the Historical Academy of Madrid. Brasseur (1814-1874) had travelled extensively in the New World, had held an administrative post in Maximilian's short-lived Mexican empire, and had written many works including historical novels under the pen-name of "de Ravensberg" and an introduction to a book by that gifted old charlatan Waldeck, of whom more later.

When Brasseur found Landa's "Mayan alphabet" he was overjoyed, thinking he had the Rosetta Stone, the key, to Mayan writing. He tried at once to read the Troano Codex (one of the halves of the Tro-Cortesianus) by this alphabet with the help of an uncontrolled imagination.

The result was an incoherent description of a volcanic catastrophe, beginning: "The master is he of the upheaved earth, the master of the calabash, the earth upheaved of the tawny beast (at the place engulfed beneath the floods); it is he, the master of the upheaved earth, of the swollen earth, beyond measure, he the master . . . of the basin of water."

In going over the manuscript Brasseur came across the following pair of symbols:

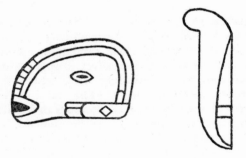

which he was otherwise unable to account for. Now, if you compare them with Fig. 5 you will see that the one on the left bears a faint (but only faint) resemblance to Landa's "M" and the other to his "U." By a magnificent conclusion-jump Brasseur inferred that these symbols spelled the name of the land submerged by the catastrophe: *Mu.*

He also, in another book, meaningfully pointed out resemblances between Plato's Atlantis and the underground empire of Xibalba in the *Popol Vuh,* a creation-myth and pseudo-history of the Mayas' neighbors the Kichés, which he had also translated and published. That was as far as Brasseur went into Atlantism.

Discovery of the Landa "alphabet" caused a natural stir among historians and archeologists, followed by a heavy disappointment when they found that trying to read Mayan by this key resulted in sheer gibberish. The efforts of other French students like Léon de Rosny to decipher Mayan writing also proved abortive. However, by diligent study over the last seventy-five years, scholars like Förstemann, Bowditch, and Morley have deciphered over a third of the Mayan gyphs. This is not enough to read off Mayan texts, but enough to give a glimmering of what they are about. Hence it is now known that the Troano Codex is no description of an eruption, but a treatise on astrology. The Dresden Codex is astronomical, and the Perezianus ritualistic. The incriptions on monuments are mostly concerned with calendric and liturgical matters.

Despite the discrediting of Brasseur's translation, his theories were further developed by two great pseudo-scientific Atlantists, Donnelly and Le Plongeon, remarkable characters apart from their contributions to Atlantism.

Ignatius T. T. Donnelly (1831-1901) was a man "with an extremely active mind, but possessing also that haste to form judgments and that lack of critical sense in testing them, which are often the result of self-education conducted by immense and unsystematic reading." Born in Philadelphia, he went into law, and in 1856 emigrated to Minnesota, where he settled in Nininger, near St. Paul, and started a small-town journal. At 28 he was elected Lieutenant-Governor of Minnesota. Thence he was sent to Congress, and for eight years, when not attending upon the nation's business, spent his time in the Library of Congress soaking up information and becoming perhaps the most erudite man ever to sit in the House of Representatives.

When finally defeated in 1870, Donnelly retired to his rambling mansion to write the first of several very successful books: *Atlantis: The Antediluvian World,* brought out by Harper in 1882, which went through at least fifty printings, the latest in 1949. He followed it with *Ragnarok, The Age of Fire and Ice,* which argued (wrongly) that the Pleistocene Ice Age was brought on by the collision of the earth with a comet, and *The Great Cryptogram,* which undertook to prove by cryptographic methods that Sir Francis Bacon had written the plays attributed to William Shakespeare.

This last theory had been suggested as a joke by Walpole in the previous century. A few decades before Donnelly's book appeared, the idea was submitted seriously by several people such as Delia Bacon, a puritanical Boston schoolteacher shocked by the notion that the author of the wonderful plays and sonnets could have been an associate of a lot of vulgar, dissipated actors. While Miss Bacon did not herself urge Sir Francis as a substitute, others like William H. Smith in England soon added this feature.

Donnelly's immense work labored to prove that Bacon had revealed himself in cipher throughout the plays; but a cruel cryptographer soon pointed out that by Donnelly's loose methods you could prove just as conclusively that Shakespeare wrote the Forty-Sixth Psalm. The 46th word from the beginning of this psalm is "shake"; the 46th word from the end is "spear"; Q.E.D.!

Despite such criticisms Baconianism grew to a major cult which survives as an active rival of the Atlantist and Ten Tribist cults. It has even branched out into heretical schools who hold that Shakespeare's works were written, not by Bacon, but by the Earl of Oxford or some other Elizabethan worthy; or who infer that Bacon wrote the works not only of Shakespeare but also those of his contemporaries Burton, Jonson, Peele, Greene, Marlowe, and Spenser. The logic is equally good in either case. The last is of course a reduction to absurdity, since Bacon could not possibly have led the active political career he did, written his own voluminous works, and also have found time to write the works of seven of the most prolific authors in the history of English literature.

Donnelly, a chubby, clean-shaven man who looked a little like his younger contemporary William Jennings Bryan, continued to lead a phenomenally active existence for the rest of his long life. He went on lecture tours and wrote several more books, including a prophetic novel, *Caesar's Column: A Story of the Twentieth Century,* which sold about a million copies. He was also elected Minnesota State Senator, helped found the Populist Party, and twice ran for Vice-President of the United States on the Populist ticket.

Donnelly thus bore the distinction of having made perhaps the greatest single contributions to two schools of speculative thought, the Atlantist and the Baconian. Atlantism had been mildly active for about three centuries, mainly among scholars of the more unworldly type. Some like Father Kircher had taken Plato's story at face value, while some considered it a fiction, perhaps, as Bartoli suggested, a satire on Athenian political conflicts. Some like Voltaire wavered, whereas some thought the story not literally true but based upon real legends with a historical foundation of fallen empires in Africa and other places.

It remained for Donnelly to convert Atlantism into a popular cult. His work expanded upon the theory, already put forward by his fellow-Americans Hosea and Thompson (and for that matter by Count Carli a century earlier) that the Mayan and other early civilizations were derived from Atlantis. Edward H. Thompson, then an undergraduate at Worcester Polytechnic Institute, had argued in an article in *Popular Science Monthly* that refugees from the sinking of Atlantis had landed in North America, spread to Lake Superior, and then been forced by the attacks of hostile tribes to migrate to Yucatán. Thompson later became an archeologist and one of the leading authorities on the Mayas, dropping all Atlantist doctrines.

Donnelly began by asserting thirteen "theses," as follows:

"1. That there once existed in the Atlantic Ocean, opposite the mouth of the Mediterranean Sea, a large island, which was the remnant of an Atlantic continent, and known to the ancient world as Atlantis.

"2. That the description of this island given by Plato is not, as has long been supposed, fable, but veritable history.

"3. That Atlantis was the region where man first rose from a state of barbarism to civilization.

"4. That it became, in the course of ages, a populous and mighty nation, from whose overflowings the shores of the Gulf of Mexico, the Mississippi River, the Amazon, the Pacific coast of South America, the Mediterranean, the west coast of Europe and Africa, the Baltic, the Black Sea, and the Caspian were populated by civilized nations.

"5. That it was the true Antediluvian world; the Garden of Eden; the Gardens of the Hesperides; the Elysian Fields; the Gardens of Alcinous; the Mesomphalos; the Olympos; the Asgard of the traditions of the ancient nations; representing a universal memory of a great land, where early mankind dwelt for ages in peace and happiness.

"6. That the gods and goddesses of the ancient Greeks, the Phoenicians, the Hindoos, and the Scandinavians were simply the kings, queens, and heroes of Atlantis; and the acts attributed to them in mythology are a confused recollection of real historical events.

"7. That the mythology of Egypt and Peru represented the original religion of Atlantis, which was sun-worship.

"8. That the oldest colony formed by the Atlanteans was probably in Egypt, whose civilization was a reproduction of that of the Atlantic island.

"9. That the implements of the 'Bronze Age' of Europe were derived from Atlantis. The Atlanteans were also the first manufacturers of iron.

"10. That the Phoenician alphabet, parent of all the European alphabets, was derived from an Atlantis alphabet, which was also conveyed from Atlantis to the Mayas of Central America.

"11. That Atlantis was the original seat of the Aryan or Indo-European family of nations, as well as of the Semitic peoples, and possibly also of the Turanian races.

"12. That Atlantis perished in a terrible convulsion of nature, in which the whole island sunk into the ocean, with nearly all its inhabitants.

"13. That a few persons escaped in ships and on rafts, and carried to the nations east and west tidings of the appalling catastrophe, which has survived to our own time in the Flood and Deluge legends of the different nations of the old and new worlds."

These theses are worth examining with some care, since they represent what you might call the centrist position on Atlantis, as distinct from the occult left-wing and the scientific right-wing. Most non-occult Atlantist books issued since Donnelly's time take essentially his position and draw heavily upon him for material. On the other hand, note that Donnelly goes far beyond Plato, who never claimed that Atlantis was the source of all civilization.

After quoting from the pertinent parts of *Timaios* and *Kritias*, Donnelly undertook to prove his theses. For one thing, he said, the lack of magical or fantastic elements in Plato's story is evidence of its truth, and the present submarine topography around the Azores corresponds with Plato's island. He argued that continents have risen and sunk thousands of feet during geological eras (true), and islands have risen and sunk in a matter of hours during volcanic eruptions (true); therefore, why could not a continent have sunk out of sight as a result of a single eruption or earthquake? Earthquakes such as that of Lisbon in 1775 had certainly had devastating effects. "We conclude," he said, "therefore: 1. That it is proven beyond question, by geological evidence, that vast masses of land once existed in the region where Atlantis is located by Plato, and that therefore such an island must have existed; 2. That there is nothing improbable or impossible in the statment that it was destroyed suddenly by an earthquake in one dreadful night and day."

Unfortunately he did not prove anything of the sort. He merely raised some inconclusive arguments to show the possibility of such a land; something quite different.

Then Donnelly set forth resemblances between many species of American and European plants and animals, and cited various authorities to show that certain plants like tobacco, guava, and cotton were not, as was generally thought, confined to one hemisphere before Columbus, but were grown in both the New and Old Worlds. The Assyrians, for example, had the pineapple. The Deluge-legends of the Jews, Babylonians, Aztecs, and others all point to the submergence of Atlantis, whose culture at its best has not been surpassed.

Donnelly thought that the Egyptian civilization blossomed all at once instead of evolving slowly, thereby showing that it was imported, and anticipated the modern ultra-diffusionists by asserting: "I cannot believe that the great inventions were duplicated spontaneously . . . in different countries. . . . If this were so, all savages would have invented the boomerang; all savages would possess pottery, bows and arrows, slings, tents, and canoes. . . ."

Like the diffusionists, Donnelly inferred that, because similar culture-traits among various European peoples can be traced to a common origin, the same was true for similar culture-traits among the peoples of the New and Old Worlds. Then he cited the occurrance on both sides of the Atlantic of pillars, pyramids, burial-mounds, metallurgy, the arts, agriculture, ships, and so on through the whole range of human culture-traits.

To show that the Old-World alphabets that descended from the Phoenician (including our own) originated in Atlantis, Donnelly printed tables in which the Latin alphabet is lined up alongside Landa's "Mayan alphabet" as published inaccurately by Brasseur de Bourbourg. The two don't look the least bit alike, but that did not stop Donnelly, who took pieces out of the Mayan glyphs and distorted them to create "intermediate forms" between the Latin and supposed Mayan letters.

Then he reconstructed the history of Atlantis by assuming that all Old-World myths like those of *Genesis* were distorted bits of Atlantean history; Poseidon, Thor, Melkarth, and other Old-World gods were Atlantean kings; the Titans of Greek myth and the Fomorians of Irish myth were Atlanteans, and so on.

Finally, he clinched his case by linguistic arguments purporting to prove that various New-World languages are closely related to tongues of the Old World. Le Plongeon (to whom we shall come) is quoted as saying that: "One third of [the Maya] tongue is pure Greek" (though actually it would be hard to find two languages less alike); A Señor Melgor of Mexico says Chiapanec, a Central American language, resembles Hebrew; the Otomi language of Mexico is related the Chinese.

Since Donnelly's formidable learning is likely to stun the average reader into taking his statements at face value, a close look at his book is needed to show how careless, tendentious, and generally worthless it is. For instance, to point out that both Europeans and Amerinds used spears and sails; that both practiced marriage and divorce; and that both believed in ghosts and flood-legends, proves nothing about sunken continents, but only that the peoples in question were all human beings, since all these customs and beliefs are practically world-wide.

Most of Donnelly's statements of fact, to tell the truth, either were wrong when he made them, or have been disproved by subsequent discoveries.

It is not true, as he stated, that the Peruvian Indians had a system of writing, that the cotton-plants native to the New and Old Worlds belong to the same species, that Egyptian civilization sprang suddenly into being, or that Hannibal used gunpowder in his military operations. Donnelly's "Assyrian pineapples" are nothing but the date-palms represented in Assyrian art. When he tried to show the resemblance between Otomi and Chinese by parallel tables of words, I don't know what he used for Chinese—certainly not the standard Northern Chinese, the language usually meant by the term. For instance he gave the Chinese words for "head," "night," "tooth," "man," and "I" as *ten, siao, tien, na* and *nugo,* when they should be *tou, ye, ya, jen* (or *ren*) and *wo.*

This common mistake about the relationship of Otomi to the languages of East Asia seems to rise from the fact that Otomi, like Chinese, has phonemic tones: the pitch at which a syllable is pronounced affects its meaning. Some speculators have inferred from this fact that Otomi must be related to Chinese or to Japanese, a surmise that is not borne out by even a slight acquaintance with these languages. Anyway, phonemic tones are nothing unique; many African languages are also polytonic.

For all its shortcomings, however, Donnelly's book became the New Testament of Atlantism, just as the *Timaios* and *Kritias* are its Old Testament, and year after year Atlantists repeat that Otomi is archaic Chinese or Japanese.

Donnelly's contemporary Augustus Le Plongeon (1826-1908) was the first to excavate Mayan ruins in Yucatán, where he lived for many years.

A sad-eyed French physician with a magnificent beard

down to his navel and a handsome American wife much younger than himself, Le Plongeon was an expert in his own curious way. Although he was familiar at first hand with the customs and speech of the Mayas, his work proved abortive, for he failed to achieve the scientific recognition he sought. He also suffered from the rapacity of Mexican officials, who in those days had the cruel custom of letting gringo archeologists dig till they found something interesting, and then confiscating their finds and harrying them from the country.

From Brasseur's attempted translation of the Troano Codex, and from some pictures on the walls of the ruins at the Mayan city of Chichén-Itzá, Le Plongeon, an even more extravagant extrapolator than Donnelly, built a romantic tale of the rivalry of the princes Coh ("Puma") and Aac ("Turtle") for the hand of their sister Móo, Queen of Atlantis or Mu. Coh won, but was murdered by Aac, who conquered the country from Móo. Then as the continent foundered Móo fled to Egypt, where she built the Sphinx as a memorial to her brother-husband and, under the name of Isis, founded the Egyptian civilization. (In cold fact the Sphinx is probably a monument to King Khafra of the Fourth Dynasty.) Other Muvians had meanwhile settled in Central America, where they became the Mayas of history.

Le Plongeon incorporated these fantasies in several books. When his small volume *Sacred Mysteries Among the Mayas and the Quiches 11,500 Years Ago* (1886) appeared, serious students naturally hooted. Thereupon the enraged Le Plongeon wrote a much bigger book, *Queen Móo and the Egyptian Sphinx,* wherein he denounced the "arrogance and self-conceit of superficial learning" displayed by the "pretended authorities," meaning Brinton and other Americanists.

Like Brasseur, Le Plongeon tried to translate the Troano Codex. The result, if no more reliable scientifically, was at least more intelligible: "In the year 6 Kan, on the 11th Muluc in the month Zac, there occurred terrible earthquakes, which continued without interruption until the 13th Chuen. The country of the hills of mud, the land of Mu was sacrificed: being twice upheaved it suddenly disappeared during the night, the basin being continually shaken by the volcanic forces. Being confined, these caused the land to sink and to rise several times in various places. At last the surface gave way and ten countries were torn asunder and scattered. Unable to stand the force of the convulsion, they sank with their 64,000,000 of inhabitants 8060 years before the writing of this book."

He also derived Freemasonry and the Metric System from the Mayas, took Mme. Blavatsky's imaginary *Book of Dzyan* seriously as "an ancient Sanskrit book," asserted that the Greek alphabet was really a Mayan poem dealing with the sinking of Mu, and printed a photograph of an Old-World leopard as an example of Central American fauna. As authorities he cited people like the London publisher John Taylor ("the learned English astronomer," Le Plongeon called him) and the eccentric Scottish astronomer Charles Piazzi Smyth ("the well-known Egyptologist") who between them founded the pseudo-scientific cult of pyramidology.

According to their teachings the Great Pyramid of King Khufu at Gizeh was really built by Noah or some other Old-Testament patriarch under divine guidance, and its measurements (which they gave with incredible inaccuracy) incorporated the wisdom of the ages and prophesied the future of mankind. Although none of the pyramidologists' predictions has come true, they are still active in the business of prophecy-mongering.

The next worker of this particular vein of pseudoscience was Dr. Paul Schliemann, grandson of the great Heinrich Schliemann, the small nervous retired German businessman who founded the modern science of archeology by digging up ancient Troy and Mykenai. In 1912 the younger Schliemann, apparently getting tired of being a little man with a big name, sold the *New York American* an article entitled *How I Discovered Atlantis, the Source of All Civilization.*

Schliemann said that his grandfather had left him a batch of papers on archeological matters and an owl-headed vase of ancient provenance. The envelope containing the papers bore a warning that the envelope should only be opened by a member of Schliemann's family willing to swear to devote his life to research into the matters dealt with in the papers inside. Paul Schliemann took the pledge and opened the envelope.

The first instruction was to break open the vase. Inside he found some square coins of platinum-aluminum-silver alloy, and a metal plate inscribed, in Phoenician: "Issued in the Temple of Transparent Walls." Among his grandfather's notes he came across an account of finding a large bronze vase on the site of Troy, in which were coins and other artifacts of metal, bone, and pottery. The vase and some of the subjects were inscribed: "From the King Cronos of Atlantis."

Schliemann gushed: "You can imagine my excitement; here was the first material evidence of that great continent whose legend has lived for ages. . . ." He went on to advance the usual arguments, taken without credit from Donnelly and Le Plongeon, for a common origin of New and Old-World cultures in Atlantis. Like Le Plongeon he claimed to have read the Troano Codex—in the British Museum, though it was in Madrid all the time. The story of the dunking of Mu contained therein he corroborated by a 4000-year-old Chaldean manuscript from a Buddhist temple in Lhasa, Tibet, of all places, which told how the Land of the Seven Cities was destroyed by earthquake and eruption after the star Bel fell, while Mu, the priest of Ra, told the terrified people that he had warned them.

Schliemann promised to reveal the full story of his discoveries in a book that would Tell All about Atlantis. He ended significantly: "But if I desired to say everything I know, there would be no more mystery about it." Alas, the book never appeared; nor were there any further revelations; nor did the owl-headed vase, the Chaldean manuscript, and the other priceless relics ever see the light of scientific investigation. Queried about the matter, Heinrich Schliemann's collaborator Wilhelm Dörpfeld wrote that so far as he knew the elder Schliemann never displayed any special interest in Atlantis and had not done any original work on the subject. The evident fact that the whole thing was a hoax has not stopped Atlantists from quoting the younger Schliemann as an authority, sometimes confusing him with his grandfather.

The last and gaudiest blossom on this particular branch is the late James Churchward, a small wraithlike Anglo-American, who in his younger days wrote *A Big Game and Fishing Guide to North-Eastern Maine* for the Bangor and Aroostock R. R., and in later years called himself "Colonel" and claimed to have traveled widely in Asia and Central America (where he was attacked by a flying snake). In his seventies Churchward burst into print with *The Lost Continent of Mu* (1926) and other Mu books published subsequently. Deriving his ideas mainly from Le Plongeon and Paul Schliemann, Churchward expanded upon them by assuming two sunken continents, Atlantis in the Atlantic and Mu (corresponding to the occultists' Lemuria) in the Central Pacific, where for geological reasons we can be reasonably sure there never has been a continent and never will be one.

Churchward shared the favorite obsession of the occultists that there was once a universal esoteric language of symbols which the ancients used in recording their secret wisdom, and that by staring at ancient symbols long enough an intuitively gifted person can conjure their meanings out of his inner consciousness and thus recover forgotten historical facts. Now, the ancients did use many symbols, just as we do with our flags and trademarks. But, unless one knows a culture intimately, one cannot tell whether some bit of antique decoration symbolized anything or was just put there to look pretty. If you think you can interpret symbols subjectively, try your skill on a page of written Chinese, without knowing in advance how to read that language. A Chinese ideograph is merely a conventionalized picture—exactly the sort of thing Churchward claimed to be able to interpret.

As an example, Churchward asserted that the rectangle stood for the letter M in the Muvian alphabet, and therefore for Mu itself. As the ordinary brick is entirely bounded by rectangles, you can see that he had no trouble in deriving everybody and everything from Mu. Moreover he misquoted Plato ("in Plato's *Timeus Critias* [sic] we find this reference to the lost continent: 'The Land of Mu had ten peoples.'" etc.) and printed nonsensical footnotes reading "4. Greek record." or "6. Various records." When he print-

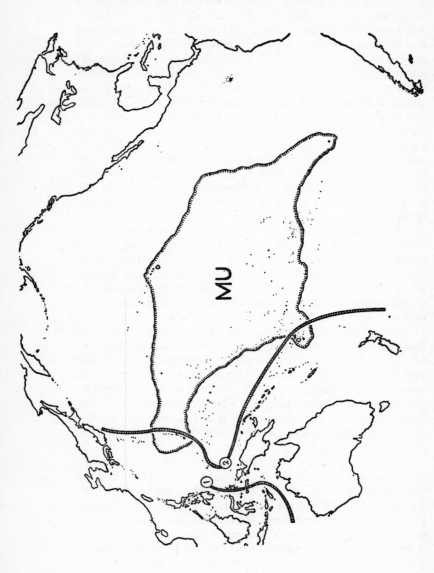

Fig. 6. MAP OF THE PACIFIC OCEAN, showing (1) Wallace's Line, (2) the Andesite Line, and Churchward's lost continent of Mu.

ed a table of forty-two Egyptian hieroglyphs, only six of them were even remotely correct.

Churchward said that he based his theory upon two sets of "tablets." One of these appears to exist, being a collection of objects found in Mexico by an American engineer named Niven. The objects look to the uninitiated eye like the flattened figurines which the Aztecs, Zapotecs, and other Mexican tribes made in great numbers for religious purposes; but to Churchward they are tablets, and their bumps and curlicues Muvian symbols conveying esoteric meanings.

The other set is more recondite: the "Naacal tablets ... written with the Naga symbols and characters" which a friendly temple priest showed Churchward in India. That is, in one book he tells of seeing them in India, and in another book in Tibet. By a lucky coincidence Churchward had just been studying the "dead language" with which these tablets were inscribed, and hence could read their account of the Creation and of the submergence of Mu.

From these sources Churchward learned that Mu was a large Pacific continent, stretching from the Hawaiian Islands to the Fijis and from Easter Island to the Marianas; low and flat because mountains had not yet been invented, and covered with lush tropical vegetation. In its days of glory Mu supported sixty-four million souls, divided into ten tribes and ruled by a priest-emperor called the Ra. While Muvians came in several colors, the Whites dominated the rest. They not only possessed a high civilization, but also practiced the pure Aryan monotheistic religion, which Jesus Christ later tried to revive. Savagery had never existed, for Churchward, who had no use for the "monkey theories" of science, held that man was specifically created, fully civilized, in the Pliocene.

Mu sent out colonies under the guidance of its priests, the Nagas or Naacals. Some of these emigrants went to Atlantis via the inland sea that then occupied the Amazon basin, while others settled in Asia, where they built a great Uighur Empire 20,000 years ago. (History, by the way, does know a real Uighur Empire—which however rose in the tenth century A.D. and fell in the twelfth, and so has nothing to do with Churchward.)

Then, 13,000 years ago, the "gas-belts," great caves underlying much of the earth, collapsed, letting Mu and Atlantis sink and making mountains in the other continents. The surviving Muvians, crowded on to the small islands of Polynesia, took to eating each other for want of other nourishment, and not only they but most of the Muvian colonies fell to the level of savages.

Like Le Plongeon, Churchward took the *Book of Dzyan* seriously, though he did not approve of it. In fact he referred to it as "nonsense" and "the writings of a disordered brain, wandering about in a fog." Perhaps he was jealous, though he need not have been. After all, several of his own bizarre brainstorms are still sold, while the few good books on Atlantism, like those of Bramwell and Björkman, have long since gone out of print.

No fervent believer in Mu, it seems, will give up his belief merely for the sake of a few facts. Thus Churchward's pseudoscientific masterpieces have begotten progeny, in the pamphlets published by Dr. Louis R. Effler, who flies about the world looking for the Muvian spiral symbol, and in the Alley Oop comic strip, whose hardboiled hero is a skin-clad native of a dinosaur-infested Mu. And in 1947 F. Bruce Russell, described as a "retired Los Angeles psychoanalyst," announced that he had found mummies eight to nine feet tall, from the lost continent of Mu, near St. George, Utah. Evidently Mu, despite anything we can do about it, marches on.

CHAPTER III

THE LAND OF THE LEMURS

Our fabled shores none ever reach,
No mariner has found our beach,
Scarcely our mirage now is seen,
And neighboring waves of floating green,
Yet still the oldest charts contain
Some dotted outline of our main. . . .
Thoreau

MENTION of Lemuria in the last chapter leads us on one hand to the science of paleogeography—the geography of the earth during past geological ages—and on the other to the occult wing of Atlantism.

The name "Lemuria" originated as follows: After the Darwinian revolution had provided scientists with a solid framework on which to hang the history of the earth and its life, the closing decades of the last century saw a great surge of activity in the fields of biology and geology. Gaps between geological eras were filled, former transgressions of the sea over the land were mapped, and the pedigrees of horses, octopi, and other life-forms were traced.

During the 1860's and 70's a group of British geologists, including Stow and Blanford in India and Griesbach in Africa, noted striking resemblances between certain formations in India and South Africa. William T. Blanford pointed out the similarity between the rocks and fossils of a deposit of the Permian Period in Central India, in a tract called Gondwana, and a corresponding deposit in South Africa. Gondwana, meaning "Land of the Gonds," is named for a forest tribe that ranges the tract and which once had the unpleasant custom of slowly torturing people to death in magical rituals to make their crops grow. The Permian Period was the last division of the Paleozoic Era, just before

the Mesozoic Era or Age of Reptiles: a cold dry mountainous period when the highest forms of life were rather
nondescript lizard-like reptiles. For those unfamiliar with
geological periods I have prepared a table in Appendix D.

Blanford and his colleagues inferred that South Africa
and India were once connected by a land-bridge that included Madagascar with its peculiar mammals, the Aldabra
Islands with their giant tortoises, the great Seychelles Reefs,
and the Maldive and Laccadive Islands. These islands and
reefs form the tops of an immense submarine mountain-
range that winds like a sea-serpent from South Africa to
the tip of India.

These observations came to the notice of the Austrian
paleontologist Neumayr and the German biologist Haeckel.
In his *Erdegeschichte* (1887) Neumayr published the first
known attempt at a paleogeographical map of the world,
showing how he thought the world looked in Jurassic
time, that is, in the middle of the Age of Reptiles. It included a great "Brazilian-Ethiopian Continent" from whose
southeast corner extended an "Indo-Madagascan Peninsula"
corresponding to Blanford's Permian land-bridge.

Ernst Heinrich Haekel was then performing the same
service for science in Germany that Thomas Huxley was
doing in England—loud and belligerent advocacy of the
revolutionary theories of Darwin. Haeckel seized upon the
Indo-Madagascan land-bridge to explain the distribution
of lemurs, creatures looking like a cross between a squirrel
and a monkey, standing below the true monkeys on our
family tree. Lemurs abound in Madagascar, and are also
found here and there in Africa, India, and the Malay
Archipelago. If, thought Haeckel, the Madagascan land-
bridge had endured from the Permian to the Jurassic, why
not into the Cenozoic Era or Age of Mammals?

In a burst of exuberance Haeckel went on to suggest
that this sunken land might be the original home of man,
since the finding of fossil forms intermediate between men
and apes had not really begun at this time. (Now over a
dozen such forms are known.) Then the English zoölogist
Philip L. Sclater suggested the name "Lemuria" for this
bridge. The name has stuck, although it now appears that
Haeckel was probably wrong about Lemuria's lasting into

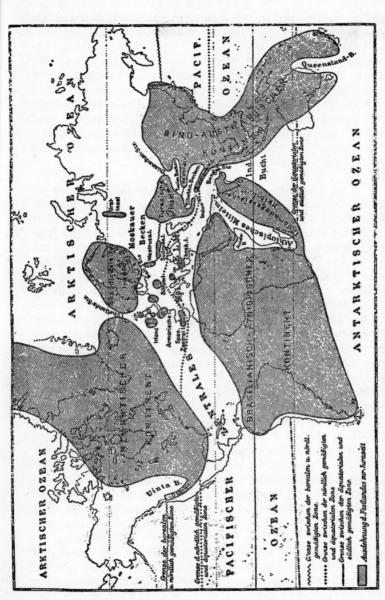

Fig. 7. THE FIRST WORLD PALEOGEOGRAPHICAL MAP, from Neumayr's *Erdegeschichte* (1887) showing his hypothetical Brazilo-Ethiopian continent and the Afro-Indian peninsula later called "Lemuria."

the Cenozoic, and most paleontologists think they can account for the distribution of lemurs without it.

Other investigators suggested that Lemuria was a remnant of a much larger and earlier continent, which they called Gondwanaland and thought once reached three-quarters of the way around the world in the Southern Hemisphere with a gap in the Pacific. These hypothetical continents, however, have little to do with Atlantism, for even if they existed they broke up long before men evolved. Nevertheless they have been exploited by occultists and Atlantomaniacs for their own purposes.

The greatest of modern occultists, the successor of Simon Magus and Cagliostro, was Helena P. Blavatsky, the founder of Theosophy. At the time she entered the current of Atlantism, in the 1870's, she was a fat middle-aged Russian woman living in New York City. She was the estranged wife of a Russian general, and had been successively the mistress of a Slovenian singer, an English businessman, a Russian baron, and a merchant from the Caucasus living in Philadelphia, and had made her living as a circus bareback rider, a professional pianist, a businesswoman, a sweatshop-worker, and a Spiritualist medium. Altogether she had led a pretty lively career, though in later years she undertook to gild refinèd gold and paint the lily by inventing an even more remarkable past wherein she was a persecuted virgin who travelled the wide world in search of occult wisdom.

Mme. Blavatsky took as her occult partner Henry Steel Olcott, a shrewd American lawyer who left wife and sons to live with her. Theosophy really got started when the pair moved to India, where Mme. Blavatsky learned to combine her considerable knowledge of Western magic and occultism with a wide and inaccurate smattering of East Indian philosophy and mythology. She led a fascinating and turbulent existence, and kept a hold on a sizeable body of followers even after she had been exposed in many chicaneries.

In 1882 she was dazzling a pair of well-connected Anglo-Indian dupes, the newspaper editor Arnold P. Sinnett and the government official Alan Octavian Hume, by delivering letters she said were written by her "Master"

Koot Hoomi, but which, as handwriting analysis later showed, she wrote herself. In these letters she was slowly feeling her way to that stunning occult cosmogony that she later advanced, a vast synthesis of Eastern and Western magic and myth about the seven planes of existence, the sevenfold cycles through which everything evolves, the seven Root Races of mankind, the seven bodies that each of us carries with him, and the Brotherhood of Mahatmas who run the world from headquarters in Tibet by sending out streams of occult force and bustling about the world in their astral bodies.

Mme. Blavatsky had picked up the Lemuria theory in the course of her reading and incorporated it, along with Atlantis, into her own gaudy cosmos. She dropped a few hints on these subjects in her Mahatma letters. Subsequently her doctrines took final if wildly confused form in her *chef d'oeuvre, The Secret Doctrine,* which she wrote in Europe after a couple of accomplices who assisted in the production of her thaumaturgic feats betrayed her, forcing her to leave India.

This huge work is supposedly based upon the *Book of Dzyan,* of which her Mahatmas showed her a manuscript copy, written on palm-leaf pages, in the trances in which she and they visited one another. The book, we are told, was originally composed in Atlantis in the forgotten Senzar language. *The Secret Doctrine* consists of quotations from the *Dzyan* and Mme. Blavatsky's lengthy commentaries thereon, interspersed with passages of occult gibberish and diatribes against "materialistic" science and "dogmatic" religion. The *Stanzas of Dzyan* begin:

"1. The Eternal Parent, wrapped in her Ever-invisible Robes, had slumbered once again for Seven Eternities.

"2. Time was not, for it lay asleep in the Infinite Bosom of Duration.

"3. Universal Mind was not, for there were no Ah-hi to contain it.

"4. The Seven Ways of Bliss were not . . ."

Presently the Universe begins to awaken: "The last

Vibration of the Seventh Eternity thrills through Infinitude. The Mother swells, expanding from without, like the Bud of the Lotus. . . ."

After various cosmic events, described in this opaquely iridescent language, life appears on earth: "After great throes she cast off her old Three and put on her new Seven Skins, and stood in her first one." (The earth seems to be conceived as a sort of cosmic strippeuse.) "The Wheel whirled for thirty crores more. It constructed Rûpas; soft Stones that hardened, hard Plants that softened. Visible from invisible, Insects and small Lives. . . . The Water-men, terrible and bad, she herself created from the remains of others. . . . The great Chohans called the Lords of the Moon, of the Airy Bodies: 'Bring forth Men, Men of your nature. . . .' Animals with bones, dragons of the deep, flying Sarpas were added to the creeping things. . . ."

Without going into the elaborate Theosophical world-plan of multiple planes of existence, chains of planets following each other from plane to plane like the horses on a carousel, and other vagaries, we are told that the history of the earth runs thus: Life evolves through seven cycles or "Rounds," during which mankind develops through seven Root Races, each comprising seven sub-races. The First Root Race, a kind of astral jellyfish, lived on an Imperishable Sacred Land. The Second, a little more substantial, dwelt in the former arctic continent of Hyperborea. The Third were the apelike hermaphroditic egg-laying Lemurians, some with four arms and some with an eye in the back of their heads, whose downfall was caused by their discovery of sex. (Mme. Blavatsky took a poor view of sex, at least after she got too old to enjoy it herself.) The Fourth Root Race were the quite human Atlanteans. We are the Fifth, and the Sixth will soon appear.

Hyperborea, like Atlantis, is derived from ancient Greek geographical speculations. The Hyperboreans were supposed to live in the Far North, either on an island or on the mainland of Europe or Asia. The usual locale was the northern coast of Asia, behind the imaginary Riphaean Mountains. Never having been there, the Greeks imagined that the Arctic was a fine place with a balmy climate, where men lived a thousand years:

On every side are the dances of maidens and sounds of
the lyre
Circling forever, and notes of the flute; as they revel in
gladness,
Crowned is their hair with the bay-leaves of gold; and no
sickness afflicts these
Fortunate people, or age always hateful. . . .

So far as is known no Greek had ever visited the Far
North until in Plato's time the city of Massalia (our Mar-
seilles) sent Pytheas to scout Northern Europe to see where
their trade-goods were coming from. Pytheas coasted as far
as the mouth of the Rhine, saw Britain, and heard of an
island far to the north called Thulë. This was probably the
Shetlands or Orkneys, though some have argued for Norway or Iceland. Beyond Thulë, he was told, one could
not go, for there the land, sea, and air no longer existed
as separate elements, but were mixed together in a kind of
cosmic pudding, with the consistency of a jellyfish. Perhaps
somebody had merely been giving the intrepid Massiliot a
hyperbolic description of an arctic fog: "so thick you could
cut it with a knife!"

Hyperboreans, it was said, worshipped Apollo, and a
wizard-priest named Abaris worshipped so well that the god
in gratitude gave him a golden javelin on which he flew
about the earth as on a witch's broomstick. He visited
Greece where he stopped a plague at Sparta by his magic,
then went on to Italy where he studied occultism under
Pythagoras before returning to Hyperborea.

But to return to the mysterious Madame and her book:
The Secret Doctrine, I grieve to say, is neither so ancient,
so erudite, nor so authentic as it pretends to be. For when
it appeared, the learned but humorless old William Em-
mette Coleman, outraged by Madame Blavatsky's pretensions to Oriental learning, undertook a complete exegesis of
her works. He showed that her main sources were H. H.
Wilson's translation of the *Vishnu Purana;* Alexander Win-
chell's *World Life; or, Comparative Geology;* Donnelly's
Atlantis; and other contemporary scientific and occult
works, plagiarized without credit and used in a blundering
manner that showed but skin-deep acquaintance with the

subjects under discussion. She cribbed at least part of her *Stanzas of Dzyan* from the *Hymn of Creation* in the old Sanskrit *Rig-Veda,* as a comparison of the two compositions will readily show. Coleman promised a book that should expose all of H.P.B.'s sources, including that of the word *Dzyan.* Unfortunately Coleman lost his library and notes in the San Francisco earthquake and died three years later, his book unwritten.

Madame Blavatsky's lost-continent doctrines seem to be based largely on the works of Donnelly, Harris, and Jacolliot. Her contemporary Thomas Lake Harris was a poet, ex-Universalist preacher, and associate of the early Spiritualist leader Andrew Jackson Davis about the middle of the century. Harris soon split with Davis and set up his own cult, first in Chautauqua County, New York, and later in California. His cult had a lurid history, for Harris (like the notorious Purnell of the House of David) denied sex to his followers but enjoyed it himself.

Harris's voluminous writings, like those of H.P.B., comprised a synthesis of the occultisms of India and the West, with multiple planes of existence, a fantastic account of life on other worlds and in the many heavens and hells surrounding these worlds, the sex-life of angels, and interviews with distinguished dead persons like Galileo. The earth, it seems, is the only planet where moral evil exists. There once was another, Ariana, but God impatiently shattered it into meteors. The human race has gone through a succession of Golden, Silver, and Copper Ages, in the course of which it was guided for a time by Adepts from Venus. Evil, especially the sorceries practiced in Atlantis, finally brought on a series of catastrophes which submerged Atlantis and other places, and which are remembered as the Biblical Flood.

H. P. B.'s other Atlantist source, the prolific but unreliable French writer Louis Jacolliot (1837-1890) made a collection of Sanskrit myths during a sojourn in India and popularized these in his books when he got back to France. According to him the Hindu classics tell of a former continent called Rutas in the Indian Ocean, which sank beneath the waters. Jacolliot interpreted these myths as referring to a former *Pacific* continent, embracing all the

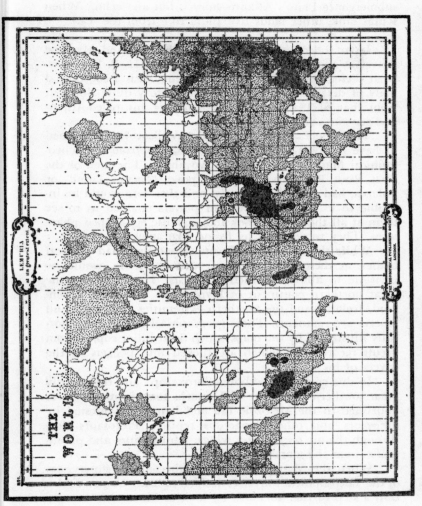

Fig. 8. THE THEOSOPHICAL LEMURIA at its greatest extent, according to Scott-Elliot. Lemuria occupies much of the southern hemisphere with an extension into the North Pacific. Dark splotches represent mountainous areas. Remnants of Hyperborea appear in the extreme North.

Polynesian islands, where civilization began and of whose submergence Plato's Atlantis story is but an "echo." When Rutas sank, other lands like India rose from the sea.

Later Theosophical writers like Sinnett and Madame Blavatsky's successor Annie Besant clothed H. P. B.'s somewhat skeletal account of lost continents with a substantial body of detail. They tell, however, a story quite different from that of Plato, altogether ignoring his paleo-Athens and its war with Atlantis.

Thus the English Theosophist W. Scott-Elliot claimed to have received the following revelation from the Theosophical Masters by "astral clairvoyance": The men of the First Root Race, living in the polar "Sacred Land" or Polarea, had astral bodies only, and therefore would be invisible to us. If our clairvoyant powers enabled us to see them at all they would look like gigantic egg-shaped phantoms. They reproduced by fission like amebas.

While the men of the Second Root Race had physical bodies, as these were made of ether they, too, would be invisible to all but trained occultists. They lived in the great northern continent of Hyperborea, of which Greenland, Iceland, and the northern extremities of Europe and Siberia are remnants. These geographical details are colorfully illustrated by the maps which Scott-Elliot copied from occult records and which are tucked into pockets in the covers of his books.

In due course Hyperborea "broke up" and the equally great southern continent of Lemuria took form. Lemuria flourished in the Mesozoic Era and hence was infested by dinosaurs and other dangerous reptiles. Pterosaurs flew through the air on leathery wings; ichthyosaurs and plesiosaurs swam the marshes.

When the Manu, one of the supernatural supervisors of the Theosophical universe, decided to bring human evolution a step further forward, he took as his model the apelike creatures that had already evolved on other planets. The Manu's first attempt resulted in jelly-like things with soft bones who could not stand up, but in time the structure of their bodies hardened. These primitive and dimwitted Lemurians were hardly more attractive than the reptiles with which they shared the earth.

From some unnamed source the author quotes a description of a Lemurian: He was between twelve and fifteen feet tall with a brown skin, a flat face with a protruding muzzle, and small eyes set so wide apart that he could see sideways as well as forward. He had no forehead, but was furnished with a third eye in back, which among us is still represented by the pineal gland in the brain. His long limbs could not be completely straightened, and he had huge hands and feet, the heels of which stuck out so far to the rear that he could walk equally well backward and forward. He wore a loose robe of reptile skin, carried a wooden spear, and led a pet plesiosaur on a leash.

By a strange coincidence, within the last decade scientists have found in southeastern Asia the bones and teeth of enormous early-Pleistocene ape-men, *Gigantanthropus* and *Meganthropus,* of whom the largest specimens seem to have been bigger than a full-grown 500-pound gorilla. These monsters match the Theosophical Lemurians in size if not in much else. Of course this does not prove that the Theosophists knew what they were talking about, any more than Plato's mention of the "Outer Continent" shows that he knew of America. It is perfectly possible to hit upon a scientific fact by pure chance, though random guesswork is not recommended as a method of scientific research.

While originally egg-laying hermaphrodites, the Lemurians began to learn about sex during the period of their Fourth Sub-Race, and by their Fifth Sub-Race were reproducing their kind as we do. Being stupid things, they interbred with beasts, the products of this perverted union being the great apes. This sin so revolted the Lhas, the supernatural beings who according to the cosmic plan were supposed at this stage to incarnate in human bodies, that the Lhas refused to do their duty.

Accordingly other beings, from Venus which had already developed a high civilization, volunteered to take the Lhas' place. These "Lords of the Flame" guided faltering humankind to the point where the Lemurians became capable of individual immortality and reincarnation. The Venerians also taught the Lemurians the arts of keeping fire, metallurgy, weaving, and agriculture. By the time the

Lemurians reached their Seventh Sub-Race they looked fairly human. Their descendants on earth today are such primitive peoples as the Lapps, the Australian aborigines, and the Andaman Islanders. Their language is the direct progenitor of Chinese, albeit the Chinese *people* are descended from the much later Turanian race.

Towards the end of the Mesozoic Era, Lemuria, like Hyperborea before it, began to break up by the sinking of its various parts, while the peninsula that curved into the North Atlantic grew into Atlantis. At the same time the Fourth Root Race, the Atlantean, appeared. The first Sub-Race of this Root Race, the Rmoahals, moved from the remnants of Lemuria to Atlantis, though some of them stayed behind and interbred with the surviving Lemurians, the resulting half-breeds looking like American Indians with blue skins.

The first Rmoahals, black-skinned men ten or twelve feet tall, settled on the southern coast of Atlantis and fought endless wars with the Sixth and Seventh Sub-Races of the Lemurians. Organized warfare was invented at this time, though the Lemurians had previously indulged in desultory raiding and murder. With the passage of ages the Rmoahals grew shorter; some migrated to northern Atlantis where their skins became lighter, though they were twice driven back to the tropics by glacial advances. The Crô-Magnons, a stalwart race from Europe's Old Stone Age, were their direct descendants. It seems there is a minor glacial period every 30,000 years and a major glacial period every 3,000,-000; one of the latter came on during the Rmoahal period.

The next Sub-Race, the Tlavatlis, were a hardy reddish-brown people, not quite so tall as the Rmoahals, who originated in an island off the west coast of Atlantis, where Mexico is now. These migrated to the mountainous region of Atlantis, whence they eventually spread out over the continent and dispossessed the Rmoahals. Whereas the bestial Lemurians and the childish Rmoahals were incapable of self-government, the Tlavatlis had attained the point of choosing chiefs or kings by acclamation.

The next Sub-Race, the Toltecs, ushered in the great period of Atlantean glory, in the early part of the Age of Mammals. They were redskins, a mere eight feet tall and handsome. (Mrs. Besant, who liked to go Scott-Elliot one better, made them twenty-seven feet tall with bodies of rock-hard consistency). They discovered the principle of hereditary monarchy, and for thousands of years their kings ruled them wisely because they kept in touch with the supernatural Adepts, as the legendary Roman king Numa Pompilius is supposed to have obtained advice from the nymph Egeria.

Unhappily the Toltecs degenerated after 100,000 years of this splendid culture. They resorted to sorcery and phallic worship and used their great psychic powers for personal aggrandizement. "No longer submitting to the wise rule of the Initiate emperors, the followers of the 'black arts' rose in rebellion and set up a rival emperor, who after much struggle and fighting drove the white emperor from his capital, the 'City of the Golden Gates,' and established himself on his throne." The white emperor took refuge with a tributary king, while dynasties of sorcerers like the "demon king" Thevatat created and worshipped elemental spirits with bloody rites.

At this time the next Sub-Race, the Turanians, appeared and fought with the Toltecs. The newcomers were a lawless, turbulent, cruel, and brutal lot, ruthless and irresponsible individualists who to increase their population for warfare practiced complete sexual promiscuity. Their descendants, the Aztecs, carried on their tradition of cruelty. At this time also, about 800,000 years ago, a catastrophe caused much of Atlantis, including the part ruled by the sorcerers, to sink beneath the ocean, reducing it from a real continent to a large island, while islands increased in size on their way to becoming the present continents, Asia and the rest. The Turanians migrated to Asia, where on the plains of Tatary they gave rise to the more civilized and psychically gifted Mongolians, the Seventh Sub-Race. Just how these races started is left vague.

The Fifth and Sixth Sub-Races, the Semites and Akkadians, also came into being now. The former, who originated in the northern Atlantean peninsulas that now comprise Ireland and Scotland, were a discontented, quarrelsome, energetic folk living under a patriarchal social scheme and constantly raiding their neighbors, especially the law-abiding Akkadians. After another catastrophe, 200,000 years ago, reduced Atlantis to two Atlantic islands, big northern Ruta and small southern Daitya, a dynasty of Semite sorcerers ruled the City of the Golden Gates in Daitya while Toltec sorcerers reigned in Ruta.

About 100,000 years ago the Akkadians drove the Semites from Atlantis. The Akkadians, enterprising colonizers with strong legal and commercial instincts, also settled the Levant; the Basques represent them today. Then another subsidence 80,000 years ago submerged Daitya and reduced Ruta to about the size of France and Spain combined. It was this island, properly called Poseidonis, of whose final submergence Plato wrote. Banning, another occult Atlantist, tells us that when Poseidonis sank in 9564 B.C., the world assumed its present shape, but not permanently, for one of these days the continent of the next Root Race (which Banning calls by the repellant name of "Numerica") will rise from the Pacific. Then in the remote future still another continent ("Nulantis") will appear, comprising the South Atlantic Ocean and parts of the adjacent continents.

Before each of these catastrophes the initiate priests, warned by occult means, would lead a migration of the worthier sections of their people to new lands. Thus it came to pass that before the subsidence of 200,000 years ago, the Occult Lodge founded the Divine Dynasty in Egypt and built the two great pyramids at Gizeh, which later generations mistakenly attributed to Kings Khufu and Khafra of historic times. Subsequently Egypt was flooded during the submergence of Daitya, but the people, forewarned, had fled to Ethiopia whence they repopulated their land when the water subsided. Their records were also preserved, having been stored in the pyramids. The dunking of Poseidonis sent another earthquake wave over Egypt, ending the Divine Dynasty, but Egypt nonetheless recovered.

The Manu, seeing in the intellectual powers of the Semites the best future prospects for human development, led a chosen band of these folk to Central Asia where they evolved into the Aryans — the Fifth Root Race who include the modern Hindus and Europeans. Scott-Elliot drops only a dark hint about the place of the Jews in this Semite-

Aryan scheme, to the effect that they "constitute an abnormal and unnatural link between the Fourth and Fifth Root Races." The tale of migrations and intermixtures out of which came the modern racial makeup of mankind is set forth in such detail as to be impractical for us to try to follow out.

Scott-Elliot goes on to describe life in Atlantis. Under the Toltec emperors the Atlanteans were subject to a collectivistic despotism like that of the Inca Empire of Peru—which indeed was derived from Atlantis. The emperor owned everything, and ruled through a squad of viceroys, under whose guidance the peasantry practiced agriculture. The viceroys collected each crop, set aside a part for the government and another for the priesthood, and divided the rest among the masses. The system worked so well that Atlantis knew no poverty until in the days of decadence the ruling class became selfish and oppressive and the system broke down.

The Atlanteans raised wheat, brought from another planet by a Manu, and other grains such as oats which were crosses between wheat and earthly plants. The greatest feat of Atlantean agronomists was the creation of the banana. They domesticated animals resembling the modern tapir, leopard, llama, and wolf, and for meat and leather kept herds of half-wild Irish elk in parks. They ate vegetables, bread, milk, meat, and fish. They showed peculiar taste with regard to the last two items, preferring their fish rotten and what are to us the less palatable organs for meat. They also drank blood. However, the kings and priests, being true initiates, were vegetarians. Drunkenness once became so common that they adopted prohibition.

The Atlanteans practiced equality of the sexes, though bigamy was allowed and sometimes practiced. Their education was highly organized, but higher education was furnished the élite only. The masses were not even taught to read and write, but were confined to vocational training. The élite wrote on metal sheets and duplicated their writing by a process like mimeographing. As artists they were indifferent painters, fond of garish colors, but good sculptors and superb architects who built gigantic structures. A Toltec house always had an astronomical observatory at-

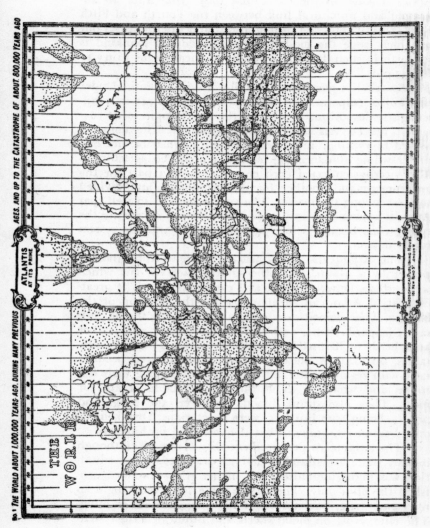

Fig. 9. THE THEOSOPHICAL ATLANTIS in its prime, according to Scott-Elliot. Atlantis occupies most of the Atlantic Ocean; Lemuria takes in southern Asia, Australia, and adjacent regions.

tached. There were no shops, since all buying and selling was done in private houses.

Their sciences were highly developed. Alchemists made precious metals in any quantity wanted; astrologers advised the viceroys on forthcoming weather conditions to enable the latter to plan farming operations. In war they fought with swords, spears, bows, and gas-bombs thrown from catapults. Their aircraft were boat-shaped structures made of plywood and light alloys, and propelled by jets of the *vril*-force invented by the Victorian novelist Bulwer-Lytton in his short novel, *The Coming Race,* wherein a race of underground supermen use this invincible and mentally directed force to blast rocks and monsters. These aircraft had a ceiling of 1000 feet (the air being less dense than it now is) and a maximum speed of 100 m.p.h. Only the rich possessed private aircraft. The emperor had a fleet of aërial warships carrying 50 to 100 men each, whose method of fighting was to play their jets on one another until one upset the other and rammed it while it was helpless. They also sailed the seas in ships propelled by a similar force.

Atlantean religion alternated between worship of the Manu and sun-worship, descending in degenerate times to diabolism and the worship of statues of themselves by rich men. The City of the Golden Gates resembled Plato's city of Atlantis, with the addition of waterworks, a guest-house where strangers were boarded free, and a population of two million.

Altogether life in the Theosophical Atlantis resembles nothing so much as life on Mars as pictured in the Martian novels of Edgar Rice Burroughs.

The schismatic Theosophical leader Rudolf Steiner added details about the psychology of the Lemurians and Atlanteans in his *Lemuria and Atlantis.* Steiner, a tall bellowing Austrian with a fine collection of university degrees, was a power in European Theosophy until in 1907 he broke with the parent organization over the issue of Mrs. Besant's deification of the young Hindu Krishnamurti. He seceded with 2400 followers and formed a new cult, the Anthroposophical Society, with headquarters at Dornach, Switzerland, where he reigned until his death in 1925.

The Lemurians, according to Steiner, possessed such feeble mental powers that while they could visualize things, they could neither remember their mental images nor use them for rational thought. They therefore had to get along on "instinct" or "indwelling spiritual force." Still, they had more control over nature than we have because of their great development of will-power, by which they could even lift great weights. Lemurian education aimed at increasing this will-power by training young Lemurians to bear pain.

Along with individual souls, the Lemurians developed the rudiments of speech towards the end of their racial history. Before they became divided into two sexes they had but little sense of sight; for a while they managed with a single eye. As long as the soul dominated the body, the soul being neuter, the body was also sexless, but at last the increasing density of matter forced sex to develop. Even so, for a long time after this interesting innovation, sexual intercourse was looked upon not as a pleasure but as a sacred duty. Lemurian women, being more spiritual than the men and much given to mystic trances, were the first to develop a sense of right and wrong.

The Atlanteans likewise could not reason or calculate, but they had the advantage over their predecessors of possessing fine memories for mental images. Atlantean education, in fact, consisted chiefly of training the memory to hold images, so that an adult Atlantean had a vast store of them available and when confronted with a problem solved it by remembering a precedent. When facing a novel situation, however, he had either to experiment blindly or to give up. Under these conditions the Atlanteans were far from progressive.

The Atlanteans were masters of the "life force" of growing things, such as a grain of wheat, and by this force they operated their aircraft, for instance. They also (especially the Rmoahals) grasped the magical power of words and used the occult energies of these words to accomplish work on the material plane. For example words could heal wounds or tame wild beasts. The Atlantean was more "natural" and instinctive in his powers and conduct than we are; he had absolute control over his physical forces, and his towns grew according to natural laws like an organism. The Toltecs achieved the power of conferring upon their descendants their individual collections of "life-pictures," thus furnishing the latter with ready-made equipment to guide their actions throughout life.

When the Semites appeared, men were losing control over the life-force and selfish individualism had become rampant. The Semites accordingly developed the power of reasoning as a counter to these changes, and along with reason a conscience. Their descendants the Aryans developed these faculties still further.

Since the time of Helena Blavatsky, Atlantis and other lost continents have become standard fixtures in bodies of Western occult doctrine. Similar syntheses of Eastern and Western occultism, complete with Atlantis, Lemuria, an eight-dimensional universe, reincarnation, pyramidology, and the like have been proffered by the learned contemporary occultist Manly P. Hall and by the late hillbilly clairvoyant-diagnostician Edgar Cayce. While an American physician-turned-occultist, W. P. Phelon, enrolled people in his Hermetic Brotherhood of Atlantis, Luxor, and Elephante (into which he said he had been inducted by initiates in Egypt) the notorious English wizard Aleister Crowley, who liked to be known as "the wickedest man in the world," included a society of Atlantean Adepts among his many gudgeon-traps.

To continue, R. Swinburne Clymer, who published the second edition of Phelon's book and who runs a Rosicrucian Foundation in Pennsylvania rivalling the better-known AMORC, assures the readers of his own voluminous works that his society preserves the original, authentic Atlantean fire-worship of which all modern creeds and sects are but corrupted descendants. According to Clymer, the Atlanteans' downfall occurred when, becoming overconfident as a result of their scientific and magical accomplishments, they tried to visit God uninvited. This rash act, by upsetting the balance of natural forces, caused the sinking of the continent. Remnants of the Atlanteans, nevertheless, still preserve the ancient wisdom in their hideouts in Yucatán.

Atlantis has of course been contacted through the spirit

world, though there is little consistency among the widely
varying accounts of Atlantean language and culture obtained from this source — in fact, about as little as among
mediumistic accounts of the spirit world itself. In the early
years of this century J. B. Leslie interviewed Atlantean
ghosts through a Spiritualist medium and published the
information thus obtained in a huge 805-page book that
included tables of Atlantean letters, numbers, and musical
notations, and there have been other revelations of this
sort as well.

Meanwhile, other occultists moved Lemuria from the
Indian Ocean (where Mme. Blavatsky at least had the
grace to leave it) several thousand miles to the Central
Pacific, where Jacolliot had located his lost land of Rutas.

Another contemporary of H. P. B., John Ballou Newbrough, after a successful career as a gold-miner, physician,
dentist, and Spiritualist medium, wrote an *ersatz* Bible
called *Oahspe*. This book is full of such extraordinary mis-
information as that Thoth founded the Muhammedan religion, and unfulfilled prophecies to the effect that all men
would soon leave their present religions to join Newbrough's pacifistic, vegetarian Jehovists. Newbrough claimed
that the book was of angelic origin and that he wrote it by
automatic writing.

Oahspe includes a map of the earth in antediluvian
times with a large triangular continent called "Pan" filling
most of the northern Pacific. The author printed a Panic
dictionary and a Panic alphabet, the latter consisting mostly
of little circles with patterns and pictures in them. His doctrines are explained in more detail by his follower Wing
Anderson, who runs a Kosmon Press in Los Angeles and
is on record as having predicted that Franklin Roosevelt
would be succeeded by a fascist dictator in 1940, that
there would be an American civil war by 1944, and that
Hitler would be swept away by domestic revolution before
he got around to attacking the United States.

Man, according to Newbrough and Anderson, appeared
72,000 years ago when angels materialized on earth and
interbred with a species of seal-like animals, the A'su, to
beget the I'hins. These in turn crossed with the A'su to

produce the Druk, who crossed with the I'hins to engender the Ghans—us. The African and Asiatic peoples are the remnants of the Druks. Anderson gives a table showing the proportion of angelic blood carried by each of these breeds, though he seems to have gotten his fractions mixed up. Pan or Mu disappeared 24,000 years ago, but is soon to rise again from the Pacific, and will be inhabited by the Kosmon race, formed by the amalgamations of all the present races. The Millenium will begin in 1980.

A writer named Frederick Spencer Oliver (not to be confused with Frederick Scott Oliver the novelist) contributed to the Pacific-Lemuria concept with a tiresome occult novel, *A Dweller on Two Planets* (1894) which he wrote under the name of "Phylos the Tibetan." In this story the narrator tells how he met his Master, a Chinese named Quong, on Mount Shasta in Northern California. Quong not only cured the narrator of his anti-Chinese prejudices but also tamed bears and pumas by a word (as St. Francis was supposed to have done with a wolf) and inducted the narrator into an order of sages who preserved the wisdom of the ancients at their Shasta headquarters.

These magi took him on a tour of the planet Venus in his spiritual body, and also taught him to remember his previous incarnations. He thus learned that when he was Zailm Numinos of Atlantis he had risen by hard work and good luck from a poor miner's son to prince of the realm, and had done very well until he got involved with two women at once. The main cities of Atlantis or Poseid were Idosa, Terna, Marzeus, Corosa, Numea, and Caiphul. The land prospered under an elective monarchy, the Rai or emperor being chosen by an aristocracy of priests (*Incala*) and scientists (*Xioqua*). The Atlanteans were of course very scientific, having aircraft (*vailxi*) and television.

This tale influenced Edgar Lucien Larkin, an elderly occultist who for some years before his death in 1924 ran the Mount Lowe Observatory in California — not to be confused with the nearby Mount Wilson Observatory. Whereas the latter is a great scientific institution, the Mount Lowe Observatory was operated as a tourist attraction by the Pacific Electric Railway in connection with their Mount

Lowe Inn. Larkin showed visitors the stars through a small telescope until in the 1930's the telescope mechanism broke down and the Inn burned.

Larkin asserted that he had spied upon the Lemurians of Mount Shasta through a telescope, learning that a thousand of them lived in a "mystic village" built around a great Maya-style temple. Occasionally, he said, they appeared in neighboring towns, clad in long white robes, polite but taciturn, to buy supplies (mostly sulphur, salt, and lard) which they paid for with gold nuggets. Every midnight they celebrated their escape from Lemuria with ceremonies that flooded the mountain with red-and-green light. However, they did not welcome visitors, and those who tried to penetrate their retreat either failed to find it or disappeared.

Eight years after Larkin's death, one Edward Lanser made a feature story of the Shasta Lemurians for the *Los Angeles Sunday Times,* wherein Lanser himself claimed to have seen the transcendental fireworks of the Lemurians from a train passing Mount Shasta. A clipping of this story excited the learned Scottish mythologist Lewis Spence to make much of it in his *Problem of Lemuria* . . . but we shall come back to Spence's theories later.

Occultists and pseudo-scientists still exploit the Shasta Lemurians and their nightgowns despite the fact that campers and state forest officials wander freely over Shasta all the time without meeting these interesting people. W. S. Cervé, for instance, credits them with aircraft and English accents. Cervé's book, published by Clymer's enemies the Rosicrucians of San Jose, combines Lemuria and Atlantis with the Jewish-Indian theory and Wegener's hypothesis of continental drift. The Mayas, it seems, were descendants of Atlanteans and Lemurians, whereas all the other Amerinds were derived from the Lost Ten Tribes. Cervé's Lemurians had a bump in the middle of their foreheads — a telepathic organ.

Finally the late Guy Warren Ballard, alias Godfré Ray King, founder of the I AM cult, claimed to have met his personal Master, Saint Germain, on Mount Shasta. Saint Germain seems to be a Mahatma remotely derived from the Comte de Saint-Germain, a slippery eighteenth-century European occultist and industrial promoter. Ballard, who graduated from selling stock in imaginary gold mines to old ladies, into occultism, picked elements of his grotesque mythology from Oliver's book, from Theosophy, from Christion Science, from Rosicrucianism, and from the Swamis, and reduced the resulting mishmash to the mental level of those comic-books whose covers show a muscular hero in a ballet-suit tearing a battleship asunder with his bare hands.

Ballard told how his Master showed him the sacred headquarters of the ruling Brotherhood in the Teton Mountains, where he saw their vast hoard of wealth and was shown magical movies of his former lives. He was delighted to learn thus that he was a reincarnation of George Washington and his wife of Joan of Arc. And of course he learned about Atlantis, Lemuria, and other vanished civilizations including one in the Sahara Desert and one (ruled by a king named Casimir Poseidon) in the Amazon River region.

For those who wish to turn Atlantism to practical account, there is or was a Lemurian Fellowship of Milwaukee, which offered correspondence courses in the "Lemurian Cosmo-Conception." The Fellowship, while following Oliver and Churchward in their concept of Lemuria, made a point of Lemuria's having been ruled by an élite minority that qualified for "citizenship" by seven years' study of occultism. The downfall of Lemurian civilization came about when this minority emigrated to China, Yucatán, and Atlantis, leaving the government in the unqualified hands of the proletariat.

Now, however, Lemuria is again rising from the Pacific. The civilization that will come into being there will be of the original Lemurian type. Its "citizens" will at the start be chosen from graduates of the correspondence courses of the Fellowship, which is even now planning the super-cities that will rise on this new continent. So, by joining the Fellowship now, you can get in on the ground floor.

These vagaries illustrate the motives and sources both of Atlantism and of occultism in general — the urge to distinction by some easy, indirect route; the assumptions of the wisdom of the ancients and the hidden hierarchy that rules the earth; the yearning for a real utopia somewhere, some time. And, sometimes, the desire to make a fast dollar.

Such doctrines are the product of the occult animus rein-
forced by a fertile imagination, wide if indiscriminate
reading, and disregard of logic, experience, and factual evidence. Not even much imagination is needed, since the
volume of occult literature is enormous, even if not familiar
to the average reader. One can easily concoct an esoteric
doctrine with Atlantist features by lifting sections from
the works of his predecessors, as Mme. Blavatsky and Scott-
Elliot used Donnelly.

Occult Atlantism shows the double attitude of occultists towards science. On one hand they would like to cut in
on the enormous prestige of science, and hence quote whatever scientific or near-scientific dicta they think will support
their ideas. They say the Mayas' "structures are scientifically
admitted to be older than those of Egypt," and "the great
scientific explorations have revealed that there is such a
continent" as a Pacific Lemuria. (Science admits nothing
of the sort.) Or they quote people like John W. Keely, a
nineteenth-century perpetual-motion crank, as "distinguished scientists," or cite obsolete science, or refer to some
authentic scientist who has gone off the deep end in a field
wherein he is not expert, as Piazzi Smith did in Egyptology.
No matter; followers of cults make no distinction.

On the other hand, since the scientific method destroys
occult doctrines, occultists try to meet this difficulty by a
frank rejection of the usual standards of knowledge. Material evidence, they say, is worthless; Truth is to be
dredged out of the inner consciousness by mystical intro-
spection. These methods are "more scientific than even
modern science," whose "great physiological discoveries"
are "no better than cobwebs, spun by her scientific fantasies
and illusions," and "little reliance can be placed on such
external evidences." On the other hand, occultism is "Gno-
sis," based upon infallible "Akashic Records," and those
who disagree with it are merely "well meaning, but grevi-
ously [*sic*] misinformed Individuals who have based their
theories upon superficial archeological observation and
theological speculation, not upon INHERENT KNOWL-
EDGE, CLAIRVOYANT VISION, or DIVINE REVELA-
TION."

If you insist upon evidence, there are plenty of records

of lost continents, but these are concealed in the Fourth
Dimension, or in secret underground libraries in Tibet open
to qualified occultists only.

The occultists, in fact, live in that dream-world of
early adolescence, where a boy becomes a pirate chief and
a girl Marie Antoinette by a pure act of imagination. The
romantic settings in which these beings exist are created,
complete with colorful if inaccurate detail, by the child-
hood faculty of conjuring up solid-looking eidetic images.
But every private paradise of that sort requires some sacrifice
from those who enter; in the case of the imaginary worlds
of the occult, the abandonment of reason. And without
reason nobody has yet figured a way to find that hair that,
perhaps, divides the false and true.

So, if we really want to get to the bottom of the lost-
continent problem, we must leave the occultists to their
dream-worlds — very pretty, but not for us.

CHAPTER IV

THE HUNTING OF THE COGNATE

In some green island of the sea,
Where now the shadowy coral grows
In pride and pomp and empery
The courts of old Atlantis rose.

In many a glittering house of glass
The Atlanteans wandered there;
The paleness of their faces was
Like ivory, so pale they were.*
Masefield

ATLANTISM ranges from the doctrines of the occultists at one extreme, through various pseudo-scientific or semi-scientific speculations, to the efforts of *bona fide* historians and scientists to find a basis for Plato's story at the other. While the occultists, with their Atlantean wizards and bisexual Lemurians, offer much more colorful accounts than Plato, they are hardly to be taken seriously by reasonable people. Arguing with them is rather like wrestling with the giant jellyfish *Cyanea*: the substance is too soft and slippery to grasp, and there is not even a brain to stun.

I therefore turn from them with some relief to the Atlantists who claim to be scientific in their approach. For, if one claims the dignity of the mantle of science, one must expect to be judged by its criteria, just as the Amerind who claimed the status of a warrior by braiding his hair in a scalp-lock accepted the risk that another brave would take up both the implied challenge and the scalp that went with it.

---

* From John Masefield: *Story of a Roundhouse*, copyright 1912 by the Macmillan Co. and used with their permission.

Outside the magical fraternities, Atlantism is a fairly small but durable cult. There once flourished, for instance, an Atlantean Research Society with headquarters in New Jersey, and in the 1930's a similar Danish group issued Atlantean stamps and money. About the same time somebody approached the late George W. Vaillant, a leading expert on the Aztecs, for help in organizing an expedition to seek Atlantis at the bottom of the sea with a diving-bell. And at the moment an Atlantis Research Centre is issuing publications in London.

One of the most active of such groups, the French *Société d'études atlantéens,* was formed in 1926 with Roger Dévigne as president and Paul le Cour as secretary. The club soon broke into a conservative faction under Dévigne who tried to keep the proceedings scientific, and a radical faction under le Cour, who seceded, set up his own *Amis d'Atlantis,* and picked quarrels with non-Atlantist French scholars like the archeologist Reinach. For some years le Cour published *Atlantis,* a bi-monthly magazine largely devoted to the occult. His group went on *piqu-niques atlantéens* talking a strange jargon of their own and wearing Atlantist emblems in their buttonholes, and finally broke up a meeting of their rivals at the Sorbonne in 1927 by throwing tear-gas bombs into a discussion of ancient Corsica.

Atlantism is active enough to support the publication of books that appear every few years exploiting and expanding the hypotheses of Donnelly and his kind. In a recent and typical example, Braghine's *Shadow of Atlantis,* the author starts off appropriately with a frontispiece that is a photograph of a pair of forged antiques, passed off as Atlantean relics. Then the author, though he repudiates the occult inspirations of the Theosophists, goes on to make a fantastic series of flat and ridiculous misstatements of fact: that all Mayan statues have beards, that the Breton language is unconnected with other European tongues, that the extinct aurochs or European wild ox was the same as the American bison, that camels were never native to North America, that the remains of elephants have not been found in America north of Florida, that there were no lions in Classical Greece, that the Otomis speak ancient Japanese, that the biographer of the famous wizard Apollonios of

Tyana was "Theophrastes," that the great pyramids of Gizeh and Teohtihuacán were built many thousands of years B.C. . . . and so on; mistakes of the sort that would be cleared up by a little selected reading and a visit to a couple of good museums. You believe Colonel Braghine at your peril.

Much of the book is devoted to the alleged cryptic inscriptions and the lost Atlantean cities of Brazil. Now, the treasure-cities of the Brazilian wilds have been one of the leading will-o'-the-wisps of speculative paleology ever since the *conquistadores* wandered the length and breadth of this vast land seeking in vain for El Dorado. The British Atlantist Wilkins wrote a whole book on these supposed relics of a higher civilization. Half a century ago the Krupps of Essen spent a half-million dollars of their steel fortune on an expedition to the Matto Grosso to look for such a city.

Then in 1925 a retired British army officer, Lt.-Col. Percy H. Fawcett, set out upon a similar errand with his twenty-year-old son and another young man named Rimell and vanished. Fawcett, an experienced explorer and a forceful character with a mystical turn of mind, thought that the Brazilian jungle concealed the ruins either of Atlantis itself or of one of its daughter cities. Several other parties went in search of Fawcett without success during the following decade. But his disappearance remained a mystery until in 1951 Orlando Vilas Boas, an Indian agent of the Brazilian government, coaxed the Calapalo Indians of the upper Xingú basin into admitting that they had killed Fawcett and his companions because they did not like the way Fawcett had treated them.

More recently a French gold-prospector, Apollinaire Frot, spread seductive rumors of his archeological discoveries in Brazil, declaring with typical pseudo-scientific fervor that the results of his investigations were so striking that he feared to publish them. He died with his findings unpublished, for like sea-serpents, psychic phenomena, and Schliemann's owl-headed vase, the Brazilian Atlantis seems to withdraw and vanish at the approach of qualified scientific investigation.

But to return to Atlantist books: An even odder specimen in this garden of queer literary flowers was the *Atlantis: die Urheimat der Arier* (*Atlantis: the Original Home of the Aryans*, 1922) of Karl Georg Zschaetzsch, who asserted that the original Aryans, blond and virtuous vegetarians and teetotalers, were the people of Atlantis. Like Carli before him and Velikovsky after, Zschaetzsch claimed that Atlantis was destroyed by the collision of the earth with a comet. The only survivors of the sinking were Wotan, his daughter, and his pregnant sister, who took refuge in a cave among the roots of a giant tree beside a cold geyser. Wotan's sister died in childbirth and a she-wolf suckled her infant. The blood of these noble Nordics became mixed with that of the non-Aryans of the mainland, and their degenerate descendants resorted to meat-eating and to the loathsome practice of fermenting liquors, invented by a non-Aryan girl witch named Heid (like Wotan, from Norse mythology).

By "racial memory" the author "proved" that these events underlay the legends of the Christmas star, the world-ash Yggdrasill and the spring of Mimir beside it, Romulus and Remus, and almost any other myths you care to mention — thus Eve's Edenic apple was "really" Heid's applejack. The Greeks were an Atlantean colony and Zeus one of their early chiefs. All of which went to prove that Herr Zschaetzsch was himself a descendant of Zeus, as you can easily see from his name!

This sort of fantastry, so common in Atlantism, explains why the dispositions of historians and scientists tend to come unglued at the mere mention of Atlantis. Part of their phobia is derived from their experiences in giving popular lectures, at the end of which somebody is sure to stand up and start an argument on Atlantis, waving a copy of Churchward and asserting that it must be so because it says so right here in print.

However, if it is possible to study such emotionally loaded subjects as sex and religion in a detached and scientific manner, we should be able to investigate the Atlantis question in the same spirit.

You can treat the problem of Atlantis in any of several ways. You can, like the occultists, swallow Plato's story whole and expand it by inspired extrapolation. Or you can take it as mainly true, but try to rationalize it by leaving out the supernatural elements. Or you can search for a real ancient culture corresponding to Plato's confederacy, though not necessarily on a sunken island. Or you can investigate the possibility of Atlantic islands and land-bridges from the point of view of geology or biology. Or, finally, you can study Plato's dialogues simply as fiction, and look for their sources as you look for the inspirations of any famous piece of fiction.

Since few people nowadays would believe in the god Poseidon and his ten sons, some rationalization is necessary in any treatment of Atlantis. Donnelly provides an example of the rationalist group. But once you start altering the story in the interests of plausibility or drama there is no telling where you will stop. As the acute T. H. Martin said: "many scholars, embarking upon the search for Atlantis with a more or less heavy cargo of erudition, but with no compass other than their imagination and caprice, have voyaged at randon. And where have they arrived? In Africa, in America, in Australia, in Spitsbergen, in Sweden, in Sardinia, in Palestine, in Athens, in Persia, and in Ceylon, they say."

As a result Atlantis has been identified with all sorts of places (see Appendix C for a complete list) and its submersion with all sorts of catastrophes. It didn't happen when Plato said it did, they say, but thousands of years earlier or later. It wasn't in the Atlantic, but in the Arctic, or South Africa, or New Zealand. It didn't sink, but was merely damaged by a hurricane; Poseidon was not a god but a mortal king. Its civilization was not of the Classical type described by Plato, but either a super-scientific machine-age civilization or a high-grade stone-age culture like that of the early Mayas. And so on.

Now, while some of these points may be well taken, you cannot change all the details of Plato's story and still claim to have Plato's story. That is like saying the legendary King Arthur is "really" Queen Cleopatra; all you have to do is to change Cleopatra's sex, nationality, period, temperament, moral character, and other details and the resemblance becomes obvious.

Moreover, while lost civilizations and submerged lands are perhaps possible, there is no compelling reason for connecting them with Plato, who after all knew nothing about what happened in the Miocene Period and had never heard of New Zealand. While it is hard to determine the exact limits of the knowledge of an antique sage like Plato, who is no longer here to be quizzed, we can be as sure of the foregoing as we can be of anything.

And Plato might, you know, have hit upon a description of real events, without knowing what had actually happened, by sheer chance. It has often happened in the history of human knowledge, as when Demokritos hit upon the theories of the atom and of the evolution of life by inspired guess. In fact the French anthropologists Mortillet and Verneau took just that view of Atlantis: while (contrary to most geologists) they believed in a North Atlantic land-bridge persisting down to the Pleistocene, they did not think this land-bridge had anything to do with Plato's story.

Therefore, let us beware of enthusiastic rationalizers of Plato's tales and other ancient myths, since such people, starting with the same story, come up with scores of different interpretations, all of which can obviously not be true at the same time.

Non-occult Atlantist theories fall roughly into two groups: those that call for a great catastrophe, a subsidence or flooding, and those that do not. Let us take the former group first.

In 1803 Bory de Saint-Vincent published his *Essai sur les îles fortunées et l'antique Atlantide,* wherein he set forth the conventional Atlantis story. Like Donnelly later, he assumed that Atlantis was much more than Plato said—the original home of civilization, in fact. There had been two subsidences. The first, the result of volcanic action in the Mediterranean, reduced the island so that the survivors had to become aggressive imperialists to find living-space. After their defeat by the Athenians, the rest of their continent sank also, leaving only the islands of the Canaries, Madeiras, and Azores as remnants.

In a plausible-looking map Saint-Vincent showed Atlantis in its heyday. It was located immediately west of the bulging part of West Africa, where the Sahara and the Sudan run down to the ocean; in fact almost touching the African coast at Cape Bojador and again at Cape Verde. The sea inclosed between Africa and Atlantis and these two straits he called the "Lac Tritonide," that is, Lake Tritonis, which plays a part in the story of the Argonauts and in Diodoros's account of the African Amazons. The northeast corner of Atlantis he labelled "Hespéride ou Pays des Atlantes"; the southeast, "Pays des Gorgones"; and the coastal region of Atlantis between "Pays des Amazones." The present Canary Islands mark the site of the original Atlas range, where lay the Gardens of the Hesperides, the snow-topped volcanic peak of Teyde on the island of Tenerife being the original Mount Atlas: a plausible idea, for the steaming 12,000-foot Pico de Teyde does answer the Classical descriptions of Mount Atlas much better than do the mountains of Morocco.

Saint-Vincent got some of his ideas from a French botanist, J. P. de Tournefort, who about a century before suggested that Atlantis had been washed away by the out-rush of water from the Mediterranean when, as a result of an earthquake, the level of that sea rose above that of the Atlantic, and the outflowing water carved the Isthmus of Gibraltar into a strait. Tournefort in turn derived this idea from Strato of Lampsakos, who back in Hellenistic times asserted that the Black Sea had become connected with the Mediterranean by overflowing into it and the Mediterranean had done likewise with the Atlantic.

Others such as the naturalist Buffon, a contemporary of Saint-Vincent, thought that Atlantis had been washed away by water flowing in the opposite direction, from the Atlantic into the Mediterranean, which theretofore had been mostly dry land. Modern geology tends to support this view. During the last advance of Pleistocene ice, say some geologists, the Mediterranean was a great low plain with a pair of large lakes separated by the ridge connecting Italy and Sicily with Africa. During the warm interglacial periods apelike men hunted pigmy elephants and hippopotami in Sicily. With the final melting of the ice-caps the ocean rose until, about 15,000 years ago, it broke through the Isthmus of Gibraltar and filled the Mediterranean basin to its present level, forcing thousands of men and millions of animals to flee. A catastrophe? The origin of Plato's story, with the Mediterranean Valley playing the rôle of Atlantis? Suggested but not proved.

As the solution to the Atlantis problem, the theory has objections. The geologists do not agree: Déperet, for instance, thinks the Ibero-African connection broke down before the Pleistocene. The Gibraltar region is built of limestone and shale, and while these rocks are not the most resistant, they are much more so than mere soil. Therefore the isthmus would not give way all at once like the St. Francis dam in Los Angeles, but would erode slowly as the salt river from the ocean cut its way into the rock. And when you compare the size of the tub with that of the faucet, you can see that filling up the Mediterranean must have taken many years. Moreover, if the sea did flood the basin thus, the dispossessed people were not Plato's cultivated Atlanteans, but skinclad Neolithic hunters, and there is no evidence that preliterate men can hand down a truthful account of a historical event by oral tradition alone for 10,000 years, or even a tenth of that period.

Other suggestions have called for even more drastic map-changes. For instance in 1926 Claudius Roux argued that in post-glacial times most of North Africa was under the sea, and the mountains of Morocco and Algeria constituted a peninsula on which Atlantis throve. Eventually the land rose, or the sea fell, and the seas and lagoons dried up leaving the present deserts and salt marshes.

Roux tried to buttress his theory by citing the Glozel finds, a group of pebbles and clay tablets inscribed with marks vaguely like writing, found on a farm near Vichy, France, in 1926 in association with paleolithic objects. Then it transpired that H. C. Rogers, a professional faker of antiques, had made the Glozel objects and planted them on the farm. Before the hoax was exposed it had fooled even the distinguished Reinach, and had led to eccentric theories of European prehistory, a libel suit, and a fine riot at a meeting of the College of France.

The American engineer R. A. Fessenden offered an even more spectacular theory of map-changes; like the eighteenth-century French Atlantist Delisle de Sales before him, he located Atlantis in the Caucasus. De Sales said the Atlanteans migrated from the Caucasus, then one of the few islands in the world-covering ocean, to the true Atlantis in the middle of the Mediterranean, including much of Italy and North Africa.

Fessenden began with sweeping assertions to the effect that "Primitive man was very literal minded," with no imagination, which is not necessarily so. He tried to show that when Greek myths spoke of events in the "Mediterranean" they really meant the Black Sea, and that the "Atlantic Ocean" of the myths was really a great shallow sea, the Asiatic Mediterranean or Sarmatian Ocean, which once covered most of Russia. This body of water Fessenden called the "Ocean of Atlantis."

Fessenden argued that he observed a curious gap in the Greek myths: they said nothing about the western half of the Mediterranean. His explanation was that a whole block of place-names had been displaced from regions east of the Black Sea, where they belonged, to the Atlantic region. Consider, for instance, that the name "Iberia" was applied both to the Spanish peninsula and to a region in the Caucasus. Mankind, he concluded, evolved in the Caucasus, built a high civilization there which was wrecked by a great flood, and then spread in all directions. Subsequently the land rose and the Sarmatian Sea dried up, leaving the Black, Caspian, and Aral Seas as remnants. While his argument was based on dubious geology, bad mythology, impossible linguistics, and Aryan-race nonsense, it was plausible enough for Wells to give it a friendly nod in his celebrated *Outline of History*.

The Orientalist Joseph Karst exploited the idea of an eastern Atlantis in an even more spectacular manner, arguing for two Atlantides, one in northwest Africa and one in the Arabian Sea. There was, he said, an even earlier center of cultural radiation in central Asia, which Karst, like Fessenden, believed to have been submerged down almost to historic times. The eastern Atlantis is Homer's Ogygia; it is also the land of Kangha in the Iranian poet Firdausi's *Book of Kings,* and the Peshotanu who rules Kangha is Plato's Poseidon.

Karst based his arguments mainly on linguistic

grounds: fantastic linguistics that classified languages into nonsensical groups like "Uralo-Basque" which included such unrelated tongues as Dakota, Chinese, and English. An "Ibero-Ethiopian race" invented civilization in Ogygia-Atlantis, which sent forth three waves of migrants before it sank altogether: the first including the Chinese, the Sumerians, and the Otomis; the second, the Mayas and Kichés of Central America; the third, the Aztecs and the peoples of the Caucasus.

The latest version of the eastern-Atlantis theory is that of the tame scientists of the U.S.S.R. Having in recent years claimed, with magnificent disregard of fact, all the world's inventions and scientific discoveries for Russia, they have at last arrogated Atlantis for that country. It existed, they say, in the area now covered by the Caspian Sea.

Other Atlantists have advanced theories, like those of Karst, of two Atlantides, one or both of which has been submerged. Thus Duvillé proposed two centers of Atlantean civilization, one on the conventional Atlantic island and one in northeast Africa, on the site of the present state of Ethiopia, while René-Maurice Gattefossé claimed that civilization was born on an Arctic continent which he called by the old name of "Hyperborea." When this land foundered, said Gattefossé, its people spread to the North Atlantic Atlantis and to other continents, erecting megalithic monuments like that of Stonehenge as they went. The Crô-Magnon men of Paleolithic Europe (who, like the Mayas, are nearly always dragged into Atlantist arguments) were pure Hyperboreans. Afterwards the Atlanteans developed the so-called Cyclopean style of architecture, involving the use of large irregular blocks without mortar, found in Bronze Age ruins in Europe.

According to Gattefossé, the "War of the Titans" of Greek myth is a recollection of periodic changes in the earth's axis, causing world-wide alterations in climates and coast-lines. Zeus, Poseidon, Atlas, and Hesperos were all Atlantean kings. Under Poseidon the Atlanteans learned to domesticate animals, and fought the Gorgons and Amazons who had a matriarchal government of which traces are still found among their descendants the Berbers. The author tries to trace the migrations of all branches of the human stem by skull-shapes, assuming all long-headed people to be closely related, and likewise all broad-headed folk. This is absurd, since the cranial index is only one of many characteristics to be considered in classifying people, and used by itself leads to such curious results as lumping the Swedes, Eskimos, and Hottentots together.

The idea of a far-northern Atlantis had been advanced by Rudbeck in the seventeenth century and by Bailly in the eighteenth. Hermann Wirth more recently put his sunken continent in the Arctic, calling it "Thulë" after the island Pytheas heard of in his travels. Like Churchward, Wirth thought he had the key to the profoundly psychological sacred symbolism of primitive man, so that by tracing the symbols of primitives the world over he could reconstruct the pre-history of man. For instance a pair of circles one above the other, connected by a short line, represents the year. Wirth believed that the last survivors of his arctic civilization were the new extinct Sadlermiut Eskimos, descendants of the Thuleans who flourished between 25,000 and 12,000 B.C., contemporary with the Crô-Magnon men; and that their culture, while high, was non-metallic. They spread to Europe, Asia, and the Americas, splitting into the present racial types as they went, and even migrating as far as New Zealand.

While most Atlantists are vague as to *why* continents sink, one of them has more than accounted for the phenomenon: Hanns Hörbiger, author of the Cosmic Ice Theory, the *Welteislehre* or WEL as it came to be known. Hörbiger was an Austrian inventor, a gadgeteer who bragged that he never calculated anything. When people doubted him he would shout:

"Instead of trusting me you trust equations! How long will you need to learn that mathematics is valueless and deceptive?"

Hörbiger told his followers that as a boy, while looking through a telescope at the moon and the planets, he had suddenly realized that they were made of ice. Furthermore, a dream revealed to him that the sun's gravity extends out only three times as far as Neptune. As a young engineer the sight of waterlogged soil exploding with puffs of steam as molten iron ran over it gave him the final clue for building a cosmos almost as colorful as Mme. Blavatsky's and even more alarming.

The universe, it seems, is a sort of cosmic steam-engine. Interstellar space is full of rarefied hydrogen and water-vapor. (True to a small extent.) When a small star, moving through this damp mixture, loses its heat, it picks up ice. When a larger star pulls it into itself it stays buried in the large star, a vast ball of ice, for millions of years while the ice melts. Finally the small star turns to steam, the resulting explosion throwing pieces of star-material into space. These condense into planets while the oxygen of the giant star, also blown out, unites with the hydrogen of space to make a ring of ice-particles ("bolides") outside the planets. The Milky Way is not, as astronomers assert, the farther stars of our galaxy, but the bolides of our Solar System.

As the hydrogen of space slows them up, these bodies all spiral inwards towards the sun. Because the smaller particles are retarded the most they spiral in most rapidly, passing the larger ones. Thus as the ice-blocks of the Milky Way fall towards the sun, most of them are picked up by the gravitational fields of the outer planets, wherefore the latter and even our own moon have acquired ice-coats hundreds of miles thick.

When an ice-block reaches the sun it causes another explosion which fills the space around the sun with fine ice-particles. The inner planets Venus and Mercury, moving through these, have also become sheathed in ice; the earth is luckily just far enough from the sun to avoid icing. Large planets catch smaller ones spiralling past them and make moons of them, eventually drawing them into themselves.

Both the capture of a satellite and its subsequent fall to the surface of a planet impose great stresses upon the planet. Its gravitational pull causes floods and earthquakes until the moon settles into a stable orbit. Then all is serene until the satellite, retarded by the hydrogen of space, nears its primary, whereupon its gravity draws the oceans into a belt or bulge around the equator, drowning the equatorial regions but leaving the polar lands high and dry.

Finally when the satellite approaches within a few thousand miles, gravitational forces break it up. The fragments shower down on earth; the oceans, released from the satellite's pull, flow back towards the poles, exposing tropical lands and submerging polar territories. All human catastrophe-myths can be fitted into this scheme somewhere, be they the Biblical Flood, Atlantis, the Revelation of St. John, or the Norse Ragnarök.

The earth has had at least six moons before the present one, captured and destroyed in the Algonkian, Archeozoic, Cambrian, Devonian, Permian, Mesozoic, and Cenozoic Eras. Since man was civilized in the Cenozoic or Tertiary Era, he remembers the fall of the Cenozoic moon, whose approach submerged all the tropical lands save a few highlands like Peru and Ethiopia 250,000 years ago. The capture of the present moon, the ex-planet Luna, submerged Atlantis and Lemuria. When Ezekiel speaks of "Tyre" or the author of *Revelation* of "Babylon," they mean Atlantis. The breakup and fall of Luna will probably end life on earth.

Hörbiger sold this doctrine to a German schoolteacher and amateur astronomer named Fauth, who in 1913 published Hörbiger's book *Glazial-Kosmogonie*. Although World War I interrupted the progress of the cult, Hörbiger appeared after the war with a long white beard and a publicity-machine that flooded Europe with books, pamphlets, and threatening letters. Soon his followers numbered millions, the more fanatical of whom interrupted meetings of learned societies with yells of:

"Out with astronomical orthodoxy! Give us Hörbiger!"

Hörbiger wrote one correspondent: "Either you believe in me and learn, or you must be treated as an enemy." Another Austrian, Hans Bellamy, was converted to Hörbigerism by recollecting that as a boy he had suffered a recurrent nightmare about an oversized exploding moon, and popularized the Cosmic Ice Doctrine in English.

Although Hörbiger died in 1931, his movement continued active for years. With the rise of Hitler the Hörbigerites identified themselves with the racial philosophy of the new rulers: "Our Nordic ancestors grew strong in ice and snow; belief in the World Ice is consequently the natural heritage of Nordic Man." But the Nazis, who bought so many queer ideas, remained unresponsive to this one.

Unfortunately for Hörbiger's theory the planets are not made of ice (though some probably do have ice-coats); if they were they would look and act differently. (When a critic pointed out that Saturn's density is *less* than that of ice Hörbiger muttered something about a "gravitational shadow," which he never explained.) Nor is the Milky Way composed of "bolides," as telescopes have long since proved; nor do stars and planets work the way he said. The moon is receding from us instead of approaching. In fact, calculations show that all Hörbiger's assertions are flatly impossible in the light of all present knowledge. But then, he always claimed one could prove nothing by equations.

An even madder theory of periodical catastrophes was brought out more recently by Immanuel Velikovsky, a Russo-Israeli physician and amateur cosmogonist whose publishers stirred up an extraordinary hoopla in 1950 to sell his book. The theory is somewhat like those put forth in the eighteenth century by the speculative mythologist Gian Rinaldo Carli, who thought that Atlantis had sunk when a comet struck it in 4000 B.C., and in the early twentieth century by Zschaetzsch as related *supra*.

According to Velikovsky, many centuries ago, as a result of a planetary collision, Jupiter erupted and spat out a comet. This comet, travelling in an eccentric orbit, skimmed the earth about 1600 or 1500 B.C., close enough to cause a worldwide catastrophe with earthquakes, tidal waves, and all the rest. This grazing contact is remembered as the opening of the Red Sea in *Exodus* and other mythological convulsions.

Fifty-two years later this comet grazed the earth again, which debacle is remembered as Joshua's feat of commanding the sun to stand still. These approaches of the comet not only stopped the earth's rotation but started it spinning again in the opposite direction, changed the position of its poles and their inclination to the ecliptic, and changed the earth's orbit so that the year was lengthened from 360 to

365 days. The gaseous hydrocarbons of the comet's tail showered down, partly as petroleum (whence modern oil-wells) and partly as the sugary *manna* eaten by the Hebrews in Sinai.

Having lost most of its tail, the comet went on the graze Mars which, jostled out of its course, also made destructively close approaches to the earth in the eighth and seventh centuries B.C. before settling into its present orbit. The comet meanwhile settled down to become—the planet Venus!

To buttress this startling proposition Velikovsky has dug up mountains of ancient lore: the Babylonian 360-day calendar; Hindu references to the "four planets"; and an immense mass of quotations from the myths of the Hebrews, Egyptians, Norse, Aztecs, and others. When they don't fit he gives them the Prokroustes treatment: As Plato's sinking of Atlantis is dated 9000 years before Solon's trip to Egypt (about 590 B.C.) Velikovsky fits the yarn into his framework by knocking a zero off Plato's figure and dunking Atlantis 900 years before Solon.

Why, if these events happened in historic times, don't the records describe them in clear language? Because, says Velikovsky, the human race suffered "collective amnesia." Mankind's subconscious was so terrified by these occurrances that it decided to forget all about them!

Despite the impressive build-up, however, Velikovsky neither establishes a case nor accounts for the success of the Copernicus-Newton-Einstein picture of the cosmos which he undertook to supersede. Some of his mythological references are wrong (for instance he uses the Brasseur "translation" of the Troano Codex); the rest merely demonstrate once again that the corpus of recorded myth is so vast that you can find mythological allusions to back up any cosmological speculation you please. The Babylonians left clear records of observations of Venus 5000 years ago, behaving just as it does now. References to a "four-planet system" probably mean that early astronomers overlooked Mercury, not a conspicuous object even at maximum elongation. And the 360-day year was simply a case of priestly inaccuracy, gradually corrected as knowledge grew.

Moreover the theory is ridiculous from the point of

view of physics and mechanics. Comets are not planets and do not evolve into planets; instead they are loose aggregations of meteors with total masses less than a millionth that of the earth. Such a mass — about that of an ordinary mountain — could perhaps devastate several counties or a small state if it struck, but could not appreciably affect the earth's orbit, rotation, inclination, or other components of movement. If a mass big enough to reverse the earth's rotation hit the earth, the energy of the impact would completely melt both bodies; if it didn't hit it wouldn't change the earth's rotation enough to matter. And the gas of which a comet's tail is composed is so attenuated that if the tail of a good-sized comet were compressed to the density of iron, I could put the whole thing in my briefcase!

As for "collective amnesia," that is like saying you're being followed by a little green man whom you can't see because he vanishes every time you turn your head to look for him. You can't very well disprove such an assertion, but neither are you obliged to take it seriously.

You may wonder if I have been selecting samples from the lunatic fringe of Atlantism in order to discredit all Atlantist theories. I assure you I have not, because the Hörbigers and Churchwards constitute the main body of Atlantism, while the "fringe" consists of the few sane and sober writers on the subject.

Among these is (or was) the Scottish mythologist Lewis Spence, who once edited an *Atlantis Quarterly* which ran for five issues before succumbing to monetary malnutrition. Between thirty and forty years ago Spence published a number of popular books on the myths of Mexico, Egypt, and other lands, wherein he showed sound skepticism towards the more eccentric conceptions of human prehistory, like those of Le Plongeon.

Spence's *The Problem of Atlantis* (1924) while less critically acute than his earlier books, is still about the best pro-Atlantis work published to date. The author undertook to prove:

"(1) That a great continent formerly occupied the whole or major portion of the North Atlantic region, and a considerable portion of its southern basin. Of early

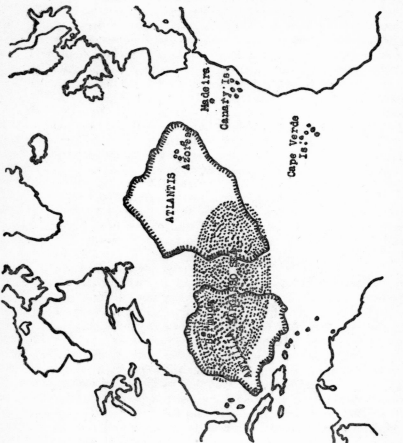

Fig. 10. MAP OF THE ATLANTIC OCEAN, showing Atlantis and Antillia (according to the theories of Lewis Spence) and the Sargasso Sea.

geological origin, it must, in the course of successive ages, have experienced many changes in contour and mass, probably undergoing frequent submergence and emergence.

"(2) That in the Miocene (Late Tertiary) times it still retained its continental character, but towards the end of that period it began to disintegrate, owing to successive volcanic and other causes.

"(3) That this disintegration resulted in the formation of greater and lesser insular masses. Two of these, considerably larger in area than any of the others, were situated (a) at a relatively short distance from the entrance to the Mediterranean; and (b) in the region of the present West India Islands. These may respectively be called Atlantis and Antillia. Communication between them was possible by an insular chain.

"(4) That these two island-continents and the connecting chain of islands persisted until late Pleistocene times, at which epoch (about 25,000 years ago, or the beginning of the post-glacial epoch) Atlantis seems to have experienced further disintegration. Final disaster appears to have overtaken Atlantis about 10,000 B.C. Antillia, on the other hand, seems to have survived until a much more recent period, and still persists fragmentally in the Antillean group, or West Indian Islands."

Spence rationalizes away most of the historically or scientifically improbable parts of Plato's account. Thus, he says, Atlantis need not have disappeared in twenty-four hours of quakes and storms, but as a result of a series of earthquakes over many years. Plato's Atheno-Atlantean war is probably a fiction based upon some local Greek war. Plato's metallurgy is unsound, metals not having come into use until thousands of years after the disappearance of Atlantis. Still, even if the Atlanteans were a stone-age people and not the source of all civilization, they had a considerable culture.

For examples of this culture Spence refers to the Crô-Magnon race of stone-age Europe, stalwart cave-men whose remains are found mostly in southwestern France and northern Spain. The Crô-Magnon people, named for the village near which the French archeologist Lartet first found their relics, were remarkable in two respects. First, they belonged to a physically impressive type of the White Race; the men averaged over six feet, with big brains, massive chins, wide cheek-bones, beak noses, beetling brows, and high foreheads, so that they looked a little like Norwegians and a little like Blackfoot Indians. Their skulls showed the unusual combination of a very long cranium with a very broad face. Second, they drew and painted astonishingly realistic pictures of game animals on the walls of their caves, and carved them on their tools and utensils; the cave-pictures being a kind of sympathetic magic to assure success in hunting.

During the final retreat of the glaciers about 25,000 years ago, the Crô-Magnon men replaced, probably in no gentle manner, their brutish Neanderthal predecessors. The Crô-Magnon period is divided into three stages, based upon types of archeological remains: the Aurgnacian, Solutrean, and Magdalenian. Most of the Crô-Magnon art belongs to the first and last of these. Other races invaded Europe during Crô-Magnon times also: a Negroid people in the Riviera and a Mongoloid People in Central Europe. After the Magdalenian stage, the Azilian-Tardenoisian peoples overran Europe: crude artists but makers of bows and fine stone fishhooks, who, again probably for magical reasons, painted multitudes of pebbles with spots, stripes, and simple patterns.

From their distribution around the shores of the Bay of Biscay, Spence inferred that the Crô-Magnons must have come from the west, especially since their remains had not been found farther east. He thought there were three waves of Atlantean invasion: the Aurignacian, the Magdalenian, and the Azilian, and cited respectable authorities like Henry Fairfield Osborn for the supposition that both the unique Basque language and the remarkable Berber-speaking stone-age Guanches, who inhabited the Canaries when Spain annexed them in the fifteenth century, were relics of the Crô-Magnons. (Some anthropologists in fact claim to see traces of Crô-Magnon blood in the people of the Dordogne Valley of France, who are tallish, rugged-looking, and brown-skinned, with abundant black hair and the peculiar Crô-Magnon skull-form.)

The Azilians, said Spence, arrived about 10,000 years

ago, about the time given by Plato for the sinking of Atlantis. Significantly they were buried facing west. They probably founded the civilizations of Egypt and Crete, the later Iberians being their descendants. The type of town planning they developed in Atlantis was reflected in the later plans of Carthage and Knossos. The Osirian religion of Egypt was developed by them, as is shown by its numerous parallels in Western European folklore; for instance Druidism and the story of King Arthur contain Osirian elements. Keltic legends of unattainable islands and submerged cities all point to Atlantis.

Like most Atlantists, Spence derived the Mayan culture from Atlantis. He got around the usual chronological difficulty (the fact that the Mayan civilization only appeared a little before the Christian Era) by supposing that the westbound Atlantean refugees stopped over in Antillia for some thousands of years, and only moved on to Yucatán when the latter island sank in its turn. He corroborates this migration by various Amerind myths, such as the Aztec legend of Quetzalcoatl, the ruddy, bearded foreigner who was said to have led a band of black-robed civilizers about Mexico spreading enlightenment until a pair of sorcerers fed him a homesickness-potion that sent him hurrying across the sea on a raft of snakes. Spence thought Mayan culture must have evolved elsewhere because "when the Maya civilization first manifests itself upon Central American soil, it does so not in any simple or elementary manner, but as a full-blown culture, with a well-defined art, architecture, religion, and system of hieroglyphic writing."

Spence also cited geologists like Gregory, Scharff, Hill, Neumayr, Suess, and Termier in support of the theory of an Atlantic continent or a transatlantic land-bridge lasting till comparatively recent times—until the later Pleistocene anyway. And he mentioned as significant the curious habit of the lemmings of Norway in swimming out to sea in great masses, as if looking for a land that is no longer there; failing to find it they circle aimlessly until they drown.

In subsequent books, *Atlantis in America, The History of Atlantis,* and *The Problem of Lemuria,* Spence added details to his theories, which however are by no means so strong as they look at first sight. Furthermore Spence's own critical sense seems to have declined with the years, until he has ended up publishing books like *Will Europe Follow Atlantis?* (1942) arguing that God sank Atlantis for its sins and will do likewise to Europe if it doesn't reform.

But to go back to Spence's earlier and more lucid period, he talked a great deal about "a world-intuition regarding the existence of a transatlantic continent," which "has its origin in folk-memory"; "we are dealing with a great world-memory, of which Plato's story is merely one of the broken and distorted fragments. . . ." The only trouble with this talk of world-intuitions and folk-memories is that there is no reason to suppose that such things exist. A memory is simply an electrical phenomenon in some-body's brain, and there is no known way by which this memory can gets into that person's germ-cells and thence into the brains of his descendants. He must either tell it or write it, and memories that must be passed on orally have a way of shrivelling into unrecognizable figments of folklore in a few generations.

Spence's ideas on scientific method remind us a little of those of the occultists: "Inspirational methods, indeed, will be found to be those of the Archaeology of the Future. The Tape-Measure School, dull and full of the credulity of incredulity, is doomed." "Analogy is the instrument of inspiration, and if wielded truly, is capable of extraordinary results." "The day is passing when mere weight of evidence alone, unsupported by considerations which result from inspiration or insight, can be accepted of the world. . . ." However, science has achieved its results by these very "tape-measure" methods, while those who have studied questions like Atlantis with "inspiration or insight" have come up with hundreds of conflicting hypotheses.

Spence also relies heavily upon myth and tradition: ". . . all tradition has a basis of fact"; ". . . tradition, if carefully employed, is a document of equal sanction with anything in black letter. . . ." "There is no myth which has not its background of reality." Perhaps Spence, like Lewis Carroll's Bellman, believes that "what I tell you three times is true." He even considers legends more reliable evidence of past changes in the earth's surface than geology, despite the obvious fact that to change a story he has heard is one of the easiest things for a man to do, while the rocks stay much the same from age to age.

Even if we confine ourselves to Spence's facts, they turn out less impressive than they seem at first. For one thing, Crô-Magnon culture has now been found in the East—in Palestine. Furthermore, despite all these assertions that the cultures of Egypt, Yucatán, and Peru sprang into existence suddenly without a slow transition from primitive culture, modern archeology has disclosed the gradual evolution of all these cultures from more primitive levels. You can, for instance, trace the growth of Egyptian culture from the neolithic Merimda people, who wore animal skins, lived in mud houses, and farmed in a crude manner, to the highly civilized men of the Fourth Dynasty. You can in your mind's eye watch them substitute clay pots for stone pots, add handles to their pots, learn to bake their pots in kilns, and lastly, in the early dynasties, adopt the potter's wheel and the practice of glazing.

Such wrong ideas about the sudden appearance of cultures arise because the relics of the later stages of a past culture tend to be more numerous than those of its early stages. As the later people often made more, bigger, and more durable houses and artifacts than their ancestors, and as these later things have not been lying around so long, the more recent relics are less likely to have been destroyed by fire, rust, termites, or vandals. For instance, if you tried to reconstruct the history of iron armor from the exhibits in the Metropolitan Museum of Art in New York alone, you might conclude that iron armor came into use suddenly in the fourteenth century in a highly developed state. But as we know from art and history, iron armor was made many centuries before 1300; the earlier pieces, however, have almost all rusted away to nothing or were turned in to the smiths for scrap.

Another reason is that people who finance archeological expeditions have an understandable weakness for showy results, and hence tend to concentrate on highly developed cultures at the expense of the more primitive.

Moreover Spence's pro-Atlantis geologists are either pre-1900 and hopelessly out of date, or belong to a minute minority. The Norwegian lemmings do not migrate in the way he says; they become starved by overpopulation, start off in any direction, swim rivers and, when they meet the sea, mistake it for another river to their undoing. The Swedish lemmings try to swim the Baltic in the opposite direction. The Crô-Magnons did not, as Spence says, have a civilized culture with domesticated horses and cows; they had a very primitive culture like that of the Australian blackfellows, and hunted wild cows and horses. The feathered war-bonnet of the Plains Indians was not derived from the Crô-Magnons by way of the Mayas, but was devised by Indians of the Sioux group from whom it spread to the neighboring tribes . . . and so on. As for the Osirian relationships of King Arthur, this legendary king is wrapped in such a fog of ignorance and speculation that to try to connect him either with Osiris or with Atlantis is to explain one unknown by another—*ignotus per ignotum.*

Even the mythology in which Spence is supposed to be an expert is not above criticism. To support the theory of a Central Pacific Lemuria, he makes much of Polynesian flood-legends, though according to his fellow-mythographer Andersen deluge-legends are exceedingly rare in Polynesia. Spence quotes as an example the Hawaiian tale of the ark in which Nuu and his wife and three sons escaped the Flood, without noticing that Nuu is obviously Noah and the tale is of Christian missionary origin.

So much for the leading Altantists, occult and secular, and their doctrines. While I have pointed out a few mistakes in these beliefs in passing, there are a lot of themes that come up again and again in Atlantism — the rise and sinking of continents, the origin of the Mayas, the nature of myths, and the relationships of languages — that cannot be gotten rid of so easily. What does science really say on these and related subjects? Be careful; nothing is easier than to say that "science says" thus-and-so. Although the Atlantists have exploited this phrase to the limit, it often turns out that "science" has said nothing of the sort; perhaps one scientist once said something of the kind, but even he might be wrong.

To begin with, we cannot accept Plato's tale as unvarnished truth; who believes the romance of Poseidon and

Kleito? True, Atlantists rationalize Poseidon into a mortal king of Atlantis. However, Poseidon is not merely one of Plato's characters, but a leading god of the Greek Pantheon. If you rationalize his cohabitation with Kleito you should also rationalize his persecution of Odysseus, his creation of a spring on the Akropolis of Athens by striking a rock with his trident, and all the other myths about him.

Moreover there is no record of Plato's prehistoric Athenian Empire. Had this great state existed, its remains could hardly have been overlooked in a country so thoroughly picked over by archeologists as Greece. True, Greece contains prehistoric ruins at Mykenai and Tiryns, but these "cylcopean" structures belong to the period immediately following the fall of Crete, roughly 1400 to 1000 B.C.

Atlantists also assert that many inscriptions in Old-World languages have been found in the Americas, and hint that in some mysterious way a Phoenician inscription in Brazil corroborates Atlantis. However, these "Phoenician inscriptions" usually turn out to be something other than was supposed. Dighton Rock in Massachusetts had its scratches attributed to Phoenicians, Druids, Persians, Trojans, Hebrews, Libyans, Romans, Norsemen, Welshmen, Chinese, and Atlanteans until Prof. E. B. Delabarre with floodlights and a camera disclosed, under the carvings of Indians and modern campers, the name of Miguel Corte-Real, a Portuguese explorer who sailed for Newfoundland in 1502 and never returned. The Grave Creek Mound stone from West Virginia was analyzed as Etruscan, Runic, Phoenician, Old British, Keltiberic, and Greek. In 1930 Andrew Price, president of the West Virginia Historical Society, jokingly wrote that it read "Bill Stump's Stone, October 14th, 1828," this hoax being inspired by an incident in Dickens's *Pickwick Papers*.

As for the South American rock-carvings in which Braghine, Frot, and others have professed to see Phoenician, Hebrew, and other Old-World writing, these are mostly crude pictures of men and animals which the Indians have carved, sometimes for magical reasons and sometimes for amusement. We know this because some of them still carve them.

Atlantists are much given to linguistic arguments to

prove that the Mayas or some other New-World people have close relatives in the Old World. The method is the same as that of those who try to show that the Lost Ten Tribes of Israel are the Irish, the Aztecs, or the Burmese. They find a word in one language, and a word with a similar sound and meaning in another language, and conclude that they are applying the science of linguistics. In this manner they have "proved" that Aztec is Indo-European; that Chinese is related to Egyptian; that Algonkian is related to Latin; that Kiché is related to Berber; that Carib is related to Hebrew, and so on until they have proved that virtually every language is related to every other.

In a sense they probably are; that is, they are all descended from the grunts of the same Pliocene ape-man. But when we say languages are "related" we mean that they were separated so recently that they still resemble each other, as the examples listed above do not.

You see, the study of linguistic relationships is much more complicated than simply finding a pair of similar words. By the laws of chance there are almost sure to be a few apparent cognates (that is, genetically related words like English *water* and German *Wasser*) in any pair of languages, related or not. Thus "ten" is *dix* in French and *disi* in Hottentot; "search" is *examine* in English and *eggämen* in Tuareg; "time" is *hour* in Szechuanese; and so on *ad infinitum*. The reason for such pseudo-cognates is that most languages have but twenty to fifty phonemes (significant sound-units) and at least several thousand words, so that some such resemblances are inevitable by sheer luck.

To find real linguistic relationships, you must not merely take isolated words, but consider all the words in certain classes, such as numbers, colors, family relationships like "father," parts of the body like "head," natural categories like "water," and so on. Let us do a little experiment along these lines to shed light on the Atlantist assertion that Maya is half Greek and half Hebrew. These are the first ten numerals in several New-World and Old-World languages:

English	Greek	Hebrew	Chinese	Japanese	Mayan	Otomi
one	hen	'echadh	i (or yi)	hitotsu	hun	da
two	dyo	shĕnayim	er	futatsu	ca	yojo
three	tria	shĕlōshāh	san	mitsu	ox	tiu
four	tettara	'arbā'āh	sz	yotsu	can	coojo
five	pente	chămishshāh	u (or wu)	itsutsu	ho	guitta
six	hex	shishshāh	lu (or liu)	mutsu	uac	dato
seven	hepta	shibĕ'āh	tsi (or chi)	nanatsu	uuc	yoto
eight	oktō	shĕmōneh	ba	yatsu	uaxac	giato
nine	ennea	tish'āh	giu	kokonotsu	bolon	guito
ten	deka	esārāh	shï	tō	lahun	detdta

English and Greek are obviously related, but no other pair of languages on the table shows any apparent resemblance. This is not final "proof," of course, but a mere sample to show that the languages in question are not so alike as Atlantists often claim. For a thorough analysis of linguistic relations you need a much longer list of comparative words, and also a study of phonology, inflection, and syntax.

When the Amerind languages are so studied, Eskimo proves to be slightly related to the languages of Eastern Siberia, which is natural enough. Otherwise, however, the Amerind languages show a wide diversity among themselves, and *no* clear relationship to any Old-World tongues. Although some of them have grammatical features faintly reminiscent of the Ural-Altaic languages like Turkish, which spread out from Central Asia, many of them display strange and unique characteristics; for instance some like Nootka do not distinguish between nouns and verbs. All of which implies that the Amerinds have been separated from the Old World for thousands of years, and that during most of that time they were split into little isolated groups of hunters and food-gatherers. Such a society cannot support the density of population necessary for civilization, as they would soon sweep the country bare of edible plants and animals, and be forced to split up to avoid starvation.

Such considerations never bother the Atlantists, however, who seize upon any chance resemblance of names or other words to prove their case. By their methods I could, for instance, "prove" that the Amerinds are the descendants of colonies from ancient Greece: I could derive the Croatan Indians from Crotona in Italy, the Cherokees from Kerkyra, the Chilkats from Chalkis, the Mandans from Mantinea, and the Aleuts from Eleusis. Why not?

As for the actual origin of the Amerinds in general and the Mayas in particular—but that requires a chapter or two to itself.

CHAPTER V

THE MAYAN MYSTERIES

> . . . hieroglyphics old
> Which sages and keen-eyed astologers
> Then living on the earth, with labouring thought
> Won from the gaze of many centuries:
> Now lost, save what we find on remnants huge
> Of stone, or marble swart; their import gone,
> Their wisdom long since fled.
>
> *Keats*

W HEN the Spaniards arrived in the Americas in the early sixteenth century, they found several well-developed civilizations. Of these the most advanced, at least in some ways, were the Mayas of Yucatán, who ever since have played a large rôle in Atlantism. For one thing, the Mayas alone of the New-World indigenes had a real system of writing, though the Aztecs of Mexico had a more primitive system of picture-writing like cartoons without words, and the peoples of the Inca Empire of Peru possessed the *quipu*, a device of knotted cords to help people remember things.

After the disappearance of Landa's treatise on the Mayas, these people dropped out of the consciousness of Europeans for about two centuries, partly because of the inaccessibility of their harborless country and partly because the conquest of the Aztecs by Hernán Cortés was better publicized. Many Aztec picture-manuscripts survived the missionaries' book-burnings, and historians like the hispanified Aztec, Fernando de Alva Ixtlilxochitl, wrote treatises in Spanish on Aztec culture and history. Most of these last, however, were buried in Spanish archives in the form of unpublished manuscripts. There were also a few little-known descriptions in Spanish of the Mayan ruins at Palenque. Altogether the Middle American civilizations were so thoroughly forgotten that the eighteenth-century Scottish historian Robertson denied that any New-World people was ever "entitled to rank with those nations which merit the name of civilized," and insisted that *conquistadores* who wrote of Indian cities and palaces must have been romancing.

In the history of any branch of science you will often find that at the beginning an enthusiastic Frenchman performs strenuous if amateurish work and publishes theories which, though wildly wrong, interest more sober students who lay the groundwork for the real science. Du Chaillu performed this office for Viking history, and likewise the rediscovery of the Mayas was begun by that singular character Jean Frédéric, Count de Waldeck (1766-1875): soldier, artist, explorer, courtier, revolutionist, and archeologist.

Waldeck commenced his adventures when at the age of nineteen he went with Le Vaillant's expedition into the unknown interior of Africa. Subsequently he became an adventurer in the French Revolution, and later a soldier and a naval officer under Napoleon. Then he took an active part in Spanish and Latin-American revolutions. In 1821 his imagination was fired by the sight of Mayan ruins in Guatemala, and on his return to London he illustrated the first modern book on the Mayas.

This book came to the attention of Lord Kingsborough, he of the Jewish-Indian theory, who sent Waldeck back to Central America with a job as mining engineer and a commission to draw American antiquities and seek traces of the Lost Ten Tribes. Although Waldeck, in his sixties, had already had enough adventures for three men, he was not slowed up yet.

During the next decade he scrambled over jungle-matted ruins, drawing them as he thought they should have looked. His beautiful drawings, alas, were "a strange mixture of inaccuracy, unjustified restoration, over-drawing, and exaggeration." Not satisfied with inserting his young Mexican mistress *sans* clothes into many of the pictures, he put things in them that were never there at all, such as four statues of men in Phoenician-style headdress holding up the front of the Temple of the Magicians at Uxmal.

Although Waldeck's fairy godmother had neglected

to give him a sense of critical accuracy along with his other gifts, nevertheless his first book, *Voyage pittoresque et archéologique dans la province de Yucatan* (1838) won him a medal and a pension from the French government. He went on to marry, at eighty-four, a seventeen-year-old girl by whom he had a son, to publish his second book at 100, and finally to drop dead at 109 just after turning to look at a pretty girl on the boulevards of Paris.

While Waldeck's books proved of little permanent value, a copy of the *Voyage pittoresque* interested John Lloyd Stephens, a successful American lawyer and globe-trotter who had already travelled extensively in Europe and the Near East. The ebullient Stephens arranged to get himself a diplomatic mission to Central America, and departed with his friend Frederick Catherwood to explore. Catherwood, an English artist, had already travelled about Muslim countries in Turkish dress drawing the antiquities of Egypt and the Holy Land, and at this time, in 1839, was the proprietor of a set of panoramas of the Holy Land in New York City, which he had painted from the drawings made in his travels.

This pair made two exploring-trips to the Mayan cities, and collaborated on two books, *Incidents of Travel in Central America* and *Incidents of Travel in Yucatán*, the real pioneer works in Mayology. At that time, when learned circles were still speculating about an Egyptian, Hebrew, Roman, Norse, Chinese, or other Old-World origin for the Amerindian civilizations, Stephens caused something of a stir by proclaiming that the Mayan cities had been built by the same brown slant-eyed people who still farmed the featureless plain of Yucatán.

Other travellers and scientists followed Stephens and Catherwood: Sir Alfred Maudslay made casts of the carvings on Mayan monuments for the Victoria and Albert Museum; Brasseur de Bourbourg unearthed the abridgment of Landa's *Relación* and tried unsuccessfully to decipher Mayan writing by means of it; Augustus Le Plongeon excavated Mayan ruins and concocted fantasies about the fall of Mu; and in the 1880's Förstemann in Germany and Goodman in California independently solved the Mayan system of numerals.

Hence during the last century a vast amount of work has been done on the Mayan civilization, so that today it is fairly well known—not so well as that of ancient Egypt, perhaps, but better than, let us say, the mysterious pre-Aryan civilizations of India.

However, there remain a number of mysteries (or pseudo-mysteries) about the Mayas. There are many things we should like to know about them but don't, and others that we do know about them but that pseudo-scientists and occultists prefer to believe we don't. For example, who were the Mayas; whence came they; what was their history; what is the secret of their writing?

Nearly every member of the Atlantist cult has propounded answers, sometimes very remarkable answers indeed, to these questions. For instance a decade ago R. B. Stacey-Judd published a huge book, *Atlantis, Mother of Empires,* with lavish illustration and a fabulously expensive format, devoted largely to idealizing the Mayas, exaggerating their accomplishments, and whitewashing their sins. He denied with partisan warmth that this noble people, at least under the Old Empire, could have practiced war and human sacrifice. Unhappily, recent discoveries prove beyond doubt that the Old Empire Mayas did engage with zest in both these practices.

What are the facts about the Mayas?

When you want to learn about a people, the first logical step is to look at them. The Mayas are a short, strongly-built people with small hands and feet, very broad heads, prominent cheek-bones, and straight coarse black hair, very sparse on the face and body. They often display the bluish "Mongolian spot" at the base of children's spines and the "epicanthic fold" that gives a slant-eyed look. Like many Mexican Indians they have a tendency to retreating chins and foreheads; in pre-Columbian days they accentuated the latter tendency by binding their children's heads.

They also wore jade plugs in their ears, were tattooed all over with green ink, and painted themselves red and black, so that altogether the Spaniards thought them the ugliest little men they had ever seen. No doubt the Mayas thought the Spaniards terrible-looking too. Visitors to their country have described them as friendly, good-natured, honest, industrious, stoical towards pain and hardship, and as intelligent as most people.

Evidently the Mayas are physically one more Amerind tribe, belonging to the Mongoloid or Yellow Race, though like many Amerinds they run to darker skins and bigger noses than such Old-World Mongoloids as the Chinese. Although Spence calls the Mayas the "American Crô-Magnons," they are as different in body from the towering, jut-jawed Crô-Magnons as any men could be. Nor is there any reason to trace their prominent convex noses back to the Armenoid racial type of the Near East, as Hooton and Gladwin have tried to do, for equally aquiline noses are common among many Amerinds like the Blackfeet, and Armenoid ancestry would probably have brought with it other Armenoid characteristics like the vast curly beards you see on Babylonian statues.

The next step is to listen. The Maya language, or rather family of languages, occupies a large area in Yucatán and neighboring regions, interspersed with Spanish and non-Mayan Indian languages. The Maya family comprises a group of typical Amerind tongues, showing, however, only the faintest resemblances to the other native languages of Middle America, and (Le Plongeon to the contrary notwithstanding) none at all to any other languages on earth. Mayan is simple and regular in its inflections, with a tendency (like English) towards monosyllables and homophones. Although it may have changed quite a bit since Columbus, this vital tongue is still the prevailing language of Yucatán, with its own books and newspapers.

When we speak of the Mayan culture or civilization we really mean two cultures: the so-called Old Empire in Guatemala, and the New Empire in Yucatán. The term "Empire" is misleading in this connection, since nowhere in Middle America was there anything, so far as we know, like the strong centralized governments implied by the term "Empire." Cortés's opponent, the Aztec "Emperor" Montezuma II, was not a hereditary despot of the European type, but an elected tribal chief, of limited powers, whose people had established a precarious ascendency over some of their neighbors.

The Old Empire (whatever we choose to call it) arose some time between 400 and 100 B.C. and flourished for a thousand years, as we know from the dates on Mayan monuments. Between 700 and 1000 A.D. the Mayas of Guatemala ceased erecting monuments, and most Mayologists believe that they abandoned their cities at that time, though a few think they continued to dwell in them.

Many causes have been suggested for this presumed abandonment: earthquakes, pestilence, climatic changes, war, and social decadence. Perhaps the most convincing explanation is that of Morley and some others: that the Mayan agricultural system contained the seeds of its own downfall.

The Mayas have always farmed by cutting and burning a few acres of jungle at a time to plant their maize with a dibble or pointed stick. While this system produces an easy living for a few years, the yield then drops off sharply, forcing the farmer to slash and burn a new patch. The old patch, instead of reverting at once to jungle, is covered by tough grasses with which the Mayas, lacking plows, draft-animals, or even digging-tools, could not cope.

In time they thus converted the neighborhoods of their cities into great grassy plains, until each farmer had to go so far to reach his plot that life near the city was no longer practical for him. Eventually the jungle returns and the cycle begins again, but meantime the Old Empire Mayas, perhaps egged on by priests convinced that they had offended the gods, migrated north and built a whole new constellation of cities in Yucatán. This New Empire, as it is called, was still a going concern when the Spaniards arrived in 1511.

The Old Empire was peculiar in being a purely Neolithic civilization, no trace of metals having been found in its ruins save those dating from the end of the period. Its rise cannot be dated definitely, since civilizations don't spring into being in a single day. However, the oldest Mayan date that most experts agree upon is that of the monument called Stela 9 at Uaxactún. According to the most generally accepted system of correlation of Mayan dates, the date on this object corresponds to 328 A.D. Apparently earlier dates exist, like those of the Leyden Plate

(320 A.D.), the Tuxtla Statuette (162 A.D.) and the La Venta Stone (31 B.C.) but there is some question as to whether these dates count from the same base date as the later Mayan dates.

Anyway, the Old Empire certainly does not go back many thousands of years B.C., as Atlantists have claimed, even though the Mayas probably carved their dates in wood for some centuries before they took to chiseling them laboriously in stone. Although similar speculations have been voiced about the cliff-dwellings of Arizona, studies of the tree-rings of the timbers used in these structures shows that they were built, not 10,000 or 20,000 B.C., but well after the beginning of the Christian Era, when the Mayan civilization had already arisen and Egypt was thousands of years old.

As to the culture attained by the Mayas, they can best be compared to the Egyptians of about the First Dynasty, before the pyramid-builders. While it is easy to exaggerate the Mayas' accomplishments, they did pretty well considering that their country was poor in minerals and that it lacked animals suitable for beasts of burden.

Furthermore the Old Empire did not, any more than the Egyptian civilization, spring into existence suddenly. It rose from a ruder culture sometimes called the Archaic or Middle Culture: a preliterate agriculture system, lacking writing, cities, and other appurtenances of civilization, but still rising like a plateau from the hunting and food-gathering cultures around it.

Of the other two main native American civilizations, the Andean and the Mexican, most archeologists believe the former to be about the same age (within a few centuries) as the Mayan. The Mexican civilization of the Toltecs and Aztecs is younger, beginning around the middle of the first millenium A.D., perhaps in answer to cultural influence radiating from the Mayas.

Among the less advanced cultures, the apartment-house culture of the Pueblo Indians of the U. S. Southwest is about the same age as the Mayan Old Empire, while the burial-mound cultures of the Ohio and upper Mississippi Valleys are even younger than the Mexican. The mounds were still in use when the Whites arrived; in fact some contain European trade-goods. Strictly speaking there was no race of "Mound Builders"; several tribes built mounds for a variety of reasons over several centuries. A Professor Haebler first asserted, wrongly, that the mounds must be of vast age because they must have been built by sedentary farming peoples, not realizing that the Indians of those areas *were* sedentary farmers, and thus giving rise to another fantasy about the Atlantean Mound-Builders.

Mayan art and architecture were highly original, with remarkable achievements and even more striking limitations. In the Old World, despite the existence of many civilizations, artistic methods and concepts have diffused until the art of each civilization has affected those of all the rest, so that despite regional differences the civilized art of the Old World is in a sense one. Mayan art, growing in isolation, shows no Old-World influence, which makes it hard for Westerners to grasp. Waldeck's mistakes were partly due to the fact that he could not free himself from the prepossessions of an Old-World artist.

Mayan architecture, like that of Egypt three or four thousand years earlier, began by developing stone structures in imitation of the existing wooden houses. Mayan architects evolved a style of massive ceremonial buildings and chiefs' houses (the *polloi* continued to live in their thatched huts) with walls, of stone and concrete faced with stucco, so thick that there was little room left inside. Never having discovered the true arch, they used false arches and corbelled vaults. That is, they let each course of stone overhang the one beneath until the two sides met at the top. Finally they crowned their concrete roofs with ornamental walls called "roof-combs."

Their structures included astronomical observatories, ball-courts in which they played a kind of cross between basketball and soccer, dance-platforms, vapor-baths, shrines, reviewing-stands, stadiums, city walls, causeways, and pyramids comparable in bulk, though not in height, with those of Egypt. However, the pyramids of the Mayas and Aztecs have nothing to do with those of Egypt, which were built several thousands of years earlier and moreover evolved from tombs, while the New-World pyramids evolved from temple platforms.

The Mayas based their agriculture on maize, with a variety of subsidiary plants such as beans and pineapples. They domesticated dogs, turkeys, deer, and bees. The Amerinds in general made up for their lack of success in taming animals by a highly developed plant husbandry; they actually domesticated more species of plants than the people of the Old World. They also showed astonishing plant-breeding ability. Thus maize was so modified by breeding that scientists still disagree as to which wild grass it is derived from. The South American Indians also domesticated a variety of plants, some like oca that have never been transplanted to other parts of the world.

When the Mayan maize-culture was working properly it gave the Mayas plenty of spare time, which they put in on their great construction projects, religious ceremonials, and art-work. They even carved jade, one of the hardest stones, without metal, using reeds and strings for drilling and cutting, with sand as an abraisive and water as a lubricant. They began to use gold, silver, and copper about the time of the rise of the New Empire, at first for ornaments, though later they learned to make copper knives. Like the Aztecs, however, their priests continued to open up their victims with knives of obsidian (volcanic glass) since the gods are conservative in such matters. Copper-working may have diffused north from Peru; the Andeans at the time of the Conquest had gone beyond simple copper-hammering to casting in bronze.

On the whole, the New-World civilizations showed themselves rather backwards technologically compared with those of the Old World. As with other Amerinds, the Middle Americans' bent was artistic and religious rather than technological or administrative. Among most Amerind peoples, for instance, the tribe relied entirely on custom to keep wrongdoers in order, lacking formal law and machinery for punishing criminals. Only a few Amerind peoples, such as the Iroquois and the Incas of Peru, displayed much talent for government.

Those who argue that the Mayas must have obtained their culture from elsewhere, for instance from Atlantis or

Egypt, cite resemblances between the Mayan civilization and those of the Old World. Such arguments, however, concentrate upon likenesses and ignore differences, which are so profound as to make the resemblances look petty and accidental. So, now that we have discussed the things the Mayas had, let's consider those they did *not* have:

The plow. The Mayas' only agricultural implement was the dibble. Poke a hole in the ground, drop in a corn, stamp the hole shut, and you have the Mayan system of planting.

Metal tools. Except for the copper knives that came into their culture late, they were innocent of these too. They fought with stone-pointed spears (and, later, arrows) and stone-edged sword-clubs.

The wheel. Although the wheel appeared in Sumer about 3000 B.C., there is no sign of wheeled traction in the New World. Nor, according to all but one of the experts, did they have the potter's wheel. (Mercer says they did.) Some Amerinds had the spindle-whorl or flywheel (mistaken by some early investigators for the true wheel) and the Aztecs made toy animals that ran on little clay rollers wrapped around wooden axles socketed in the toys' legs. Why they never took the obvious step of developing these toys into wheeled vehicles is as great a mystery as why Old-World peoples stamped impressions with seals for thousands of years before anybody thought of the printing-press.

Old-World food-plants. In general the New and Old Worlds had no food-plants in common. Now, it is incredible that if Atlanteans had colonized both Mexico and Egypt, they should have taken wheat only to Egypt and maize only to Mexico. The few apparent exceptions to this rule do not necessarily contradict the principle. The coconut-palm and the gourd, for instance, occur in both hemispheres, but bear buoyant seeds that were probably spread by drifting over the oceans. In other cases, where similar plants were domesticated in both hemispheres, as with the fig, the cottonbush, and the strawberry, the plants belong to different species.

Old-World epidemic diseases. The voyages of Columbus and his colleagues brought smallpox from the Old

World to the New for the first time, and yellow fever in the reverse direction; the hemispheres had no epidemic diseases in common. Siphylis, usually considered an American disease brought back to Europe by Columbus's ships, is a disputed case, but it does not count because it is not strictly speaking an *epidemic* disease. These, to thrive, need the density of population found in the towns and villages of agricultural societies. Since hunting and food-gathering people are too thinly spread for epidemics, the common ancestors of the New and Old-World peoples must, when they parted company, have been in the hunting or food-gathering stage.

Old-World domestic animals. The Amerinds had no Old-World domestic animals except the dog, whose taming in the Old World goes back to the Neolithic hunting cultures. If the Amerinds' ancestors had had the horse or pig or chicken before they left the Old World, there is no reason why they should not have brought them along, especially since many Amerind tribes eagerly adopted these creatures when Europeans brought them to the New World.

Old-World calendars. The Mayan calendar differs drastically from Old-World calendars. Whereas the latter are based upon a year of twelve months of about 28 days each, with a few extra days to make it come out even, the Mayan calendar is based upon a year of eighteen months of 20 days each.

Old-World writing. Donnelly and others to the contrary notwithstanding, there is not the slightest resemblance between the Egyptian hieroglyphic writing, or any other Old-World script, and Mayan writing. Although the Mayan system is based upon principles much like those of Egyptian hieroglyphics, the signs and the language are entirely different. Hence our knowledge of Egyptian writing has been of no help in the decipherment of Mayan.

The Mayas also lacked the bellows, glazing, kiln-dried bricks, stringed musical instruments, the true arch, and the rake. Evidently the Mayas' ancestors arrived in their later homes unburdened by food plants, animals other than the dog, writing, or calendars — that is, as savages in the hunting or food-gathering stage of culture.

Some Atlantists argue that the Mayas might have been

ordinary primitive Amerinds converted to civilization by a boatload of refugees from Atlantis. However, in the light of history, such conversion is unlikely. Most primitives are suspicious and conservative, reluctant to adopt foreign ways. If they let strangers settle among them at all, it is usually at the price of the strangers' dropping their own customs for those of the primitives. In fact the newcomers are lucky if they are not eaten, as happened to the first Spaniards in Yucatán. Thus the Norse settlers in Greenland, instead of influencing the Eskimos, were eventually absorbed by them; and the Spanish priest Jeronimo de Aguilar, found by Cortés after he had lived among the Mayas eight years, had not only failed to influence the natives, but had almost forgotten Spanish. Although primitives may sometimes pick up a culture-trait from visitors when it fits easily into their own pattern, they usually submit to wholesale cultural change only as a result of long contact with or overwhelming pressure from a more advanced culture.

Now, to return to the characteristics of the Mayan civilization: The Mayan religion was much like other early agricultural polytheisms, with a multitude of departmental gods and an elaborate calendar of observances. Among their chief gods were the sky-god Itzamna and the rain-god Chac, who was more or less the same as the storm-god Kukulcan. Kukulcan in turn was equivalent to the Kiché Gucumatz and the Aztec Quetzalcoatl. Like Quetzalcoatl, Kukulcan had culture-hero functions. The story that he came from the west has led some historians to suppose that he was a real Mexican chief who conquered Yucatán under the New Empire. In the later stages of the New Empire the Mexicans did exert influence in Yucatán, but whether a real Kukulcan led them is one of those things than can probably never be settled.

Quetzalcoatl-Kukulcan has received disproportionate attention as a result of the Spanish friars' belief that he was the Apostle Thomas who had travelled about the Americas preaching. Needing some such theory to square the existence of the Amerinds with the New-Testament statements about the Apostles' being sent unto the ends of the earth, they studied the New World carefully for traces of the Apostles' passage. When they found that some tribes used the cross as a symbol (actually of the four winds or of the four cardinal directions) they were sure they had found a trace of Christian influence.

To justify the Conquest, the friars in Mexico made much of a blond god who civilized his people and then departed across the sea, having promised to return some day. In truth, however, Quetzalcoatl is always shown as black-haired and black-faced in the Aztec manuscripts, when he is not shown in the form of the masked wind-god Ehecatl. It has also been surmised, with just as little foundation, that he was a Norseman who had wandered south from the Vinland settlements. Actually the Amerinds of North and Central America have a widespread myth of culture-heroes who came from across the sea, often called "white," probably meaning white-clad or white-painted. Among the Yucatecan Mayas the white strangers are the Chanes, who landed from scaly boats at Vera Cruz, taught the people the civilized arts, and founded Chichén Itzá.

We can be quite sure these alleged civilizers weren't Norsemen, because when the Norsemen actually arrived, they found the myth already in circulation. Thorfinn Karlsevni's men captured a pair of Indian boys who, when they learned to speak Norse, told their captors that they understood that across the sea lay a land "where people went about in white clothes, uttered loud cries, and carried poles with banners fastened to them." Probably this is just one more family of culture-hero myths proving nothing historically, like the group of Old-World stories about the archer (William Tell, Egil, Palnatoki, etc.) compelled by a tyrant to shoot an object from his son's head.

The gloomy Mayan religion was closely involved with the Mayan agriculture. It entailed a good deal of human sacrifice, though perhaps not on the grisly scale of the Aztecs, who killed as many as 20,000 people at the dedication of a single temple. The Mayas, like the Aztecs, ripped open victims and tore out their hearts; they threw virgins down sacred wells, shot other victims with arrows, mounted victims' heads on huge skull-racks next to their temples, and ate selected victims at ceremonial feasts.

Mayan religion was closely connected with Mayan

astrology, a well-developed pseudo-science that made much of the revolutions of the planet Venus, and the Mayan calendar, which in turn was involved with the Mayan systems of numbering and writing. Since no Rosetta Stone for deciphering Mayan writing has ever been found, Mayologists have had to interpret written Mayan by the laborious process of trial and error. During the last century they have learned to read the numerals and calendric signs, the names of many of the gods, and some common nouns. The Mayas seemingly did not write verbs.

The Mayan written language was mainly ideographic —that is, a word was represented by a conventionalized picture, as in Chinese writing. In addition there were some phonetic elements, mostly standing for whole syllables, and the same glyph might be used either in the phonetic or in the ideographic sense. The precise extent to which the Mayas used phonetic elements in their script is a subject of dispute. Such writing may be called ikonomatic or rebus writing.

A Mayan glyph normally included two or three of these picture-elements wrapped around one another in a complicated fashion. Besides, the glyphs differed widely according to the scribe who was writing them, and according to whether they were painted on paper or chiseled on stone. Moreover a glyph might have two radically different forms: a "normal" form and a "head-form" that had only certain "essential elements" in common. If the normal form had, let us say, the essential element of three dots, the head-form would comprise one of those grotesque heads in profile that haunt Mayan art, with the three dots marked on the cheek or ear. The reader was supposed to be able to pick out the essential elements in either kind of glyph. In addition the Maya scribes sometimes used "full-figure forms" in which the head of the head-form was provided with a body, and "beast-forms" wherein the essential elements was combined with a picture of a snake or other animal. Altogether they could hardly have made their system more cryptic if they had deliberately set out to do so.

The Mayas, like the ancient Gauls, used a vigesimal system of numbers (with a base of 20 instead of 10). They had two systems of writing these numbers. One was ideographic like our Arabic system, with the usual head-forms and other complications, but without the place-value concept. Thus it had a series of symbols for the numbers from 0 to 19, and another set for multiples of numbers: 20, 400, 8000 and so on.

The other set of Mayan numerals, like our Roman system, used simple elements repeated; a dot represented 1 and a bar 5. But this system, like our Arabic system, used place-value; multiples were donated by stacking the symbols vertically, just as we string them out horizontally. For instance, the number 53 in this system would be:

• •

• • •
▬▬▬▬▬
▬▬▬▬▬

Fifty-three is $(2 \times 20) + (13 \times 1)$. In the first place, at the bottom of the stack (corresponding to the right-hand digit in an Arabic numeral) the Mayas put the symbol for 13: two bars and three dots. In the second place, above the first, they put the symbol for 2 — two dots. The Mayas' greatest accomplishment was inventing symbols for zero several centuries *before* the Hindus invented ours.

You have probably heard of the Mayan calendar, whose system of leap-year corrections was more accurate than that of the Julian calendar that preceded the modern Gregorian calendar. Most of the existing Mayan writing is in fact concerned with keeping track of dates, a more difficult task among people of that level of culture than among us who have the Naval Observatory to help us.

During the Old Empire the Mayas employed a complicated system of recording the dates they chiselled on their monuments. They wrote out each date, not in one calendar, but in three. They recognized a period of 260 days, for which the Mayan name was probably *tzolkin,* divided up into shorter periods. Then they had a 365-day period, the *haab,* divided into 18 periods of 20 days each, with five extra days to make it approximately equal to the true solar year.

By writing the date of a given day according to both
the haab-calendar and the tzolkin-calendar, the Mayas gave
a combination that repeated itself only once in 52 haabs.
Then to make sure there was no mistake, the Mayas also
wrote the date in the "long-count" calendar, a system of
recording the number of days that had elapsed since the
date 4 *ahau* 8 *cumhu,* which on our calendar would be
about 3113 B.C. Atlantists sometimes jump to the conclusion that the Mayan calendar must have been actually
put into use at that time, but it is much more likely that
4 *ahau* 8 *cumhu* represented some event in Mayan mythology. Many Old-World calendars also start from some mythical date, like the orthodox Jewish calendar which counts
time from the Creation of the World in 3761 B.C. Morley
believes that the Mayan calendar was devised, either by a
single man or by a committee of priests, some time in the
fourth or third century B.C.

To make things a little more complicated, the "long
count" was recorded, not in a straight vigesimal system, but
in a modified vigesimal system: 20 days made one *uinal*;
18 *uinals* made one *tun*; 20 *tuns* made one *katun,* and so on
up by multiples of 20 thereafter to the great cycle of
2,880,000 days. We need not go into the lunar calendars
and other complications.

Unfortunately the Mayas stopped using the "long
count" several centuries before the Spaniards arrived, so
that while we can locate dates accurately enough with relation to one another during the thousand-odd years that the
long-count system was in use, we can only roughly correlate this system with our own. By using Mayan astrology,
history, archeology, and other evidence, Mayologists have
worked out several correlations between the two calendars,
according to which Stela 9 at Uaxactún is dated all the
way from 203 B.C. to 584 A.D. Three of these systems, by
Goodman, Martínez Hernández, and Thompson, differ
among themselves by a maximum of only five days, and are
therefore lumped together as a single system. The Goodman-
Martínez-Thompson system, according to which Stela 9 is
dated 328 A.D., is now the most generally accepted.

The surviving Mayan literature consists of many inscriptions on temples and monuments, and three of the

original Mayan books; plus some writings in the Mayan and Kiché languages using the Spanish alphabet. The three old codices are all religious, calendric, or astrological, the Dresden Codex being a prophecy covering a period of 34,000 years to the end of the world. The inscriptions contain similar matter, though some of them might possibly contain brief notices of historical events as well.

The Spanish missionaries in Yucatán quickly adapted their own alphabet to the native languages so that they could print prayers and hymns for their converts. The Indians found this system so much easier than their own writing that they soon dropped the latter and began writing their own languages in the Spanish alphabet. They wrote a number of village chronicles, perhaps transcribed from old native histories before these were destroyed, known as the *Books of Chilan Balaam* (or *Chilam Balam* or *Balan,* depending upon the dialect) of which about a dozen have survived. "Chilan Balaam" means literally "interpreter jaguar," the title of certain oracular priests of the pagan priesthood who were responsible for historical records, prophecies, and the like.

These books consist of brief notices, wrapped in obscure and enigmatic language, of plagues, tribal wars, the deaths of chiefs, and so on. For instance the best of them, *The Book of Chilan Balaam of the Village of Mani,* begins: "This is the order of the katuns since the four katuns during which the Tutulxiu left their home and country Nonoual to the west of Zuiua, and went from the land and city of Tula, having agreed together to this effect."

In addition, the southwestern neighbors of the Mayas of Yucatán, the Kichés, wrote a work in Romanized Kiché, the *Popol Vuh,* which is mostly mythological. It tells of the creation of the world by the wind-god Hurakan, and a great feud between the gods Hun-Ahpu and Xbalanque on one side and the giant Vukub-Cakix and his family on the other. The fourth and last book of the *Popol Vuh* is devoted to the mythical progenitors of the Kichés, giving lists of battles and the genealogies of chiefs, which, while they may contain scraps of fact, don't agree with the scanty information from other sources.

Evidently the Mayas developed a historical literature, all of which has now been lost, at least in its original form. On the other hand it seems likely that their history never got beyond the stage of annals like those that monks kept in European monasteries during the Dark Ages. The latter read like this:

> 709. Hard winter. Duke Gottfried died.
> 710. Hard year and deficient in crops.
> 712. Great flood.

Better than nothing, but hardly to be compared with Herodotos or the *Books of Kings*.

However, by piecing together the statements of the *Books of Chilan Balaam* and the accounts of the Spanish clerics, we can get a good idea of Mayan history during the New Empire. (There are no surviving historical facts about the Old Empire, beyond what we can infer from archeology.) After the Mayas built their new cities in Yucatán around 1000 A.D., a confederacy of the three strongest clans, the "League of Mayapan," dominated the peninsula for a time. These clans were the Xiu or Tutulxiu, the Cocom, and the Chel. About 1194 a Cocom chief, Hunac Ceel of Mayapan, imported mercenary soldiers from Mexico to make war against Chac Xib Chac of Chichén Itzá, chief of the Xius, and made his clan supreme in the land. Subsequently the Xius revolted under Ah Xupan Xiu and destroyed Mayapan, after which there was bitter warfare amongst the clans until the Spaniards came. The Spaniards were preceded by several years of plague and bad weather.

The Spaniards quickly overran the country by the terror their guns and horses inspired and by the atrocities they committed to cow the natives. Then the Mayas, fierce and wily fighters when aroused, revolted and drove the Spaniards out. It looked as though old confederacy would be revived, but the hatreds engendered by the long feud proved too bitter. The Cocom chief lured a party of Xius, including the son of the Xiu chief, into Zotuta with protestations of friendship, and when they were asleep in the guest house set fire to the house and burned them all up.

This act touched off a terrific war between the Xius and the Cocoms, and when the Spaniards came back in 1537 the Xius went over to them, received baptism, and helped them conquer the other tribes. In 1541 the Spaniards and Tutulxiu together destroyed the army of 70,000 Cocoms in a last great battle.

The Mayas really had no chance; if these Spaniards had not conquered them somebody else would have. Given the great advantages in military tools and techniques that the Old World had over the New, any small, well-equipped, and resolutely led European army could have destroyed any of the major American states, as in fact happened in Mexico and Peru. The nature of tribal society with its small and mutually hostile communities made it impossible for any great number of Amerinds to unite against the White menace. Some always went over to the Whites to get even with their local enemies, just as many Greek city-states went over to the invading Persians, oblivious to their own fate once the outsiders gained control.

Finally the Middle Americans, while brave enough, had no notion of European strategy, tactics, and imperialistic methods of conquest and exploitaion. They fought to capture prisoners to sacrifice to their gods, and after each side had caught its share they went home like gentlemen until the time came for another battle.

The Mayas' downfall may have been hastened by their priests' habit of issuing gloomy prophecies, such as:

> What time the sun shall brightest shine,
> Tearful will be the eyes of the chief,
> Four ages yet shall be inscribed,
> Then shall come the holy priest, the holy god.
> With grief I speak what now I see.
> Watch well the road, ye dwellers in Itzá.
> The master of earth shall come to us.

The priests' object may have been to keep their own people in a state of apprehension to make them easier to control. But as a result, when the Spaniards came, although some Mayas fought like demons, others sighed "This must be it" and stolidly awaited the end.

Magical and pseudo-scientific cults have subjected the Mayas to their attentions. During the 1930's Harold D.

Emerson published a magazine, *The Mayan,* in Brooklyn,
devoted to "spiritual enlightenment and scientific religion."
It mixed Theosophy, Atlantism, a smattering of Mayan
archeology, and general guff. Then a society called "The
Mayas" sells correspondence courses in occultism from San
Antonio, claiming that their leader received his initiation
into the transcendental mysteries from a Mayan priest in
the jungles of Central America.

Still, the Mayan mysteries on examination turn out to
be not so mysterious after all. The Mayas are neither
morons nor Atlantean supermen, but a likeable and quite
human people. If some of their ancient customs were not
what we should consider good, neither were those of our
own ancestors. The purpose of all this discussion of the
Mayas is to show that, while we don't know many things
about them, there is nothing in what we do know to suggest any great mystery in their background. The same factors that raised other civilizations — challenge-and-response,
the workings of heredity to produce occasional geniuses,
and the self-accelerating nature of technological progress—
raised them from their primitive condition.

No Atlantis is necessary to account for the Mayas; in
fact, their known history will not fit the lost-continent
schemes of the Donnellys and Spences. They arose, not in
immemorial antiquity, but about the beginning of the
Christian Era. Their ancestors could not have been civilized
before they reached Middle America. And finally, the
Mayas were racially, linguistically, and culturally entirely
distinct from the Egyptians and other Mediterranean peoples, so that any theory based on a supposed common origin
of the Mayan and the Phoenician, Egyptian, or other
Mediterranean civilization falls to the ground. While these
arguments do not in themselves disprove the existence of
Atlantis, they certainly knock one of its main props out
from under it.

Chapter VI

WELSH AND OTHER INDIANS

Across the seas of Wonderland to Mogadore we plodded,
  Forty singing seamen in an old black barque,
And we landed in the twilight where a Polyphemus
    nodded
  With his battered moon-eye winking red and yellow
    through the dark!*

Noyes

MANY and strange have been the theories advanced to account for the origin and cultures of the American Indians. These theories involve not only sunken continents but also extraordinary voyages, astounding trading-enterprises, and amazing feats of converting primitives to a civilized way of life. The Jewish-Indian theory, already mentioned, was not the only one. Joseph Smith, the founder of Mormonism, set forth a variant of this in his *Book of Mormon,* a tedious and puerile pseudo-Biblical fabrication which he claimed to have copied from golden plates lent him by the angel Moroni. The angel thoughtfully also provided a pair of magical spectacles, the Urim and Thummim, which translated the inscriptions on the plates into English.

According to Smith, one Jared and his brother sailed to America from the Old World at the time of the Confusion of Tongues, God having kindly provided their ships with artifical light. Their American descendants increased to several millions, but were completely destroyed, save for one lone survivor, in a war that ended with a great battle on the hill Cumorah in New York State.

Later a certain Lehi, on God's orders, emigrated from Jerusalem with his family to America in the reign of

* By permission of J. B. Lippincott Co., publishers, from Alfred Noyes: *Collected Poems in One Volume,* copr. 1906, 1934 by Alfred Noyes.

Zedekiah. Although they met the last Jaredite, they failed
to profit from his experience, for they too fell into civil
war, and in a last battle on Cumorah all the better element
were killed off. The descendants of the survivors became
the American Indians. Previously Christ had visited the
Nephites, one of the rival clans, to convert them, and had
explained that he was on his way to preach to the Lost
Ten Tribes somewhere else.

The Welsh-Indian theory appeared in 1583 in a propaganda pamphlet by Sir George Peckham, and again in 1584
in Humphrey Lhoyd's *Historie of Cambria, now called
Wales,* and was soon repeated by other authors like Hakluyt,
Purchas, and Raleigh. The English were just beginning to
get imperialistic ideas about the Americas, so that any story
indicating early British discovery of these continents had a
strong appeal to them.

The hero of the Welsh-Indian stories, Prince Madoc
ab Owen Gwynnedd, seems to have been a real twelfth-cen-
tury Welshman, an expert sailor and fisherman according
to the brief allusions to him in bardic and monkish literature. According to Lhoyd, Madoc, distressed by a war between his father and his brothers, sailed away into the
Atlantic with several hundred followers. Later he came
back, said he had found a fruitful land to the west where
he had left some of his people, collected ten more shiploads
of colonists, and went away for good. Subsequent versions
of the legend had Madoc landing in Florida, Mexico,
Yucatán, and even the west coast of South America. The
number of his ships was blown up to eighteen and the number of colonists to 3000.

A travel tale first published in 1704 under the name
of "Morgan Jones" further exploited the Welsh Indians.
Jones claimed to be a Welsh preacher who had been caught
by the Tuscaroras in 1660 while going with the Port Royal
expedition into the wilds of North America. The Tuscaroras
would have killed him but for the intercession of an Indian
of the Welsh-speaking Doeg tribe, otherwise unknown. It
was later reported that another Welshman named Griffith
was taken in 1764 by the Shawnees to visit a Welsh-speaking
tribe. The Welsh Indians, you see, withdrew westward as
the continent became better known, like the Great Salt

Mountain staying just out of easy reach. Every time the fable started to die down somebody would stir it up again; in the 1790's patriotic Welshmen sent John Evans to search North America for the Welsh Indians to reconvert them to Christianity, and Southey wrote a poem, *Madoc*, on the theme.

For a time the Mandans, a farming and bison-hunting tribe on the upper Missouri River, were identified with the Welsh Indians. Evans thought they were, and so did the artist George Catlin, who in the early nineteenth century published a big book, *American Indians*, calling attention to the Mandans' coracle-like boast and other culture-traits. The Mandans were a natural choice, being a friendly people with a slightly higher culture than their neighbors and slightly lighter skins. However, nobody knows whether the latter were due to a natural variation in pigmentation (some Amerind tribes being naturally darker than others) or to their having been in contact with Whites for a century during which their women tendered the ultimate in hospitality. Holand, the Kensington Rune Stone man, thinks they got both their culture and their coloring from the survivors of a party of fourteenth-century Scandinavians who wandered inland to the Great Lakes and, unable to go back, went native.

Be that as it may, the poor Mandans were nearly wiped out by the great smallpox epidemic of 1837 and those left alive scattered to other tribes. Thereafter the search for the Welsh Indians resumed; Brigham Young was still looking for them in the western deserts in 1854. The usual announcement of their discovery says that a particular tribe—Caribs, Mandans, Flatheads, or what-not—uses Welsh words, such as that for "cow," though it is hard to see how any Amerinds could have had a word for "cow" in pre-Columbian times when there were no cows in the Americas. At last accounts Robert Pritchard of Invermere, B. C., had identified the Kutenais of British Columbia with the Welsh Indians, as usual announcing that they use the Welsh word for "cow." The Welsh Indians seem to have been chased as far as they can go, unless they mush north to Alaska.

As if the Jewish Indians, the Mormon Indians, the Welsh Indians, and the Norse Indians weren't enough, the

Polynesian Indians have also played a part, in theory, in the populating of the New World. In the early nineteenth century Ellis asserted that the Polynesian islands had been settled by people from Peru, and in 1830 Dr. John D. Lang of Sidney, Australia, denied this on the ground that the ancient Peruvians were not a sea-going people; therefore the Américas must have been peopled across the Pacific by the Polynesians.

One comes across arguments of this kind all the time in Atlantism and the search for the Amerinds' family tree. The so-and-so's are not known to be navigators, therefore they couldn't have reached a place by water; or they could not have built such-and-such a structure because they are too primitive, and so on. Spence, for instance, trying to establish a lost continent in the Pacific, argues that the Melanesians must have reached their islands afoot because they are not seagoing. In the first place the writer often has his facts wrong (some Melanesian tribes are almost as mighty sailors as the Polynesians) and in the second, people sometimes change their way of life. Thus the Easter Islanders, originally seamen like other Polynesians, settled on a treeless island with no material for shipbuilding, with the result that when their old boats wore out they were stuck.

The Polynesian-Indian idea, which became popular among French anthropologists, is not so silly as some proposed derivations for the Amerinds. The Polynesians, a racially anomalous people, seem to be partly or largely Mongoloid (with some Melanesian or Oceanic Negro admixture) and are therefore physically rather like the Amerinds. The argument has raged for over a century between those who thought the Polynesians influenced the Andeans, those who believed the Andeans influenced the Polyensians, and those who denied any contact between them. Recenty a sextet of stalwart young Scandinavians under Thor Heyderdahl sailed a balsa craft from Callao, Peru, 4300 miles to the Tuamotu Archipelago to "prove" that noble Nordics from Peru settled Polynesia about 500 A.D. Heyderdahl did not explain how their rafts returned to Peru against wind and current—and anyway to show a thing can be done is not the same as proving that it was done.

Now, it does not seem reasonable, in view of differences of race and language, to assert that Polynesians *are* Andeans or *vice versa*. The Polynesians' own traditions, corroborated to some extent by archeology, say that they moved into Polynesia from Indonesia (they particularly mention Java) within the last thousand years, reaching Easter Island about 1300 A.D. Therefore they could hardly have founded the Andean civilization, which started some centuries earlier.

There is still room for argument, all the same, over the question of an occasional contact between the two peoples, whether this took the form of Polynesian canoe-galleys' paddling to Peru or Andean balsa rafts' wafting to the islands or both. Even some conservative archaeologists seem to be coming around to the belief that there may have been some connection, since the Amerinds' own traditions tell of such visitors from the Western Ocean.

But it does not follow that Polynesian influence started the Andean civilization, or even that it had any profound effect upon it, or that it influenced the Mayas at all, since the Mayas and Andeans were unknown to each other. Their only contact was slow cultural seepage via the almost impassible jungles and inhospitable coasts of Panama and Columbia, and while some objects were carried from one continent to the other by traders, and some techniques slowly diffused across, more complicated culture traits like the Mayan calendar failed to make the passage.

Further, the vast distance from South America to the nearest Polynesian archipelago, the Tuamotus, would have made the voyage desperately risky for even such accomplished sailors as the Polynesians. And besides, useful Polynesian traits that one would have expected to diffuse to the New World as a result of these contacts, such as the double canoe and the raising of chickens, pigs, bananas, and sugarcane, failed to do so.

Meanwhile those who, like Spence and Churchward, believe in sunken Pacific continents, have been putting the Polynesians to their own peculiar uses.

Like Jacolliot they argue that the Polynesians are the remnants of the people of a former Polynesian continent, a Pacific Lemuria. They cite the statues of Easter Island, a lonesome and infertile speck of land in the Southeast Pacific, midway between the Tuamotus and the Chilean coast, as examples of Lemurian art-work.

Now, almost as much pseudo-scientific pother has been made over Easter Island (or Rapanui as the natives call it) as over the Mayas. As with the Mayas, there is much that we do not know about the Easter Islanders. Again, the reason is not that these people are inherently mysterious, but that the evidence about them has been allowed to perish unrecorded.

The Rapanuians actually comprise about 450 brown-skinned fishermen of mixed Polynesian-Melanesian type, notorious as among the world's most accomplished thieves. Their traditions tell of their arrival in two canoes from the west under a chief named Hotu Matua about 700 years ago. They once had a higher culture than they do now, with a system of picture-writing on boards, and a religion in whose observance they made the strange statues that are their best-known culture-trait.

These statues are busts from three to 36 feet high, carved with stone tools out of soft volcanic rock. The people dragged these objects into position with grass ropes, set them up in rows around the island, and provided them with stone hats. Being topheavy the statues have now nearly all fallen down. The Rapanuians had mounted about a hundred of these busts when they abandoned the custom, leaving a lot of half-done statues and stone chisels in the quarry where all these sculptures were made. These stern, highly stylized images remind the beholder of Shelley's lines about the ruined statue of Ozymandias:

> ... Half sunk, a shattered visage lies, whose frown,
> And wrinkled lip, and sneer of cold command,
> Tell that its sculptor well those passions read
> Which yet survive, stamped in these lifeless things ...

Between the lack of resources on the island, the raids of South American slavers, and the cultural depredations of missionaires (who burned most of the boards with the picture-writing) the native culture collapsed before it had been properly studied. Hence the exact meanings of the writing and of the statues have been lost. However, the many Atlantist statements that the statues could not have been executed by the simple Rapanuians, but must have been produced by an advanced civilization, are simply untrue.

Some Lemurists even take seriously the claim of Juan Fernandez to have sighted a continent in that area in 1576, maintaining that at that time Lemuria had not wholly vanished, but was still represented by a large archipelago in the Rapanui region!

The story of the Polynesian Indians brings us to the diffusionist controversy, which has an important bearing upon Atlantis because nearly all Atlantists use diffusionist methods in their reasoning. The fact that the Atlantists and the diffusionists start with the same facts, using the same logic, and come out with entirely different conclusions, indicates that something is wrong with these methods.

When social anthropology was getting started about a century ago, the pioneer German anthropologist Adolf Bastian asserted that cultural likenesses between different peoples were due to a "psychic unity" of the human race, which caused the human mind, when faced with the same problem, always to come up with the same answer. This is of course pre-scientific psychology. The rise of Darwinism suggested to the early anthropologists that human societies evolved like animal species, from small and simple to large and complex. Separate groups, they thought, tended to evolve independently along parallel lines, and even, under similar circumstances, to develop similar forms from different ancestry, just as the tuna, the porpoise, and the mako shark, though of vastly different ancestry and inside construction, look much alike on the outside.

The evolutionary anthropologists, largely accepting Bastian's "psychic unity" theory, assumed that all human societies were going through similar stages of the same process, and that modern primitives were perfect examples of our own ancestors a few thousand years back, and if left alone would develop our own type of civilization in time. Now, there is something to these ideas, though they are not the whole truth by any means and though some nineteenth-century anthropologists carried them to unwarranted extremes.

Then in the early years of this century, some students of anthropology, mostly amateurs, reacted in an extreme manner against the teachings of evolutionary anthropology. The English psychologist William H. R. Rivers, a leader of the group, after pioneering in physical psychology, got interested in anthropology and went on expeditions to Melanesia. There he found a technique of mummifying corpses like that used in ancient Egypt. On this basis Rivers, the anatomist Sir Grafton Elliot Smith, and W. J. Perry of the University of Manchester developed the diffusionist or dispersionist theory that all civilization came from one (or at most a few) Old-World centers.

The members of this school should properly be called *extreme* diffusionists, because all anthropologists, even such anthropological evolutionists as Tylor and Morgan (who derived the Amerinds from the Tamils of India) have admitted that much diffusion has occurred. Nobody claims that all the people who use matches or guns invented them independently, though extreme diffusionists like to pretend that the "orthodox anthropologists" really believe such to be the case.

Elliot Smith, following Eduard Braun in Germany and Miss A. W. Buckland in England, traced all civilization back to Egypt. He affirmed on feeble grounds that the discovery of copper-smelting stimulated the Egyptians to develop writing, agriculture, and all the other elements of civilized culture. Smith and Perry called this early Egyptian culture "Heliolithic"; they claimed that it included sun-worship, mummification, pyramids, the swastika symbol, metallurgy, irrigation, and the custom of putting papa to bed when a child is born. Before this revolution the Egyptians, like all other men, lived like apes.

The Helioliths were supposed to have wandered all over the world looking for gold and pearls, which they valued for religious reasons, and to have incidentally founded all other civilizations, including the Mayan and Inca. Diffusionists thought you could identify such cultural influences because the distinctive Egyptiac traits were tied together in a "culture complex" that tended to be transmitted in a solid bundle or not at all.

Smith put the Heliolithic expansion in the ninth century B.C.; Perry, about 2500 B.C. Their chronology was fantastic; Smith asserted that the cultural diffusion started out from Egypt after 1000 B.C. and arrived in India in pre-Aryan times, *before* 1000 B.C. We are even asked to believe that the Mexican pyramids erected during the early centuries of the Christian Era were copied from Cambodian pyramids built 500 years *later!*

Their general treatment of facts is about as reliable as that. For instance Perry, wishing to prove that all North American cultures were degenerate copies of that of the Mayas (who got theirs from Egypt) said the Mayas had irrigation while the North American Indians did not—when the reverse is true. And he described Middle American history as one of "uninterrupted culture degredation" when it is of course nothing of the kind, but the story of the successive rise and decline of various peoples: the rise and fall of the Mayan Old Empire, the rise and partial decline of the New Empire, the rise and fall of the Toltecs, and the rise of the Aztecs. Nobody can say what would have happened to the Middle Americans if Europeans had not arrived. Perhaps the Aztecs would have fallen while the Mayas achieved another renaissance, or perhaps both would have declined while another people like the Zapotecs took up the torch of Middle American culture.

Moreover it is hard to see why the diffusionists should pick the Egyptians as the source of Heliolithic culture, since of all the great nations of antiquity they were the most stubborn and self-satisfied stay-at-homes. Except for a few coasting voyages around the Red Sea and the temporary conquest of Syria under the Eighteenth and Nineteenth Dynasties, they seldom left their favored land, leaving exploration and colonization to Cretans and Phoenicians. It is equally ridiculous to say that everybody lived like apes before the arrival of the Heliolithics, when such well-developed cultures as that of Iraq came before the rise of Egyptian civilization.

What is more, the northern peoples of Asia and North America could hardly have derived such ingenious inventions as pants, felt, skis, snowshoes, and igloos from Egypt, since the ancient Egyptians neither made felt nor wore pants, nor did they have snow to ski on or make igloos of.

Now that Elliot Smith has been gathered to his sun-worshipping ancestors, the cult is carried on by Mitchell and Raglan in England, and in America by Harold S. Gladwin, author of the recent *Men Out of Asia*. They all think inventiveness such a rare trait that no major invention could have been made more than once; that most savages are that way because they have degenerated from Heliolithic standards; and that similar culture-traits in any two parts of the world are *prima facie* evidence of diffusion.

This matter of "degeneration" is pretty much one of the personal prejudices of the anthropologist. Diffusionists apply the term to any loss of a culture-trait, like rice-culture or large stone monuments. But the change may be an intelligent adaptation to new conditions, as when the Polynesians moved to islands unsuited to rice-culture, and took to fishing and growing taro and bananas instead. And while the great pyramids which Egyptian kings built as their tombs are from one point of view an engineering triumph, from another they represent a vast waste of effort, and the Egyptians were much better off when they dropped the custom.

Gladwin has an interesting variant of the diffusionist theory of New-World culture. After he has described a series of migrations from Asia by alleged Australoids, Negroids, and other peoples (mostly guesswork on his part) he suggests that when Alexander the Great died, the fleet he was building to explore the Indian Ocean sailed away, manned by the Egyptians, Phoenicians, Cypriots, Greeks, and other people whom he had hired as sailors. They went east to India and Indonesia, picking up recruits and women on the way, becoming the Polynesians, and pushing on to South America where they became the Arawaks. Their White leaders Viracocha, Quetzalcoatl, and the rest travelled about founding all the American civilizations; hence the "fair god" culture-hero myths of the New World. The native Amerinds, says Gladwin, are mere putterers who would never have gotten anywhere without such leadership.

To show his disesteem for those who will not accept

this fantasy, Gladwin erects a straw-man who he calls Dr. Phuddy Duddy, representing the "Voices of Authority," whom he then demolishes with the vim of a Mayan *nacom* tearing out his victims' hearts. Unfortunately Dr. Duddy's stuffing seems a bit moldy. Gladwin attributes to Duddy a lot of convictions that few or no modern Americanists hold, as that the Amerinds came over in one migration, that they are all of the same physical type, that they are smarter than Old-World peoples, and so on. Also he has avoided facts like the recently found Tepexpan skull that don't fit his scheme. That is called selection of evidence, and is how the Atlantists and Lost-Ten-Tribes hunters get away with it among the imperfectly educated.

As you can see, the uninventiveness of man is a cardinal dogma of the diffusionists. Scientists have long argued about invention by preliterates. Modern science, we know, is built upon (1) philosophical speculations about the universe and (2) technological invention. These are so closely associated today that, while the garret genius still flourishes, more and more inventions are made by scientists and engineers in the laboratories of universities and corporations.

However, it was not always so. The farther back one goes the more distinct science and invention become, until in ancient times science was a matter for priests and philosophers, while nameless common men made inventions. Archimedes, a great Classical scientist, apologized for his inventions as beneath a philosopher's dignity, while on the other hand we do not even know the names of the inventors of such important devices, all less than 1500 years old, as the rudder, the windmill, the horse-collar, and the iron-casting furnace.

Although invention seems to have been a continuous process ever since the first sub-man learned to chip flint and feed fire, don't take seriously those stories of primitive life wherein a hero named Ug invents the bow, the canoe, the domestication of animals, and monogamy all in one lifetime. In real primitive life such developments are strung out over thousands of years.

Now, when two peoples far apart use a similar gadget or custom, did they invent it independently, or did they get it from a common source? Raglan and Smith assume that the latter is *always* the case, and by this convenient assumption the ultra-diffusionists trace all civilization to a single source, usually located in Egypt. On the other hand the late Prof. G. N. Lewis derived all civilization from Brazil, while others have sought it in Peru, the Ohio Valley, India, the Arctic, Atlantis, and Lemuria. (The "orthodox" or "anti-diffusionist" view, which I think is supported by the weight of the evidence, is that civilization arose more or less independently in at least four places: Iraq, China, Mexico-Guatemala, and Peru, in that order.)

But is invention so rare? The U. S. Government has for years been issuing over 50,000 patents a year, or about one patent per year per 25,000 citizens. That is not to say that you will get one invention a year from any group of 25,000 people, because there are other factors to consider: unpatented inventions, patents on very minor improvements, the contrast between the encouragement our civilization gives inventors and primitive conservatism, and so on. Still, any considerable primitive group should produce *some* inventions from time to time.

The ultra-diffusionists, in fact, show a strange and snobbish prejudice against the concept that inventiveness is a widespread human attribute that crops up among working folk and primitives as well as among scholars and scientists. Raglan for instance insists that preliterates can't invent because they have no scholars and scientists among them, though, as we have seen, scholars and scientists are not necessary for invention. Therefore, say the diffusionists, everything must have come from somewhere else—it doesn't much matter where. Laymen who read a little anthropology or Atlantism tend to agree with them, perhaps because, never having made an invention themselves, they find it hard to imagine anybody else's doing so. As Lord Raglan tartly puts it: "We are often told that the Bonga-bonga have discovered the art of smelting iron, or that the Wagga-wagga have invented an ingenious fish-trap, but nobody claims to have seen them doing it."

But then, neither has anybody seen a Heliolithic Egyptian teaching them how to smelt iron or trap fish.

Of course, if it be true that there is no *a priori* reason

to assume that a culture-trait was either borrowed or invented locally, we are left in the uncomfortable fix of perhaps never being able to learn how many culture-traits originated. While we all like simple formulas to settle such troublesome cases, such assumptions are after all mere guessing. In the cases where we do know the origin of a trait, it turns out that both invention and diffusion have played large parts in the growth of culture.

For instance, printing was independently invented in China and Germany. The crossbow was invented in Classical times by Zopyros of Tarentum; re-invented or revived in medieval Europe; and independently invented in southeast Asia. The fire-piston, a handy little gadget for lighting tinder by compressing air in a small cylinder, was independently invented in southeastern Asia or Indonesia and (in 1802) in France.

Where documentary evidence is lacking, we can make a good guess at the origin of a trait from its distribution. If the trait is found in one continuous area the chances are that it was diffused from one center. If on the other hand it occurs in two widely separated places, with no trace of its existence in between, it is likely to have been invented independently in both.

As an example the blowgun is found (without mouthpiece) in southeast Asia and Indonesia, and again (with mouthpiece) in the tropical parts of the Americas. To take the blowgun from one of these areas to the other, one would have to transport it either across the vastnesses of the Pacific, or over the great windy plains and through the hardwood forests of northern Asia and North America. Such a trek would take primitives many generations, during which time the travellers would be passing through land lacking canes and bamboos for making blowguns and poisnous plants for envenoming the darts, and where the tool itself would be almost useless.

Hence, by all reasonable calculations, the blowgun must have been invented independently in the two places. To argue that the Indonesians must have brought the blowgun to America because the blowgun is found in both places, and then that it must have been brought from Indonesia because Indonesians brought culture-traits (like the blowgun) to America, is circular logic: assuming what you wish to prove. The blowgun makes a neat case because the Malayan peoples not only devised it independently, but even, in Borneo and Celebes, provided it with a gunsight and a bayonet, developed independently of the artisans who added these accessories to guns in Europe in the seventeenth century.

True, primitives do not invent very often—or perhaps I should say that their inventions are not often adopted. In a conservative tribal environment, you see, it may be less of a problem to make an invention than to get it accepted by your fellow-tribesmen without yourself being liquidated as a dangerous innovator. Hence the same invention may have been made in tribal society many times before if finally took.

However, even in the short time that primitives have been under anthropological observation they have made some inventions. Thus in 1871 the Sioux Indians invented the central-fire primer for the rifle-cartridges and the practice of reloading empty cartridge-cases. The ghost-dance religion, launched by the Paiute Indian Wovoka in 1889, was an invention of sorts. And about 1900 a Gilbert Islander living in the Marquesas in the Pacific invented a detachable outrigger to keep people from stealing his canoe.

In short, there is nothing incredible about independent invention, even among primitives. In civilized society it happens all the time: every year the U. S. Patent Office conducts hundreds of "interferences," investigations to discover who, of two applicants for a patent on the same invention, was the first inventor. And many scientific hypotheses have been developed independently by different men, the most celebrated instance being evolution, discovered at the same time by Darwin and Wallace.

Furthermore, even when we can trace the diffusion of a culture-trait far and wide from one center, such traits spread from many different places: tobacco-smoking from North America, rice-culture from southeastern Asia, the stirrup from central Asia, the gun from Europe, and so on. Therefore there is no good reason to credit the Egyptians or any other single people with all the inventions of early civilization.

Some apparent similarities between culture-traits of far-separated peoples may be due neither to diffusion, great inventive genius, nor "psychic unity," but to the limitations of the material. If you set out to make a paddle, for instance, you soon learn that your paddle must be of a certain length and shape (within rather narrow limits) or it won't work very well. Likewise in the burial customs beloved of Atlantists and diffusionists, you can do only so many things with a corpse: bury it as among us, burn it as in India, mount it as in Inca Peru, throw it away as in Tibet, or eat it as the ancient Irish did. If you decide to keep it you will have to preserve it against decay by gutting, greasing, stuffing, drying, or other forms of taxidermy. Since most of these few alternatives were worked out long ago, the burial customs of different tribes are bound to look alike now and then by pure chance.

The ultra-diffusionists, moreover, fail to distinguish between the handing on of material objects and that of techniques. Material things obviously travel much farther and faster; guns diffused all over the world within a few centuries of their invention, according to tradition by Friar Berthold the Black in 1313. But only in a few places did non-Europeans learn to make guns for themselves. A pot can change hands in a matter of seconds, while the art of pottery requires training, which a wandering trader could probably not give even if he wanted to.

Further, if people who know how to make a given artifact think about the matter at all, they are probably unwilling to give away their knowledge to outsiders for fear of losing the commercial or military advantages that exclusive possession gives them. Thus the Chinese kept the nature of silk secret from the West for centuries until a couple of monks smuggled silk-worm eggs out in the reign of Justinian. Therefore to argue diffusion from likeness of techniques all by itself, in the absence of material objects, is backwards reasoning. A few Egyptian scarabs or Greek coins, found in undisturbed American sites, would make a much stronger case for Old-World trade with the New in ancient times than Mr. Gladwin's comparisons of culture-traits; but no such staggering discoveries have ever been made except in books by writers like Churchward.

Moreover the diffusionists are probably wrong in thinking that any collection of culture-traits called a "complex" will stick together indefinitely and can be traced all over the world. There are two kinds of culture-complexes: the logical or organic, where one trait necessarily entails the others (as the domestication of the horse implied the saddle, bridle, and whip) and the accidental or adventitious. Groups of traits of the former kind stick together, certainly, so that when the Plains Indians adopted the horse they took over the saddle, bridle, and whip along with it. Contrariwise, groups of the latter kind show only a slight tendency to stick together. Thus tobacco-smoking went all over the world shortly after the discovery of the Americas, without (fortunately) taking with it such other Amerind traits as scalping and shamanism.

Finally, the whole diffusionist controversy is somewhat unreal, since, as Malinowski once stated: "Diffusion is but a modified invention, exactly as every invention is a partial borrowing." In other words, nearly all inventions are improvements on something already invented, and people who borrow culture-traits usually modify them in the process. Where you draw the line between borrowing and invention is therefore a matter of personal preference.

Some inventions seem to be cases of what Kroeber has called "idea diffusion" or "stimulus diffusion"; a primitive hears about an invention used by others without ever seeing it in operation, and is stimulated by this news to try to develop his own version of the invention. This process is neither pure invention nor pure diffusion, but involves both; it probably accounts for most of the scripts invented among preliterate peoples in recent centuries. Thus King Njoya of Bamun, Kamerun, invented a system of writing about 1900, and while he probably got the germ of the idea from Arab or European sources, he created, not a phonetic alphabet like that of the Arabs or Europeans, but a system that was ideographic like Chinese. Likewise in 1821 the Cherokee Indian Sequoya worked out a Cherokee syllabary, and a few years later Momolu Duwalu Bukele of the Vai of West Africa either invented the script still used by his tribe or converted it from an ideographic to a syllabic system.

The inventive spark, it would seem, does sometimes flash even among primitives.

The ultra-diffusionists have been especially eager to derive the New-World cultures from the Old World, arguing that civilized Amerinds must have been connected with Egypt because they built pyramids and left mummies behind them; with China because they used jade; with India because they sculped elephants upon their monuments. All these arguments dissolve at the touch: the Peruvian "mummies" are mere dried-up corpses preserved, not by Egyptian embalming technique, but by the dryness of the Peruvian climate; the Mayan jade is from a New-World source; and so on.

The myth of the Mayan elephants was started by John Ranking's *Historical Researches on the Conquest of Peru, Mexico, Bogota, Natchez and Talomeco in the Thirteenth Century by the Mongols, Accompanied with Elephants* (1823). Ranking's elephants were the fossil remains of mammoths and mastodons, found in great numbers in the Americas. Some of these beasts were alive when the first men arrived in the New World, but became extinct long before the rise of the Mayan civilization.

The fiction was further built up by Count Waldeck, among whose scientific felonies was to draw pictures of Mayan glyphs with non-existent elephant heads in them. Here, for instance, is one of his drawings (left), with a truthful drawing of the selfsame glyph made some years later by Catherwood for comparison:

When Elliot Smith learned of Waldeck's elephant drawings almost a century later, he insisted upon taking them seriously, though more accurate reproductions of the glyphs, such as Catherwood's drawings, had long been available.

Diffusionists also pointed to the monolith called Stela B at Copán, on which are carved two creatures that do look somewhat like elephants with mahouts on their back. (See Fig. 11.) In fact when the stela was damaged by vandals some years ago, the diffusionist Mitchell in his book on the Mayas accused the anti-diffusionist archeologists of having mutilated the monument in order to suppress evidence against their theories.

But if one examines the monolith closely, one sees that the "elephants" have nostrils, not at the ends of their "trunks" as elephants should, but in front, at the roots of these organs. Furthermore they have large round eyes surrounded by feathers. Feathered elephants, as you know, are extremely rare; these are probably conventionalized macaws.

In the case of the Mayas, then, we can be reasonably sure they invented many of their culture-traits like their calendar, writing, paper, and architecture; others they got by diffusion: metallurgy probably from South America.

A similar verdict applies to other Amerind civilizations and culture-traits. The bow for instance seems to have diffused over from Asia after the beginning of the Christian Era, before which it was unknown in the Americas. Theretofore the Amerinds used a gadget called a spear-thrower: a stick about two feet long with a handle at one end and a hook or spur at the other that could be engaged in a hollow or socket at the butt-end of a javelin. The user held both spear-thrower and javelin in one hand, the hook engaging the butt of the spear, and when he threw he let go the spear but gripped the thrower so that the latter acted as an extension of his arm. The Mayas said Aztec mercenaries brought in by Hunac Ceel introduced the bow to their country.

On the subject of pottery in the New World there is evidence both for independent invention and for diffusion. The question is unsettled, and both processes may have been at work. Lastly, despite talk of the Indians' being "putterers," they created several ingenious devices like the tobacco-pipe and the hammock that were quite unknown in the Old World.

Fig. 11. THE "MAYAN ELEPHANTS" on Stela B at Copán, as drawn by Sir Alfred P. Maudslay.

Now that we have brushed away the fantasies of the diffusionists, the Ten Tribists, and the Welsh-Indian hunters, what *is* the story of the native American cultures? Well, during the Pleistocene Period the advances of the ice locked up so much water in the glaciers that the sea-level fell far enough to connect Alaska with Siberia. And this was the route followed by the first Americans from Asia.

Nobody knows just who these first immigrants were. There are several candidates: Lime Creek man who left tools of bone and stone in Nebraska, and whose discoverers thought he might have come over during the third interglacial period about 40,000 years ago; Sandia Cave man, who left leaf-shaped spearheads in New Mexico during the last retreat of the glaciers, perhaps 25,000 years ago; Cochise man from Arizona, a vegetarian who ground seeds and nuts on flat milling-stones; Abilene man from Texas, who made crude fist-hatchets like those of a Neanderthaler; and Tepexpan man, who left a skull in a Mexican alluvial plain 10,000 to 15,000 years ago. By 8000 B.C. Folsom man was hunting the extinct straight-horned Taylor's bison in North America with his peculiar spearheads grooved on the sides like a bayonet. After him Yuma man made eight-inch stone blades or spear-points with concave bases and a peculiar ripple-pattern, and after Yuma man came the historically known American peoples.

It is not quite certain what these early Americans looked like. Tepexpan man was pretty much a common Amerind, with sloping forehead, flattish nose, and wide zygomatic arches (cheek-bones). While all the Amerinds belong to the Mongoloid or Yellow race, they vary in height, skull-form, and skin-color just as Whites do. Some resemble, not the specialized Mongoloids of China with their ultra-broad heads and dish faces, but the less specialized types found here and there in Siberia, Tibet, and Indonesia, tall men combining the Mongoloid coloring, hair, and cheek-bones with long skulls and prominent noses.

In 1492 the long-headed Amerinds were mostly scattered around the margins of the two continents, while the broad-heads occupied large solid areas in the middle. This fact suggests that the earlier waves were long-headed, while the later were broad-headed like most modern Asiatic

Mongoloids. There are several fossil skulls of possible great age besides Tepexpan man, such as those from Punin, Ecuador; Abilene, Texas; and Lake Pelican, Michigan. While all of these may not be so old as their finders hoped, the chances are that some are, and all are of this long-headed Mongoloid type.

Down to about 1000 B.C. all evidence points to a thinly spread population of hunters and food-gatherers, here and there experimenting with the useful plants of which the New World afforded such a pleasant variety. No Atlanteans, no Muvians, no Egyptian Children of the Sun radiating enlightenment: just stone-age Mongoloids. During the next thousand years some tribes developed higher cultures. While dates are uncertain, one of the first such stirrings may have taken place about the middle of the first millenium B.C. in the state of Vera Cruz, Mexico, on the Gulf Coast, whence culture diffused west to the Valley of Mexico and east to Guatemala. Another such center arose in Peru. Linton suggests that the rather sudden flowering of American civilizations around the time of Christ followed the domestication of the bean, which by providing a bigger and more reliable supply of protein than could be furnished by hunting, made possible denser populations than maize, which provides starch only, could support alone.

By the beginning of the Christian Era the Amerinds had developed such adjuncts to civilization as writing and stone cities, though the latter were less cities in our sense than community centers whither the Amerinds gathered from their scattered huts for trade and ceremonies. At this time the Polynesians had not even begun their wonderful colonization of the South Pacific islands; the Scandinavians, so far from being white gods bringing civilization to the Amerinds, were a handful of miserable clam-diggers huddled on the shores of the Baltic Sea. The bow was still spreading over North America; soon the Andeans would develop metallurgy (which the Polynesians lacked) and copper would reach the Mayas and Aztecs 800 or 1000 years later, while the Andeans themselves went on to discover bronze-casting.

So much for the Amerind civilizations. The Amerinds were neither supermen nor mere "putterers," but ordinary men who did pretty well considering what they had to work with; who neither came to America completely uncultured, nor invented every last culture-trait themselves, but who still showed ingenuity without help from Atlantean en-lighteners, Egyptian gold-prospectors, Jewish refugees, Macedonian sea-captains, Welsh princes, or other Old-World cultural sources. And to use a non-scientific argument, haven't we stolen enough from the Amerinds without depriving them of their just claim to the same degree of originality that other peoples have shown?

CHAPTER VII

THE CREEPING CONTINENTS

The years will come, in the eld of the world
When Ocean will loosen his grip on things,
When the land will extend itself afar,
And Tethys new continents will disclose
Nor will Thulë then be the end of the earth.
Seneca

YOU cannot, then, prove whether or not sunken continents ever existed by comparing the cultures of different peoples, because to do so you would have to make the same assumptions as those of the diffusionists about culture-complexes, the scarcity of invention, and so on. In the first place, as I have tried to show you, these assumptions are wrong, and in the second they give contradictory results: for instance, Atlantis versus Egypt versus Brazil as the source of all civilization.

On the other hand neither can you disprove sunken continents by the evidence of cultures alone. It is safe to say that no continent with a high civilization, like that of Egypt in her days of glory, existed in the Atlantic 10,000 years ago, because it would have left relics of that age on the neighboring shores of North America, Europe, and Africa, which it did not. Still, the Atlanteans might, as Spence asserts, have been a stone-age people with a few modest accomplishments in the way of culture. The evidence of human remains does not flatly rule out Atlantis according to Spence's time-table. That is not to say that archeology *favors* the existence of these continents; it just doesn't settle the matter one way or the other.

But *do* continents rise and fall the way Plato said they did? Or, if they don't, is there any other way land

can be flooded on a vast scale so as to give the impression
of a world-wide Deluge? For the answers to these questions
we must go to geology, a science that produces rugged characters and quaint technical terms that I will try to keep
to a minimum.

The idea that present land areas were once under water
and *vice versa* is one of the oldest thoughts that men have
had about the earth they live on, and it happens to be, in
a general way, correct. Creation-myths often start with a
waste of waters which stays that way until a god drains off
the surplus water (Hebrews) or hauls the land up to the
surface with his fish-line (Polynesia) or until the Muskrat
brings mud to the top to make the first continent (North
America). As I told you in the first chapter, many Classical
writers thought long and hard about the matter; thus
Herodotos, Aristotle, Polybios, Strabo, and several others
noted the presence of the fossils of sea-creatures on land.

However, it is one thing to say that present land was
once water and present water land, and quite another to
show just what the map looked like on the morning of
January first, 100,000,000 B.C. Geology only began to grow
as a science, as distinct from a vague body of speculation
about whether rocks reproduced like animals, in the seventeenth and eighteenth centuries.

In the early years of the nineteenth century the French
anatomist Baron Georges Cuvier first arranged geological
formations in the order in which they were made by comparing the fossils found in them. Obviously life had
changed during the earth's history, and Cuvier, trying to
account for this fact and at the same time to save the Biblical creation-myth, assumed that a series of catastrophes had
from time to time destroyed all life on earth. After each of
these, he thought, God created new plants and animals,
and *Genesis* merely tells of the last of these creations.

While that theory seemed reasonable when only a few
geological periods were known, with wide gaps between
them, it no longer fitted the facts when so many fossils had
been found as to show that evolution had been continuous,
with no Cuvierian catastrophes. Darwin put this discovery
into rational form a little less than 100 years ago, and his
theory of evolution by selection of the fittest has (with

minor modifications) survived vicious attacks from the pious to triumph over all opposition.

To unravel the past history of the surface of the earth we must use every fact about the rocks and the fossils found in them that we can lay our hands on. This study is called paleogeography, or historical geology. For instance, if you want to know what a given square mile was doing in the early Triassic Period, you can tell easily enough if there is a fossil-bearing early-Triassic deposit on the surface of that square mile. You know that limestone and chalk are laid down on the bottoms of shallow seas, that sandstone is deposited by rivers on deltas and plains, and that coal is made in swamps.

But what if that area is covered by a formation of another age, or lies at the bottom of the sea? Moreover your troubles get worse as you go back in time, and formations become rarer and less well-preserved.

Still, the situation is by no means hopeless. We can learn much about the geography of the past by studying how the earth is built, how various rocks were formed, and how living things are distributed. Fossils tell us what land-bridges and water-channels existed in former times, and what sort of climate there was, and how the ocean currents flowed.

To show the opening and closing of land-bridges, the fossils of large land-animals are the best indicators, since these creatures move about actively on land but cannot cross even narrow stretches of water. On the other hand flying things like birds and insects can be blown across by a gale; small land animals may be wafted over on pieces of driftwood; and the seeds of plants and the eggs of small animals may be carried across by birds.

Thus when we find that in some past age the same species of large land-animal lived in two land areas, now separate, we know that they must have been joined either then or shortly before. In this way we know, for example, that North and South America were separated throughout the first half of the Age of Mammals (the Cenozoic Era of Appendix D). About the end of the Miocene Period the Isthmus of Panama appeared above water and let animals cross over, which is why there are armadillos in Texas and jaguars in Brazil.

Several decades ago many scientists who lacked the mass of evidence we have today, but who should have known better, made reckless guesses about former land bridges across the oceans, and the Atlantists promptly picked up these guesses as "scientific proof" of their lost continents. Gregory in England, for instance, filled the Pacific Ocean with a regular network of supposed continents and land-bridges. They used such arguments as insisting that titanotheres (rhinoceros-like beasts that throve in North America and Asia during the early Cenozoic Era) must have crossed a mid-Pacific bridge because the obvious Siberia-Alaska bridge would have been too cold. But in the first place we don't know how cold the Bering Strait isthmus really was then (the climate may have been mild) and in the second we don't know how much cold the titanotheres could stand (they may have been hairy like the mammoth).

Another builder of land-bridges, the English naturalist H. E. Forrest, tried to persuade his readers that there had once been a land-bridge across the North Atlantic Ocean, taking in Iceland. While such a bridge probably did exist once upon a time, Forrest wanted his bridge in the Pleistocene Period, the last age before the present Recent Period, late enough to furnish a basis for Plato's story. And that, alas, will not work.

Forrest based his arguments largely on the distribution of plants, creeping things, and fresh-water fish. These are slowly-evolving organisms, some of which may have been about where they are and what they are since the Mesozoic Era sixty or more million years ago, and they therefore prove nothing about the shape of the land in the Pleistocene Period. For the Pleistocene, on the other hand, the distribution of large land-animals shows a bridge from Siberia to Alaska via Bering Strait but none from Laborador to Europe.

Recently Simpson of the American Museum of Natural History has worked out a time-table of the opening and closing of the Bering bridge. This he did by comparing the percentages of species, genera, and families of animals found living at the same time in different parts of the earth during various past periods. Simpson concluded that the Bering bridge was open to land traffic during most of the Cenozoic Era, with an interruption in the middle Eocene Period, another in the middle and late Oligocene, and of course still another since the last retreat of the glaciers in the Pleistocene, which has lasted to our own time. There may have been shorter interruptions as well, but these are hard to detect.

Simpson and many of his colleagues agree that, according to fossil evidence, the main land masses have stayed in very much the same places where they are now during the Cenozoic Era, even though shallow seas have overflowed the continents and land-bridges have opened and closed. Therefore during the last 50,000,000 years at least the connections between the continents have been those that you see on the map today: Bering Strait, Panama, Suez, and some time earlier a bridge from Australia to the Malay Peninsula through New Guinea. During this time, say these scientists, there have been no direct connections among Africa, South America, and Australia, which have exchanged animals via the northern continents only.

The distribution of animals is fatal to many of the Atlantists' sunken lands. For example Lewis Spence suggested two Pacific continents, one stretching east and west from the Hawaiian Islands to the Malay Archipelago, taking in New Guinea, Celebes, and Borneo; the other reaching north from New Zealand. The first continent unfortunately lies right across one of the sharpest animal boundaries on earth: the "Wallace line" through Indonesia that divides the Oriental from the Australian region. As one goes from Borneo to New Guinea one passes from the Indo-Malayan world of monkeys, cats, buffaloes, and elephants to the vastly different world of kangaroos and echidnas. In the latter region the mammals are all (with a few easily-explained exceptions) egg-laying monotremes or pouched marsupials; in the former they all belong to the higher "placental" type to which we ourselves belong as well. The only placental mammals to reach New Guinea are dogs and pigs brought in by men, small rodents that could have floated over on the driftwood ferry, and bats that flew in. These facts show that the water-barrier of the Celebese, Banda, and Timor Seas has been where it is for a long time, probably ever since the Mesozoic Era, and Spence's continent, which would have let the beasts of these two zones mingle freely, is out of the question.

Back before the Cenozoic Era, however, the evidence regarding land-bridges does not point so clearly to any one conclusion. Furthermore some evidence does strongly suggest direct connections among the southern continents in the Mesozoic Era and earlier. These connections may have been via Antarctica (an idea favored by the late great Henry Fairfield Osborn) or by way of land-bridges across the South Atlantic and Indian Oceans. Which brings us to the story of the Gondwanaland theory.

Before going into that subject, however, we had better take time to consider what a continent is. The present picture of the earth that geologists have worked out from the action of the tides, the behavior of earthquake shocks, the things meteors are made of, and other indications, may not be entirely right, but it is surely more correct than the pictures that went before it and will do until a better one comes along.

As geologists see it, the earth consists first of the skin or crust of rock that we see. As one goes down, though, the rock gets hotter and hotter until at a depth of 50 or 100 miles it is white-hot and would be molten if it were on the surface. Down there, however, it is not a true liquid because the enormous pressure on it keeps it in a state like glass — a non-crystalline or amorphous stuff that resists quick stresses like a solid but yields like a liquid to long-continued ones. The earth's substance from there down is in fact a kind of hot glass called "magma," stiffer than steel, down to the nickel-iron core of the earth, a sphere 3500 to 4000 miles through.

We can forget the iron core and concentrate on the two outer layers: the thin crystalline crust on top and the thick glassy substratum, as it is called, beneath it. And the crust is made up in general of two main kinds of rocks: dense rocks like basalt, mostly silicon and magnesium salts, and light rocks like granite, mostly silicon and aluminum

salts, with various intermediate kinds. The heavy mag-
nesium-bearing rocks are called "sima" (silicon-magnesium)
and the light aluminum-bearing rocks "sial" (silicon-
aluminum).

Now, the rocks of these different kinds are not distributed in a haphazard manner at all; instead, the land
areas are mostly sial while the sea-bottoms are mostly sima.
The continents, in fact are really patches of sial "floating"
on a crust of sima like cakes of ice in a river. Geologists
guess these patches to be anywhere from ten to sixty miles
thick, but in any case they extend down into the sima
farther than they stick up out of it, just as ice does in water.
Granitic rocks are so characteristic of continents that they
are sometimes called "continental-type" rocks.

So close is this relationship that geologists, when they
find sial rocks under water or on an island, suspect that
here may be the place of a former large land area. On
the other hand they think that islands made entirely of
volcanic sima, like the Samoan group, must have grown up
from the sea-bottom by the outpourings of volcanoes.
They do not therefore believe such islands to be the remains of any continent, and do not take the Central Pacific
continents of Spence and other Atlantists seriously, since
the islands of Polynesia are all of this volcanic-sima type.

To tell where continents might possibly have existed,
we have to study the ocean floor. The ocean bottom presents the following forms: shallow continental shelves which
are merely the submerged edges of the continents; vast submarine plains, miles deep and monotonously smooth; and
areas of moderate depth with a broken, mountainous relief as if a mountain-range had sunk till only the tops
showed.

By studying the speed of earthquake vibrations, geologists can get a good idea of what the earth's crust under
the oceans is made of, since these vibrations travel at speeds
that vary by as much as 26% — too much for experimental
error. They have learned in this way that the great deeps are
plain sima, while the continental shelves and the rough,
moderately deep places are partly sial. The greatest areas
of deep sima are in the Central Pacific, the southern Indian,
and the Arctic Oceans. These are therefore the "permanent"

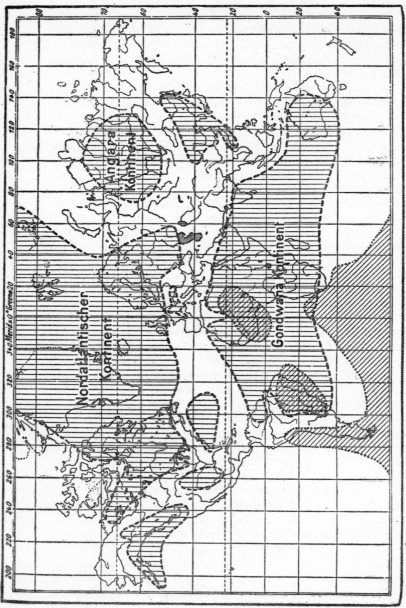

Fig. 12. SUESS'S IDEA OF THE WORLD IN THE LATE PALEOZOIC ERA, according to Dacqué's *Paläogeographie*.

oceans, where no continents are to be expected, past, present, or future.

So much for Mu, Pan, and the Theosophical Lemuria, which Atlantists have located in the Central Pacific — the unlikeliest place on earth for a continent. The great Pacific sima depression is in fact probably one of the most stable and long-lived features of the earth's surface, going back in much its present form at least half a billion years. Fig. 6 shows a map of the Pacific with the "andesite line," east of which not one pebble of continental-type rock has ever been found.

Men have naturally wondered why most of the earth's sial should be concentrated in one hemisphere centering in Europe, leaving the opposite hemisphere nearly all water, but nobody has yet given a final answer to the question. In the late nineteenth century Osmond Fisher and George Darwin suggested that the moon was an agglomeration of the missing part of the sial crust, thrown off by centrifugal force when the earth was hotter and more liquid and spinning faster. However, a few calculations showed that the moon's mass is about 37 times too great for the purpose, and moreover it has a specific gravity of 3.7 compared to 2.5 for the earth's granitic crust.

The great submarine mountain ranges (where, you remember, we might reasonably suspect a lost continent) lie in the southwest Pacific (including the Fijis and New Zealand) and in the northwest Indian Ocean. The Atlantic Ocean appears to be sima with little patches of sial here and there. Also, "continental" rocks occur in islands where the sea-bottom is thought to be made partly of sial: among others, the Fijis, the Seychelles in the Indian Ocean, and the Canaries and Madeiras in the Atlantic, the last probably representing a peninsula that once stuck out west and northwest from the Moroccan coast.

Some have even claimed that continental rocks occur in small amounts in the volcanic Azores, right where orthodox Atlantists put Atlantis. While most geologists are cautious about the geological history of the Azores, for lack of information, some concede that they might just possibly be the remains of a larger island, perhaps the size of Spain. However, those who think so are sure the island had sunk by the end of the Miocene Period, when our ancestors were still running naked through the woods and living precariously on berries and beetle-grubs.

On the question of sunken continents, modern geologists are divided into three schools: the stable-continent school, the transoceanic-continent school, and the continental-drift school. The stable-continent geologists like Matthew believe in no sunken continents; the most that happened, they say, was a series of slight movements of the existing lands to open and close land-bridges. On the other hand, the transoceanic-continent school believes big former continents like Gondwanaland possible, while the small continental-drift school holds that the continents are stable in size and shape but that they drift about the surface of the earth.

Gondwanaland and the transoceanic-continent school in general began with the Viennese geologist Melchior Neumayr (whom you may remember from the third chapter as the author of the world's first paleogeographic world map) and with the Lemurian speculations of Blanford, Sclater, and Haeckel.

Neumayr was followed by another Austrian, Eduard Suess, who in his day was the Grand Old Man of European geology, at which he worked actively for more than half a century. In the 1880's Suess expanded upon the ideas of those who had gone before him in an immense five-volume treatise on the geology of the world, *Das Antlitz der Erde* (*The Face of the Earth*). Suess thought that in the Paleozoic Era, the age of fishes and invertebrates and coal, there was one large continent in the Southern Hemisphere in addition to those in the North. This he named "Gondwána-land" after the tract in India which Blanford had described. (Do you recall the Gonds and their unattractive tribal customs?)

At the same time there were, according to Suess, two continents in the northern hemisphere. One, North America with a peninsula reaching over to Europe by way of Greenland and Iceland, he called "Atlantis"; the other, consisting of eastern Asia, he named "Angara-land" after a Siberian river. The shallow sea separating Gondwána-land from Angara-land he baptized "Tethys" after a Titaness from Greek myth, the wife of our old friend the sea-god Okeanos. Suess thought that while Gondwanaland broke up during the Mesozoic Era, the Atlantean land-bridge lasted until well into the Cenozoic — not late enough to be Plato's Atlantis, perhaps, but late enough to have affected the migrations of mammals in the Cenozoic Era.

Fig. 12 shows Suess's idea of the world during the late Paleozoic Era, according to a more recent German geologist, Edgar Dacqué. Dacqué, a strong Gondwanalander, also believed in an eccentric theory of evolution, the "Doctrine of Types." He supposed the human stock to have existed as a separate stem of evolution clear back to the Paleozoic Era, or as a sort of central trunk of the evolutionary tree, all other animals representing degenerate side-branches.

During the Jurassic and Cretaceous Periods when dinosaurs ruled the world, according to the Neumayr-Suess school, Gondwanaland gradually broke up by the sinking of its various parts. Australia and New Zealand separated first, which fact accounts for their lack of placental mammals, since placental mammals had not been invented at the time of the separation. South America went next, leaving it with the very curious assortment of mammals that lived there until it was re-connected with North America: herbivorous placental mammals like the ground-sloths, and carnivorous marsupials like those found today in the Australian region, for instance the thylacine ("Tasmanian wolf") and the dasyure ("native cat").

The last part to sink was the land-bridge connecting South Africa with India, the geologists' Lemuria, leaving Madagascar and the great Seychelles Reefs with their islands as a fossil of its former existence. Some of the curious animals found in out-of-the-way spots in the Southern Hemisphere, such as the platypus, and the giant tortoises of the Galápagos and Aldabra Islands, are thought to be in a sense survivors of the life of Gondwanaland before its breakup. Opinions differ about the cause of this fragmentation: some say parts of the continent sank into the substratum and dissolved, while some continental-drifters think Gondwanaland came apart at the seams in

the course of its drifting and left pieces like Australia behind.

Many American geologists take a position between the
stable-continent and transoceanic-continent schools. On
one hand they think there may be something to the evidence for direct migrations between the southern continents
during and before the Age of Reptiles; on the other, they
object to a vast Gondwanaland of Suess's kind because it
would displace enough water to submerge the other continents completely, which we know was not the case. The
question of how much water there used to be on the earth,
by the way, is another unsettled one. Some think the total
water in the oceans has increased in the last billion years
or so, because every time there is a volcanic eruption some
water that had been dissolved in the magma of the substratum (called by the odd name of "juvenile water")
escapes in the form of steam. On the other hand the earth
is probably losing water-molecules from the upper atmosphere into outer space, and which process is working the
faster is anybody's guess.

American geologists therefore tend to picture the connections between the southern continents as narrow necks
of land like the Isthmus of Panama rather than as broad
ocean-filling continental links, as you can see from Schuchert's and Joleaud's maps.

The American geologist Schuchert outlines paleogeographical history as follows: Not counting Antarctica, there
are seven main land-masses — North America, Northern
Europe, Eastern Asia, South America, Africa, India, and
Australia. (The geologists give these places various names
in former times: North America is "Laurentia" or "Eria"
and so on.) In Paleozoic times the land area of the world
was greater than now because there was less sea-water and
the sial continental masses were bigger. Connections among
these masses tended to run east and west instead of north
and south as they now do. You might imagine the map of
the world as a microscope-slide on which there are seven
amebas, ever swelling and shrinking and reaching out
pseudopods towards each other. Once they stretched their
tentacles east and west, where now they reach north and
south.

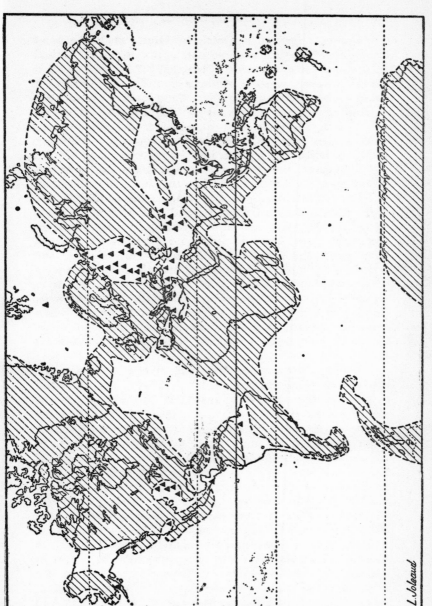

Fig. 13. WORLD MAP OF THE PENNSYLVANIAN PERIOD, showing continents connected by "isthmian links," according to Joleaud's *Paléobiogéographie.*

L. Joleaud

During times of mountain-building, when the land stood high above the sea, the northern masses were joined in a single continent called "Holarctica," which encircled the North Pole, and the southern masses were joined in a single Gondwanaland. However, these world-wide mergers of continents took place only at long intervals. The rest of the time the continents were lower, and shallow seas washed over much of their surface, sometimes reducing North America, for instance, to a mere cluster of islands.

The earth, you see, alternates between states of mountain-making and resting. During the times of mountain-building ("orogeny," geologists call it) the continents are high, climates are cold and dry, and land-bridges are open to pedestrians, whereas in the resting periods the continents are low, flat, and small, land-bridges are covered with water, and free flow of the ocean currents gives the world a mild, moist climate. These up-and-down movements of the continents do not take place smoothly, but with smaller up-and-down movements. When the continents sink, the inland seas ("epeiric seas") come and go several times, each time advancing a little farther and retreating a little less, until after tens of millions of years the land reaches its lowest point.

During these periods of sinking, great rifts open in the skin of the earth, through which lava wells out to spread over tens of thousands of square miles. You can imagine the scene: a great pancake of doughy lava hundreds of miles across, silvery-gray in the sun and glowing dull red at night where the red-hot interior shines through cracks in the cool crust, spreading slowly foot by foot over the level land, while trees in its path turn brown and go up in a puff of flame and the herds flee. The greatest of these flows are found, safely solidified, in India, northwestern Europe, the western United States, southern South America, and Australia. Similar flows take place on the ocean bottom, quietly, because more than a mile and a quarter down the water is under such pressure that it cannot boil even in contact with white-hot lava. Such flows occurred in the Mediterranean in the 1880's and 90's, for instance, without causing any disturbance at the surface of the sea. And the coolth of the sea-water soon freezes these submarine lava-flows.

When the land has sunk as far as it will go, the downward motion is reversed. Parts of the continents fold up as if pushed together from the edges, and the continents rise —again, not smoothly, but by an up-and-down movement, faster than the descent. The epeiric seas drain away and dry up, and erosion cuts the high parts of the lands into rugged mountains like the Alps. Volcanoes erupt. The climate gets cooler and dryer, with sharper changes from place to place and from season to season. In an extreme case an ice-age may ensue. In time the uplift stops and the cycle begins over.

Most geologists think we ourselves live at the end of a period of orogeny, and can look forward to 25 or 50 million years of sinking, shrinking, and warming of the lands we live on. It's a sobering thought that the human race may some day have to get along on two-thirds or even half its present land, though much of the remaining land will have a languid, even climate like that of Hawaii.

Although many spasms of orogeny have shaken the earth, we need remember only the biggest since the Pennsylvanian Period of coal and amphibians about 200,000,000 years ago. The end of the next period, the Permian, saw a great orogeny called the "Appalachian Revolution" from the mountains built at that time. At this time Gondwanaland, if it existed, reached its greatest size, and glaciers covered much of the Southern Hemisphere.

Next came the Mesozoic Era, with three smaller orogenies and a big one, the Laramide Revolution, at the end. This last upheaval built the Rockies and ended the Age of Reptiles. The Gondwanalanders think that during the Age of Reptiles the east-west links between the southern continents disappeared for good and were replaced by the present north-south links, so that a map of the world during the Laramide Revolution would look fairly modern.

The Laramide Revolution ended about 60,000,000 years ago, ushering in the Age of Mammals or Cenozoic Era. This in turn was closed by a great orogeny, the Cascadian Revolution, which built the Himalayas and brought on the Pleistocene Ice Age. The Pleistocene Period saw four advances of the ice (we live at the end—we hope—

of the fourth), the rise of man, and the disappearance of giant mammals like elephants and rhinoceroses almost everywhere except in India and Africa.

Geologists agree pretty well as to when and how cycles of mountain-building and subsidence happen. But as to what makes them occur they do not agree at all, nor yet as to how and whether continents come to "founder" and disappear. The natural buoyancy of continental rock in the sima crust makes it hard to see how such a vast floating mass *could* disappear.

The Irish geologist Joly, for example, worked out a theory of orogenic cycles based upon radioactivity. All rocks contain radioactive elements, and while the amount may be too small to detect in a given sample, it can warm the rock up a lot if its heat accumulates for millions of years without being able to escape. Joly thought that this heat melts the substratum from a stiff hot glass to a true liquid, expanding it and making the earth's diameter about forty miles greater than it is now. Since the basaltic magma of the substratum swells, its density becomes about 6% less than it was, so that the granitic continents cannot float so high in it as they did. As they sink, the epeiric seas will naturally cover parts of them. Since this expansion of the earth stretches the crust, it tears open, allowing lava to flow out on to the surface.

Then in time currents in the substratum allow the magma to cool and solidify again, whereupon the whole process is reversed and a time of orogeny occurs. Very neat and plausible, though I think the scheme has several flaws. Considering the pressures in the substratum, it is not likely that it could be truly liquefied except by a temperature so high as to melt the whole crust, and there are technical objections to some other features of Joly's theory as well.

Willis, another geologist, thinks that the continents have been connected by "isthmian links": mountain-ranges thrown up from the ocean bottom by the same squeezing forces that build mountains on land. Being made of dense sima rocks to begin with, these ranges have no natural tendency to "float" and hence in the course of millions of years sink back to their former level of their own weight.

Barrell on the other hand believes that these isthmuses were originally of light sial-type rocks like the rest of the continents. But magma, you know, is always melting and dissolving its way into the surface rocks to form "dykes" and "intrusions" of basalt. If enough of this heavy stuff invaded the isthmuses, they would be made dense enough to sink of their own increased weight, which is what Barrell thinks happened.

Since somebody can bring good, solid objections on one ground or another against all these hypotheses, however, we had better agree that nobody really knows why continents or parts of continents sink and let it go at that. No doubt a sound explanation, perhaps combining features of older theories, will be forthcoming some day.

That leaves the continental-drift hypothesis of Alfred Wegener, a professor of geophysics and meteorology at the University of Graz, Austria: a theory developed independently in the United States by Frank B. Taylor (another case of independent invention!)

Wegener, who perished exploring the Greenland icecap in 1930, said: If the continents float on the sima crust like cakes of ice in water, why can't they drift like cakes of ice? He therefore assumed a single super-continent back in the Paleozoic Era, a "Pangaea" that included all the modern continents. If Pangaea ever existed it must, like the larger Gondwanaland of Suess and Dacqué, have been a land of immense deserts, as the prevailing winds would sweep over it for thousands of miles after they had lost their moisture without being able to pick up any more.

Wegener's world maps show how all the continents can, with a little stretching and ingenuity, be fitted together like pieces of a puzzle. Pangaea, *selon* Wegener, began to break up in the Mesozoic Era, the Dinosaur Age, and its parts drifted asunder until finally in the Pleistocene Period Europe came loose from North America. Obviously we need not worry about land-bridges connecting the continents if they were once all huddled together so that they touched. And Atlantis, on this basis, would simply be North America, which Europeans would suppose to have disappeared when it had merely drifted beyond range of their ships.

The Wegener theory has several points in its favor. For instance the mountain chains along the western margins of the American continents look as if these continents had been drifting westward and had been crumpled at their leading edges. Surveyors once issued a report (not confirmed by later observations, though) that Greenland was still drifting westward at sixty feet a year.

The Wegenerites have suggested various mechanisms to explain the drift. Wegener himself thought that Pangaea was pulled apart by a combination of centrifugal and tidal forces; Daly asserted that the earth was slightly egg-shaped with Pangaea at the bulge, so that the continents were simply coasting downhill; others believe that slow convection currents in the substratum carry the continents about, like bits of scum on the surface of an overheated pot of cocoa.

However, the theory also has some probably fatal flaws. Paleontologists like Simpson object that the distribution of animals proves that Europe and North America were separated at least as early as the Mesozoic Era; a Pleistocene connection between the two, they say, is preposterous. Chaney points out that continental drift implies a much greater shift in climatic zones during the Cenozoic Era than actually took place. Others object that Wegener's jigsaw puzzle does not fit together nearly so well if we take the boundaries of the continents to be the edges of the continental shelves, as we should, rather than the shorelines. They also note that the sima of the ocean floor is stronger and stiffer than the sial rocks that go to make up the continents, and ask: how could these slabs of weak rock plow their way through the stiffer stuff in which they rest? Finally the geophysicist Lambert calculated that the forces Wegener relied upon to tow his continents about the earth were only one-millionth the size required by the job.

Schuchert thinks that the continents may indeed have drifted, but only a few hundred miles throughout geological time, and not necessarily always in the same direction. Altogether, perhaps we had better put the Wegener theory on the shelf marked "very doubtful" and leave it there for the time being.

Besides the Wegener hypothesis there have been a lot

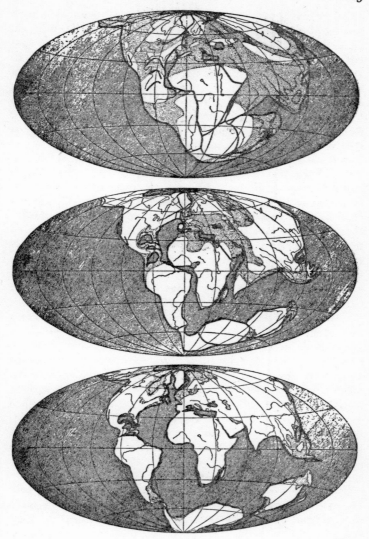

Fig. 14. THE THEORY OF CONTINENTAL DRIFT — the world (top to bottom) in the Pennsylvanian, Eocene, and Pleistocene Periods, according to Wegener's theories.

of other marginal theories having to do with geology, which for one reason or another failed to convince most geologists. Of these, perhaps the one most interesting to us is that which the French geologist Pierre Termier proposed in 1912. It had come to Termier's attention that in 1898 a cable-ship, fishing for the broken ends of a telegraphic cable 500 miles north of the Azores, brought up splinters of tachylyte in its grappling-irons from a depth of two miles. Tachylyte is a natural non-crystalline volcanic glass that ordinarily forms when lava cools in the open.

Reading a geological report on the eruption of Mt. Pelée in the West Indes, Termier noticed that tachylyte had formed there from lava that was exposed to air, but that lava that solidified under a blanket of previously hardened lava formed crystals. He then jumped to the conclusion that tachylyte can form only under the slight pressure of the atmosphere, and that therefore the cable-ship's tachylyte must have come into being in air. The answer, of course, was the sinking of Atlantis; Termier guessed that between 40,000 and 200,000 square miles had plunged 10,000 feet beneath the sea.

Schuchert soon pointed out that Termier's chemistry was at fault. Neither pressure nor the presence of air is the critical factor in the hardening of lava; instead, it is the rate of cooling, and lava that freezes under a blanket of previously hardened lava forms crystals because it cools more slowly. Both laboratory experiments and observations of underwater lava flows indicate that tachylyte forms quite as readily under water as elsewhere. Nevertheless the Atlantists picked up Termier's argument and have continued to repeat the tale of Termier's tachylyte ever since as if his theory had never been refuted, just as they have the story of the Japanese Indians of Mexico.

So much for Gondwanaland, Pangaea, and other geological lost continents. All of them are very much a matter of educated guesswork; European geologists seem to take them more seriously on the whole than their American colleagues. In this field one finds such wide differences of opinion among the people best qualified to judge that the layman might just as well write off the geological lost-continent question as unsettled.

However, most geologists agree very well on some points: That these former lands, if they existed, broke up long before civilization arose, and that they sank little by little over millions of years. While there have been great floodings of the land, there is no reason to think that a whole continent can be submerged at once. The material underneath has to go somewhere, which takes time; and great gas-filled chambers under the earth, like those of Churchward, would violate the laws of physics. A few miles down, long before you come to the substratum, you enter the zone of flow, where the pressure of the overlying crust makes the rock slightly plastic. Below that depth any cavity would soon fill up.

Thus, while a volcanic eruption may make drastic changes over a few square miles, alterations on a continental scale take millions of years. Therefore geology offers no support to Plato's stories, first, because their timing is all wrong by millions of years, and second because when land does sink, it goes down much more slowly than Plato thought it did.

But if geology doesn't actually confirm Plato's story, does it really disprove it? What do we know for sure about the rate of the rise and sinking of lands, the flooding of large areas, and the effects of earthquakes? And do these facts offer any clues as to where Plato got his ideas?

As it happens, we do know quite a bit about the drowning of lands, both temporarily and permanently. The four main causes of inundations are the flooding of river-valleys, the invasion by the sea of an area below sea-level, the tsunami or earthquake wave, and permanent sinking under water as a result of movements of the earth's crust. Let us consider them in order.

The flooding of river-valleys. Such floods, like those of the Mississippi, are disastrous enough and are probably the source of flood-legends, since civilizations usually arise in river-valleys where such floods are likely to take place. Still, the area affected by such floods is limited and the effect usually temporary, though landslides sometimes dam rivers, making small permanent lakes. Nevertheless, such floods sometimes do enough damage to make their ignorant

victims think that the whole world has been deluged.
Chinese histories state that in the reign of the Emperor
Yao, in 2287 B.C., a flood caused the waters of the Yellow
River to mingle with those of the Yangtze, making a great
inland sea. The eminent engineer Yü spent years bringing
it under control. A combination of high water and a storm
to pile it higher around a river-mouth may do marvelous
damage; for instance a storm of this sort flooded 3093
square miles of the lowlands of Bengal in 1876, killing
215,000 people.

Noah's flood may be based upon a real flood that submerged something like 40,000 square miles of the Euphrates
Valley some time between 5400 and 4200 B.C. In 1929 Woolley, digging in the valley, found an eight-foot layer of clay
with no human relics in it, below which relic-bearing layers
of rubbish began again. Above the clay the relics were of
Sumerian type only; below, of mixed Sumerian and pre-
Sumerian. Sumerian chronicles treat the flood as a historical
event and mention kings and cities that existed before and
after it. Woolley deduced that before the flood the Sumerians had invaded the land and built their cities on the
higher ground, while the more primitive pre-Sumerians
hung on in the lowlands. The flood wiped out the pre-
Sumerians but left some Sumerian cities standing, and their
inhabitants repeopled the whole land. Later evidence indicates several such floods, Iraq (according to one theory)
having had a wetter climate then.

Inundation of areas below sea-level. If the Mediterranean basin was flooded from the Atlantic as some
geologists have suggested, this flood would be the greatest
such deluge on record. For a historical case, however,
the best example is the swamping of the Zuyder Zee in
1282. This bay was dry land below sea-level until a great
storm broke through the natural dykes and let in the
North Sea, which flooded the whole area in one day. The
industrious Dutch have been busy ever since pumping the
water out again. In several other parts of the world —
in the Saraha Desert, around the Caspian and Dead Seas,
and in Australia — a connection with sea-level water would
produce similar floods. In 1906, in fact, the Colorado River
flooded a part of Imperial Valley (in Southern California)

that is below sea-level in just this way, and the dry 4000-square-mile Lake Eyre in South Australia recently filled with water all of a sudden.

Various ancient authors like Florus and Timagenes told how the Kelts and Germans of the northern coast of Europe had been driven from their homes by floods like these. Their accounts sound like the much later medieval-Keltic stories of Ys and Lyonesse. In Breton legend the city of Ys stood in early Christian times on the shores of the Bay of Trespasses. Its king, the rich and pious Gradlon (the hero of Edouard Lalo's opera *Le Roi d'Ys*) protected it from high tides by a wall and a basin to receive the overflow. Once Gradlon's dissolute daughter Dahut got drunk with her lover and, after various pranks, opened the sluice-gate in the wall with the key which she had stolen from her father; or, in another version, she accidentally opened the sluice when meaning to open the city gate to let her lover in. St. Gwenole, the founder of the first monastery in Armorica, warned his patron the king, who leaped on a fast horse and galloped off just ahead of the surf.

A similar tale is told of Lyonesse, an island supposed to have stood off the tip of Cornwall. When it sank the sole survivor was one Trevillon who, like Gradlon, rode furiously to the mainland. Cardigan Bay in Wales and Lough Neagh in Ireland also have stories of this kind connected with them. The Irish of Connemara tell a tradition of a sunken city offshore that will one day come up again, at which time Galway will in its turn be submerged. One stormy night in 1946 these folk were shocked to see a host of lights twinkling over the water where the city should be. Their fears were confirmed when a man who tried to telephone Galway was told by a fresh operator: "There's no reply; they must be all dead in there." At dawn, however, the Connemarans were relieved to see the "city," a fleet of Spanish trawlers riding out the storm in the lee of the Aran Islands, hoist anchor and sail away.

Gidon, the translator of Bessmertny's fine *Atlantis-Rätsel*, used these references to coastal floods in Strabo and other Classical writers to build his own Atlantis theory. Most geologists agree that during the greatest advances of the Pleistocene ice, England and Ireland were joined to Europe, and that at this time most of the North Sea was a low plain over which the Thames, the Rhine, and other rivers sluggishly wound their way. The plain sank beneath the sea between 25,000 and 10,000 years ago, and today fishermen sometimes dredge up stone-age tools and mammoth teeth from the North Sea bottom.

Gidon, stretching things a bit, postpones this flooding to the Bronze Age and backs up his theory by a study of the distribution of plants in northwestern Europe. He concludes that Plato's story is based upon a series of inundations like that of the Zuyder Zee, plus other elements such as rumors of Atlantic islands, and that Plato's Atlanto-Athenian war is but an echo of the migrations of the Kelts and Germans displaced by these floods. Ingenious, though hardly to be proved in the present state of our knowledge.

The tsunami. The earthquake is certainly a leading wholesale killer of men. Within the last two centuries earthquakes have slain 60,000 people at Lisbon (1755), 30,000 in Calabria (1783), over 100,000 in Messina (1908), over 100,000 in Kansu, China (1920), and over 142,000 in Japan (1923). Chinese history tells of even more destructive quakes like that of 1556 which killed 830,000, and while the figure may be exaggerated the destruction was no doubt great enough. However, there are certain things to remember about earthquakes before making a seismologist of Plato.

The great losses of life in tremblors are *not* due either to the swallowing of people by cracks in the earth or to permanent submersion of land under water. They result from (1) the collapse and burning of houses, (2) landslides, and (3) tsunamis striking a densely populated shore. The Messina quake, for example, caused enormous damage although not a first-class earthquake, because the houses of Messina were built with thick walls of loosely cemented rubble masonry with no bracing to speak of, and they all collapsed at the first shock.

As an example of what landslides can do, the Kansu disaster occurred in a region of deep valleys between hills of loess, loose soil that has been piled up by the wind. Thousands of people lived in cave-villages dug into these hillsides, and the quake made the sides of the hills slip down and bury the caves. In the Japanese quake of 1923 a landslide swept down from Mt. Hijiridake, four miles from the sea, at 100 miles an hour, wiped out the town of Nebukawa on the shores of Sagami Bay, tossed a railroad station and a waiting train into the bay, and deposited a tangerine grove where the station had been — with most of the trees still upright and growing.

There have however been many earthquakes that caused little damage although just as violent as those I have mentioned, because they took place in thinly populated regions or under the sea. The area of intense damage of even the most violent quake is quite small. Thus the Owens Valley earthquake in California (1872) was probably more severe than the San Francisco quake thirty-four years later, but caused less than 100 casualties.

But occasionally an earthquake under the sea will set up a tsunami — a Japanese word that seismologists use for the earthquake wave, which many people call by the misleading name of "tidal wave." Actually such a quake makes a series of waves, the biggest first. While a tsunami cannot usually be seen by the naked eye out at sea because of its gentle slopes, it becomes higher and steeper as it approaches the shore. If it happens to meet a gently sloping offshore bottom, or a shallow bay opening toward the direction whence it comes, it may rise scores of feet with horrid results. Usually the wave does not take the steep form of a breaker, however, but is more like a sharp sudden super-tide that rises far above normal levels. Some of the record tsunamis are:

Place	Date	Height, ft.
Lima, Peru	1724	80
Kamchatka	1737	210
Callao, Peru	1746	80
Lisbon	1755	60
Arica, Peru	1868	52
Iquique, Chile	1877	80
Krakatoa	1883	50-100
Sanriku, Japan	1896	50-100

A volcanic eruption instead of an earthquake caused the Krakatoa tsunami, when the island of Krakatoa blew up with a roar heard 3000 miles away and the wave drowned 36,500 people living on the shores of the nearby Indonesian islands. The west coast of South America is particularly liable to these disasters; the Callao wave sank nineteen out of twenty-three ships anchored in its path and carried the rest far inland.

The heights of most tsunamis are estimated from the levels at which damage is done or debris is found, since most sensible folk, seeing a tsunami coming, would not stand gauping at it long enough to make a good estimate. Also, they reach their greatest heights only along short stretches of beach where the bottom formation is of the right kind. If you are ever at a beach in an earthquake zone and hear a rumble from the sea, which then recedes for hundreds of yards leaving crabs and fish flopping: run, do not walk, for the nearest high ground. You have from five to thirty minutes.

However, even the most violent tsunami does not mean any very great change in the shape of the land, for after the water recedes the shoreline is left much as before.

Earthquakes have caused some considerable tsunamis in the Mediterranean. Classical writers described them, and Plato, who would not have clearly distinguished between temporary and permanent effects, or between a rise of water and a subsidence of land, might have gotten the idea of the foundering of continents from some such account. In fact before this book is over I shall point out the very passage that I think may have inspired him.

Plato also, you will remember, said that in the Atlanto-Athenian war the Athenian army was swallowed up in the earth by the great quake. It used to be a general belief that in major earthquakes, cracks would open in the earth big enough to gulp down a whole city. Now however it turns out that this idea is much exaggerated. Earthquakes do open cracks in the ground, but only the sort that a man might break his leg in if he were careless; nowhere nearly big enough to swallow a house, let alone a city or an army.

Permanent submersion. Either earthquakes or volcanic eruptions may change the contour of the land and lower it beneath the sea. The Krakatoa eruption destroyed an

island 1400 feet high and left water 1900 feet deep where
it had been; but the island had had an area of only four-
teen square miles in the first place. Because of the small
areas they affect, volcanic eruptions cannot be compared
with continental sinkings of the kind Plato described.

Two kinds of motion go to make up an earthquake.
First comes a vibration or quivering, generally less than
an inch in amplitude, but sometimes violent enough to
make stones leap from the ground and houses tumble about
their owners' heads, or to start landslides. The other movement is a slower but permanent displacement of parts of the
earth's crust, which may start a tsunami or, if it happens
in a city, may break water-mains and keep people from
putting out the fires that spring up when houses with
lighted stoves in them collapse.

The earth's crust, geologists tell us, is made up of odd-
shaped blocks about five miles or so across and sometimes
joined together into thirty-mile-wide super-blocks. These
blocks are always moving slightly with relation to each other
but, since they fit tightly together, they cannot slide freely
past one another along the cracks between them, called faults.
Each block along such a fault is therefore slowly bent
until its natural stiffness overcomes the friction and it
snaps back into shape. During this "snap" the grinding
together of the rocks along the fault sets up the vibrations
that make up an earthquake. The "snap" takes several
minutes and leaves the earth displaced on the two sides
of the fault, with fences broken and roads offset. Some of
the biggest known displacements are:

Place	Date	Length of fault, mi.	Max. horiz. displ., ft.	Max. vert. displ., ft.
Owens Valley	1872	40	16	23
Japan	1891	60	12	20
Alaska	1899	?	?	47
San Francisco	1906	190+	21	3
Newfoundland	1929	?	?	35

As a result of earthquakes in 1868 and 1906, the distance between Mount Tamalpais and Black Mountain, California, increased about ten feet. Evidently the geodetic

survey people have a never-ending job, for no sooner do they get the terrain accurately measured than the thing gives a shudder and a squirm and changes on them.

Luckily, while earthquakes occur nearly everywhere, they are common only in two definite earthquake zones or belts. The first of these is a horseshoe-shaped zone around the shores of the Pacific Ocean, including New Zealand, New Guinea, Japan, Alaska, California, and Chile, while the other zone branches off from the first in the East Indes and extends west to the Himalayas, Iran, and southern Europe and into the Atlantic where it peters out.

Atlantists may ask: if the sea-bottom off the coast of Newfoundland dropped thirty-five feet, and the shores of Yakutat Bay in Alaska rose forty-seven feet, each in a single earthquake, why could not one or a few quakes like that sink a continent? A fifty-foot drop could drown quite an area of low flat country like the Florida Everglades.

But the theory, like so many others, does not work in practice. The above figures are for the greatest movements only; the displacements over most of the earthquake areas were much less. For example, while the Alaska 1899 quake caused a maximum rise of forty-seven feet, the total area affected by this quake was only fifteen to twenty miles across, and only over an area six to ten miles wide did the rise exceed twenty feet. Moreover a rise in one place usually balances a sinking in another, since the blocks on each side of the fault are strained in opposite directions, and when the quake comes the bent parts snap in opposite directions.

Finally, records of such submergences in historic times show that they are all local changes. In one of the biggest, for instance, sixty square miles of the Chittagong district of Bengal sank beneath the sea in 1762. That is an area about the size of Staten Island, New York, and less than one-fifth of 1% of Ireland with its 32,000 square miles.

True, Donnelly says a quake submerged 2000 square miles in India in 1819, but this is a misleading description of what happened. The tremor in question took place in the Rann of Kachh, a flat desert of 8000 square miles east of the Indus River Delta. During the southwest monsoon this plain is covered with a thin sheet of salt water, and during the Indus flood it is covered with fresh water. The rest of the year it is dry and strewn with dazzling-white patches of salt: a weird, silent place with little life except occcasional herds of wild asses.

The quake of 1819, which enlarged the Rann a little, took place along a fifty-mile east-west fault. The land to the north rose about ten feet and that to the south sank an equal amount. A shallow permanent lake of several hundred square miles formed in the western part of the Rann, from whose waters the fort of Sindri long forlornly rose. During the following century, however, silting returned the Rann to much its original shape.

The most that we can say, therefore, is that if there were a large island of very low flat relief, nowhere more than a few feet above sea-level, it *might* be possible, by an earthquake greater than any recorded, to submerge several hundred square miles of it at one clip. But, in the first place, Plato describes Atlantis as mountainous. In the second, a few hundred square miles is but a tiny fraction of an island of continental size. And in the third, such an island would not disappear after submergence, but would remain as a shoal or bank as Plato said Atlantis did. As you can see from a relief map of the North Atlantic ocean-bottom, there is no such bank anywhere near where Plato located Atlantis.

Donnelly thought Dolphin Ridge, which makes an S-curve down the middle of the Atlantic, was a relic of Atlantis. But except for the small and steeply mountainous Azores region, nearly all of Dolphin Ridge is under two or three miles of water, and there is no known way to get a large island down to that depth in anything like the 10,000 years required. The last word on Dolphin Ridge as a site for Atlantis has probably been said by Dr. Maurice Ewing of Columbia University, who lately announced that after thirteen years of exploring the Mid-Atlantic Ridge by sounding, dredging, and lowering a camera and a searchlight down as much as 18,000 feet to photograph the bottom, he had found no trace of sunken cities.

Besides all the foregoing methods of flooding lands, two other changes will bring about results of the same sort, but much more slowly. One is the continual movement of parts of the earth's crust even when there are no earthquakes to help it along. For instance the Baltic Sea region is rising at a maximum rate of about half an inch a year. The Baltic area was pressed down into the substratum by the weight of the Scandinavian ice-cap during the Pleistocene Period, and now that the ice-cap has melted the region is bobbing up again like a piece of wood that has been pushed under water and then released. At that rate the Baltic Sea should be pretty well dry in another 100,000 years. Faster up-and-down movements occur in Japan—as much as three of four inches a year—but these are not the same over any large area; instead the islands seem to be heaving up in some places and down in others.

Ice-ages also change the shape of continents by locking up enough water in glaciers to lower the sea-level. During the fourth Pleistocene glaciation, geologists have figured, the sea-level dropped between 200 and 400 feet below its present level, connecting the British Isles with Europe and Alaska with Siberia. At present, judging from the shrinkage of the Arctic glaciers, the climate is probably warming and the sea rising. If the warmth ever gets to the point of melting the Greenland and Antarctic ice-caps completely, sea-levels will be raised something like thirty or forty feet, or even more, so that the world's seaports will all have to move miles inland. Luckily the process takes place only with extreme slowness.

Dr. Trask of the U. S. Geological Survey recently opined that the Pleistocene ice-caps withdrew so much water that there were no oceans left—just small seas, mere puddles, at the bottoms of the ocean basins. However, any such lowering would have let the placental mammals into Australia, which it obviously did not do.

Besides all the forms of change in the earth's surface that we have discused, there is the common everyday erosion of wind, wave, and rain, the sculpturing effect of rivers and glaciers, and the building and washing away of beaches. Normally these are all slow, non-catastrophic processes, even though a big storm can do considerable damage. Thus the 1938 New England hurricane kicked up forty-foot waves, and a storm in 1099 washed away the little island of Loumea off the coast of Kent, England, changing it into a tidal flat, the treacherous Goodwin Sands of sinister repute.

Now, Plato did see that erosion had been at work on the Greek peninsula, for which we must give him due credit. In fact he was one of the first writers to realize that such a process as erosion existed. Where he went wrong was in assuming that Greece must have been whittled down to its present jagged form by a series of deluges, instead of by the slight but persistent day-to-day action of wind and water.

Then what conclusions does geology teach about lost continents? That there *may* have been such continents in the Paleozoic Era, 200,000,000 years ago, when the highest life consisted of lizard-like reptiles. That these "continents" were probably not real continental masses but narrow isthmuses connecting some of the present land-masses as the Panama Isthmus does now. That these continents, if they existed, lost their shape during the Mesozoic Era, so that by the beginning of the Cenozoic Era the map of the world looked quite modern. That there *may* have been an island where the Azores are now before the end of the Miocene Period, but that it could not have been very impressive and that it probably disappeared at least five or ten million years ago. That while earthquakes and tsunamis can do vast damage, they do not plunge whole continents beneath the waves, nor is there any reason to think they could. That Plato had a vague idea of some of the effects of earthquakes and erosion, but no notion at all of the time it takes these processes to bring about large changes in the earth's surface.

By normal movements of the earth's crust, like that of the Baltic region, or by a long series of terrific earthquakes, it might be possible to submerge a low flat island of good size in 100,000 years or so. That, however, takes us back beyond the last glacial advance, and all the data of history and archeology tell us that civilization did not arise until seven or eight thousand years ago. Furthermore there is good reason to believe that traditions cannot be handed down by word of mouth for more than a few centuries and still retain enough of their original form to be recognized. Therefore the events of 100,000 years ago have nothing to do with Plato's Atlantis.

Moreover such a large flat island would have left remains in the form of shoals, which it has not done, Plato to the contrary notwithstanding. If you want to call the hypothetical land around the Azores that some geologists think may have existed 10,000,000 years ago "Atlantis," I see no reason to object.

Provided, that is, that you bear in mind that our ancestors were probably still sitting on a branch and scratching at that time, and could neither have crossed the ocean to reach this isle nor have created a culture when they got there.

Therefore, in spite of the fact that lost lands are possible, any idea of connecting them up with Plato's Atlantis breaks down on the time-factor. Geology and history operate on different time-scales: the rise and fall of a civilization is but a tick of the clock in geological time, and during such an interval lands do not change shape enough to matter. So let us, once and for all, grant decent burial to all the theories of Atlantis that assume that a great island did, as Plato says, sink beneath the sea in a day and a night of storm and earthquake.

CHAPTER VIII

THE SILVERY KINGDOM

Here in a sleep of exhaustion lay long-tested, godlike
    Odysseus;
On to the Phaiakes' city however proceeded Athena.
Once they resided in spacious Hypereia near the Kyklopes,
Who with superior power did evermore plunder them until
Godlike Nausithoös thence unto Scheria brought them and
    placed them
Distant from laboring men.

The Odyssey

To say that Plato's Atlantis "is" a sunken
Atlantic island or America or anything else in the real
world, without qualification, shows muddy thinking. Plato's
Atlantis, strictly speaking, was an idea in Plato's mind,
nothing more and nothing less. Plato embodied this idea in
the form of writing on a roll of papyrus, which writing
has been many times copied and reproduced and translated in many forms since Plato's time, and other men,
reading Plato's dialogues, have in turn had ideas engendered
in their own minds.

Of course there must have existed something in the real
world to give Plato the basis for his tale—to engender Plato's
concept of Atlantis. But this need not have been at all
identical with the Atlantis described by Plato in *Timaios*
and *Kritias*.

If I seem to be laboring the obvious, it is because some
Atlantists fail to keep this distinction in mind, to the confusion of everybody. So when I say that Atlantis "is" this or
that, I mean merely that this or that impinged upon Plato's
senses and incited him to compose the Atlantis story. And
this real thing can be simple or complex (that is, a combi-

nation of different stories, myths, or facts) and it need not look much like the final product.

Philosophers have a term *isomorphism,* meaning that a map, picture, description, or other symbol corresponds point for point with its referent (the thing it symbolizes). Now, Plato's Atlantis tale is obviously not completely isomorphic with any one real source. We have already thrown out the Greek gods and the prehistoric Athenian Empire as fictions, and now it seems that the sinking continent must go also, as just too improbable.

The possibility remains that Plato's Atlantis was based upon traditions or rumors of some far-off civilization that flourished once but later disappeared, without necessarily having been on a sunken Atlantic island. The Atlantis-in-America theory was of this kind. Then in 1675 a Swedish scholar, Olof Rudbeck, wrote a ponderous treatise in several volumes to prove that Atlantis was Sweden. He began by assuming that Plato's Atlantis and Homer's isle of Ogygia were the same. Then, from the scanty and vague sailing-directions in the *Odyssey* and some remarks by the unscientific Plutarch on the shape of the earth, Rudbeck inferred that Atlantis must have lain between the latitudes of Mecklenberg, Germany, and Vinililand, Sweden. By bending the poetry of the Viking Age to his service he proved to his own satisfaction that Atlantis was Sweden with its capital near Upsala, and was moreover the fountainhead of all civilization.

Since Rudbeck's time Atlantis has likewise been "found" in Africa, Spain, Ceylon, and practically all other parts of the world. Rudbeck's work incited the French astronomer Bailly, a distinguished victim of the French Revolution, to compose an even more extreme origin for Atlantis. Bailly in his *Histoire de l'astronomie ancienne* "developed a grandoise system of racial migrations, based upon certain recurrent errors in astronomical tables brought back by some missionaries from India," which errors he maintained "could not have been drawn up from observations made in India at all, but in Central Asia, at a latitude of 49°." Atlantis, Bailly concluded, was really Spitsbergen in the Arctic Ocean. In ancient times before the earth had cooled to its present temperature (an idea from Buffon)

Spitsbergen was comfortably warm, but its subsequent refrigeration made the Atlanteans migrate south to Tatary. Later this race of giants dwelt in the Caucasus, in the region of Caf (an imaginary mountain in Iranian mythology) and laid the foundations for all the ancient civilizations of Asia.

Being a euhemerist (one who thinks one can turn myth into history by leaving out the supernatural elements) Bailly deduced that Atlas was an astronomer-king of Spitsbergen who invented the terrestrial globe; Ogygia and Hyperborea were likewise parts of the bleak Svalbard Archipelago. He conducted a long correspondence with the politely skeptical old Voltaire on the subject, and was attacked by the Keltomaniacs who were trying to derive all civilization from the Druids.

Keltomania had begun with a seventeenth-century Irishman, John Toland, who expanded the references of Pliny the Elder and other ancient writers to the Druids' white robes and golden ornaments into a description of an occult brotherhood preserving the arcane wisdom of antiquity. The Keltomaniacs founded occult societies like the Ancient Order of Druids to exploit this craze, and perpetrated remarkable literary forgeries like the "Ossianic Poems" of James Macpherson in the 1760's and the *Barzaz-Breiz,* or songs of the Breton Bards, of Villemarqué in 1839.

For the eighteenth and early nineteenth centuries were the heyday of the speculative mythologist, when any scholar felt qualified to construct, on the basis of any ambiguous mythological allusion in the works of some jejune late-Roman poetaster or obscure Byzantine annotator, a whole new cosmos-shaking theory of the origins of mankind and of civilization. Hence the conjectures of Buffon and Saint-Vincent about the washing away of Atlantis by earthquake waves, or of Carli about its destruction by a comet.

One contributor to this stream of speculation was Francis Wilford, who in 1805 advanced an Atlanto-Druidic hypothesis according to which the British Isles were a remnant of a former Atlantic continent where the events of the Old Testament had actually taken place (and not in Palestine as most people thought). Wilford was a British officer in India. When his Indian acquaintances learned of his enthusiasm they obligingly forged and sold him a great mass of Sanskrit documents to help him to prove his contention: ancient poems and the like. Wilford wrote up his doctrine as a series of articles for an English magazine; then learned how he had been hoaxed, and caused the articles to be published with large deletions that made them practically unintelligible.

Wilford's theory was taken over by another Neodruidist: that pious, imaginative, and rather stupid though talented artist and poet William Blake. Blake incorporated the Wilford hypothesis in scattered allusions throughout his turgid and tedious mass of apocalyptic free-verse. According to Blake, King Albion led the survivors of the sinking of Atlantis to Britain where he founded the Druids.

For all this foofaraw the real Druids of ancient Britain and Gaul were neither better nor worse than most barbarian priesthoods. Like Aztec and Mayan priests they were much given to human sacrifice, despite the pathetic efforts of some modern Keltophiles like Talbot Mundy to clear them of the charge. The cult of Neodruidism flourished until driven out of business in the late nineteenth century by the competition of occultisms based upon Oriental traditions of Egypt, India, and Tibet.

Meanwhile another Atlantist school, started by Serranus in the sixteenth century, had been proclaiming that Atlantis was Palestine: the reverse of the theory of Wilford and Blake. Like Kosmas they deemed Plato's story to be but a corrupted bit of Biblical history. Baër, a Swede living in France, published a book in 1762 to prove that Atlantis was Judea, the ten kingdoms of Atlantis the twelve tribes of Israel, the Atlantic Ocean the Red Sea, and Atlas the mythical patriarch Israel himself.

During the last century North Africa has become a favorite haunt of Atlantis-hunters.

Godron started the Atlantis-in-Africa school in 1868 by placing Atlantis in the Sahara; then, following him, Félix Berlioux in 1874 claimed to have located the capital of Atlantis on the west coast of Morocco between Casablanca and Agadir, where the Atlas mountain chain slopes down to the sea. Here, said Berlioux, and not on any island, lay Plato's city of Atlantis, otherwise Kernë, the capital of the Atlantioi in the account of Diodoros the Sicilian. Once the Atlanteans had ruled a great North African empire, but were defeated in the thirteenth century B.C. by a combined Egyptian-Phoenician army while the Berber Gaetulians took their capital. The surviving Atlanteans lived on as a subject race, and traces of them persist in the form of blondness and blue eyes occasionally met in the North African mountains. And on the strength of verbal similarity alone, the Medusa and Perseus of the North African myths of Diodoros are identified with the historical Medes and Persians — a treacherous line of argument.

Kernë bobs up elsewhere in ancient literature. A little before 500 B.C. the Carthaginian admiral Hanno made a celebrated voyage down the West African coast, and by good luck a Greek translation of his report has survived. The sight of the country on fire alarmed Hanno's sailors, who did not know that the natives were burning the grass for pasturage, as they still do. The crews killed three women of a hairy race whom the natives called *gorillai* (but which were probably chimpanzees) and took their skins back to Carthage to hang in a temple.

Hanno ended his voyage at a small island which he named Kernë — either the modern Hernë at the mouth of the Rio de Oro, or Arguin at Cape Blanco 200 miles further south. Here for many centuries trade was carried on by the method used in those days for dealing with natives who feared slavers. First, the trader set out his goods and withdrew. Then the natives approached the spot and set out what they were willing to pay for each item and withdrew in their turn. The trader, who had been watching from a distance, came back and either took the price offered or declined it, and so on until a deal was reached without either party's coming within shouting distance of the other.

Berlioux's Moorish-Atlantis theory begot numerous progeny such as Pierre Benoît's colorful but rather depressing novel *L'Atlantide,* one of the best-known of the large family of lost-continent novels.

L'Atlantide recounts the adventures of a pair of French army officers who find Atlantis in the Ahaggar Mountains of southern Algeria, the country of the Tuareg, a tall truculent tribe of Berbers among whom the men wear the veil but the women do not. In the novel, the Ahaggar is ruled by an imperious young lady named Antinéa, supposedly the name of the Tuareg's legendary matriarch Tin Hinan.

Antinéa is a typical Rider Haggard heroine (or villainess), resembling his "Asika" in *The Yellow God* (from which several plot elements are borrowed) or his "Ayesha" in *She*. She is the last descendant of Poseidon and Kleito, and keeps a pet leopard named Hiram for a bodyguard. When Europeans wander into her reservation she makes them her lovers, and such is her fatal beauty that when she discards them they all commit suicide or otherwise come to a bad end. Thereupon Antinéa has their corpses plated with orichalc and mounted in niches in a red marble chamber maintained for the purpose. Benoît, like the French Atlantist Claudius Roux, assumed that the "submersion" of Atlantis was actually a rise, causing the Saharan Seas to dry up — though geology indicates that the Sahara has been land continuously ever since the early Cenozoic Era.

L'Atlantide had considerable success as lost-continent novels go. It has been published twice in English, as *Atlantida* in the United States and as *The Queen of Atlantis* in Great Britain. Moreover it has twice been made into a motion picture. About 1929 it appeared as a French silent film with Brigitte Helm, and again in 1949 as United Artists' *Siren of Atlantis*.

The latter unfortunately turned out to be a remarkably dull movie. The late Maria Montez was cast as Antinéa with a series of eye-filling gowns and a Spanish accent, while Jean Pierre Aumont, her real-life husband, gave a glassy-eyed performance as the hapless French officer beguiled by Antinéa into murdering his best friend, a brother-officer who has antagonized the temptress by resisting her lure. Although the story has considerable cinematic possibilities, the production in this case was so hammy that the final result was unintentionally funny. The picture was such a financial disaster that the following year Señorita Montez sued the producer (a man with the Atlantean-sounding name of Nebenzal) for the unpaid balance of her salary, and got a $38,ooo judgment against him.

Altogether *L'Atlantide* cannot be considered a very good novel. Though better than some lost-continent stories, it remains a not too successful imitation of Haggard, and the theme of the beautiful siren the mere sight of whom turns strong men's wills to water seems naïve in these unsentimental days.

Benoît's novel is said to have incited the volatile Count Byron Khun de Prorok to go hunting Atlantis in the Sahara. De Prorok started out as a competent archeologist who did sound work on the site of Carthage. Later he seems to have gone in for a type of exploration that, if it produced less substantial scientific returns, provided him with adventure and furnished lively copy for his books. He hinted at Atlantean traces in the Sahara and in Yucatán, and in 1925 penetrated the Ahaggar Massif. There he opened the tomb of some Tarqi dignitary and excitedly announced that he had found the bones of Tin Hinan herself — a view in which the museum experts, however, did not concur.

Meanwhile Knötel in 1893 had put the Atlantean empire in Northwest Africa, with the reservation that the Atlanteans proper were a caste of priests of the god Thoth-Ouranos-Hermes, who had come to thse parts from Chaldea.

Between 1908 and 1926 Captain Elgee in England and Leo Frobenius in Germany developed another theory, of Atlantis in Nigeria, on the West African coast a few hundred miles north of the Equator. The explorer Frobenius discovered things in Yorubaland (part of Nigeria) that convinced him he had found Atlantis, complete with elephants, luxuriant vegetation, blue-clad natives, and copper-ore. Frobenius equated the Nigerian god Olokon with Poseidon, and pointed out that the land had been the home of powerful maritime nations ever since the thirteenth century at least. He also convinced himself that the Yoruba culture contained many non-African elements such as the short bow, tattooing, number-magic, and the king's sacred parasol.

Civilization, he thought, had begun on a lost continent in the Pacific Ocean, whence it had spread to Asia and thence westward, stimulating the rise of such cultures as the Egyptian and the Atlanto-Nigerian. Tartessos in Spain was an outpost of this African Atlantis, and the Uphaz mentioned in the Bible along with Tartessos ("Tarshish") as a source of gold and trade-goods was Yorubaland itself, whose capital was at the site of modern Ilife.

The main weakness of Frobenius's theory (aside from the impossible Pacific continent) was his conviction that he could trace all the early migrations of mankind by comparison of their artistic symbolism, for the degree of resemblance between any pair of symbols, as any trade-mark lawyer will tell you, is a very subjective matter.

Following Frobenius, the geologist Paul Borchardt of Munich undertook in 1926 to find Atlantis in Tunisia, in the region of the shotts or salt-marshes that stretch westward from the Gulf of Qabès which the ancients called the Little Syrtis. The biggest of these dismal swamps, the Shott el Jerid, was probably the ancient Lake Tritonis where Diodoros located his Amazons and which played a part in the story of the Argonauts. This body of water, said Borchardt, was the original Atlantic Sea.

Furthermore Borchardt identified the ancient Mount Atlas, not with the modern Moroccan range of that name, but with the Ahaggar Mountains. He tried to correlate the names of Poseidon's ten sons as given by Plato with the names of modern Berber tribes, and thought the "Pillars of Herakles" were real temple pillars instead of the mountains flanking the Strait of Gibraltar. He deemed the mineral wealth of the shott country a confirmation of his theory, and suggested that Atlantis with its brass and orichalc, the brazen palace of Alkinoös in the *Odyssey,* and the City of Brass in the *Arabian Nights* were all one and the same. Near Qabès he found the remains of a fortress which he took to be the city of Atlantis itself; alas, it turned out to be Roman.

Undiscouraged by this debacle Albert Herrmann went hunting Atlantis in southern Tunisia. He thought he had found it in the village of Rhelissia, where he discovered traces of irrigation works pointing to a higher culture than that of the present Rhelissians. He reasoned that Plato was wrong on three counts. First, he took Herodotos's meaning for "Atlantic" instead of the alleged older meaning referring to Lake Tritonis. Second, Atlantis fell in the fourteenth or thirteenth century B.C. instead of the ninety-sixth.

Third and lastly, Solon and his priest, conversing through an interpreter, got fouled up in translating Egyptian measurements into Greek units, so that everything came out thirty times too big. With that correction Plato's Atlantis, irrigated plain and all, shrinks to modest dimentions that would fit comfortably into a corner of Tunisia.

And Herrmann went on to derive all civilization from Friesland, of which Atlantis was merely a colony in the days of Frisian glory.

One more member of the Afro-Atlantean school, Butavand, got around the lack of tangible Atlantean relics in North Africa by locating the lost Atlantis at the bottom of the Gulf of Qabès off the Tunisian coast. He assumed that the Gulf was once dry land, out to about the present 100-fathom line, until an earthquake lowered this land beneath the waters of the Mediterranean and at the same time raised the bottom of the suppositious Saharan Sea, which thereupon drained off and dried up. Perhaps the Strait of Gibraltar broke open at this time too.

Finally, one of the latest and also the most plausible African interpretations of Atlantis is that of Silbermann. After surveying the general field of Atlantist theories, he pointed out that in view of the known chronology of Egyptian civilization and the difficulty of handing down a story by word of mouth alone for more than a few centuries, Plato's dating of the rise and fall of Atlantis 9000 years before Solon's time could be dismissed right away: if such a civilization had existed, it would have been entirely forgotten long before the rise of Egypt.

Silbermann thought that the Atlantis story was originally a Phoenician account of a war with the Libyans of the shott region of Tunisia that took place around 2540 B.C. About the eleventh or tenth century, he thought, some Egyptian of Saïs made this story into a romance, placing the events "in the time of Horus," which date Plato interpreted much later as about 9600 B.C. This romance may have been forgotten for a time, but it was revived about 600 B.C. when Niku II rebuilt the Egyptian Navy and

there was a search for books about Libya. The story was
also translated into Greek for the benefit of the Hellenes of
Saïs, and one Greek version formed the basis for Solon's
account while another was preserved as the story of the
Atlantioi in Diodoros of Sicily.

These theories, ingenious though some of them be,
can hardly all be true at once. Besides their obvious mutual
incompatibility, many of them reveal the patriotic desire
of some Atlantis-hunters, from Rudbeck on, to prove their
own land the source of all civilization. While such a motive
may arouse a certain sentimental sympathy, it has nothing
to do with science.

Still, the general concept of a fallen civilization
whereof rumors helped to inspire Plato in his literary
labors is not implausible. Perhaps the leading contenders
for the honor of having served as the prototype of Atlantis
are Minoan Crete, Carthage, and Tartessos. Let us look
at them in that order.

In Classical times Crete was a depressed backwater
where Dorian landlords bullied their native tenants,
a haunt of pirates and a recruiting-ground for mercenary
archers. A web of legend surrounded it: Crete, it was said,
had once been a great sea-power ruled by King Minos, the
son of Zeus (in the form of a bull) and Europë. Hephaistos,
the divine smith, gave Minos a brazen robot named Talos
who kept strangers away from Crete by throwing stones
at them.

Once Minos prayed to Poseidon to send him a bull
from the sea, promising to sacrifice it to the god. Poseidon
furnished the bull, but Minos liked it so well that he
sacrificed another in its stead. In revenge Poseidon caused
Minos's wife Pasiphaë (sister to the enchantress Kirkë)
to fall in love with the bull. The queen prevailed upon
the skillful Daidalos, an Athenian fugitive, to arrange a
tryst with the animal. Daidalos did so, having disguised
Pasiphaë as a cow. Poseidon also made the bull so savage
that Minos had to appeal to Herakles to come and take
the thing away.

Pasiphaë in due time gave birth to a bull-headed man,
Asterios, known as the Minotaur, whom Minos shut up in

the maze that the obliging Daidalos built for him. Subsequently Minos warred on Athens because a son of his had been killed there, and in the peace treaty demanded seven youths and seven maidens each year for the Minotaur to eat. You probably heard as a child how Theseus broke up the racket by killing the Minotaur with the help of Minos's daughter Ariadne (whom he eloped with but lost to the god Dionysos) and Daidalos, who then made his famous flight from Crete with his son Ikaros.

For many centuries scholars tried to guess what historical reality lay behind the Minos-myths, especially the recurrence of the bull-motif. Most of their guesses went wide of the mark until Evans's excavations, begun about 1895, disclosed the basis for the story: the public spectacles of Minoan Crete at which young men and women performed perilous gymnastics on the horns of live bulls.

Crete flourished as a naval and commercial power for about two thousand years until it fell in the time of the folk-migrations at the end of the second millenium B.C. Almost nothing of its actual history is preserved, since we cannot read Cretan writing, and the Minos-legends are too scanty and mythical to recover history from. However, Crete was allied with Egypt during much of its history. The ruling class in the capital at Knossos lived a refined and luxurious life with startlingly modern bathrooms. The men, small dark Mediterraneans, wore corsets and loin-cloths; the women, bare-breasted big-skirted dresses that look like a burlesque version of Victorian costume.

The Cretan Empire was a federation of island cities under the dominance of Knossos. Colonies on the Greek mainland, at Mykenai and Tiryns, became so important that after the destruction of Knossos by earthquake about 1400 they assumed the leadership of the federation. Then as the power of the empire declined, its local governments were taken over by the barbarous Greeks (*Achaiwoi* they called themselves, later *Achaioi* or Achaeans) as the Germans took over the West Roman Empire piecemeal 1500 years later. Homer presumably portrays a romantic reflection of the life of this period, when barbarians who have formed a taste for the niceties of civilization have seized control of it and are exploiting its luxuries.

In the early years of this century K. T. Frost in England pointed out striking resemblances between Plato's Atlantis and the archeologists' Crete. Both were island kingdoms, sea-powers, that met sudden downfall at the hands of men from Greece. The self-centered Egyptians, he thought, heard rumors of the conquest of Crete by "Mycenian" invaders from the Greek mainland, who later appeared among the barbarian invaders repelled by Rameses III about 1190 B.C. Hearing no more of the Mycenians, whom the Achaeans had conquered in their turn, the Egyptians assumed that Crete had disappeared and invented a submersion legend to account for it.

But we cannot take Frost's theory too seriously even though Balch and Magoffin followed him in it. For one thing we don't know that the Mycenians conquered Crete. For another the Egyptians are unlikely in a matter of 600 years or so to have moved Crete clear out of the Mediterranean, enlarged it a hundred-fold, and predated it by 8000 years. Still the Cretan public works and bull-ceremonies are suggestive; perhaps they entered the Atlantis story as fragments which Plato picked up and which his subconscious wove into his Atlantean fiction. Spence explains the Minoan-Atlantean resemblances by supposing Crete to be an offshoot or colony of Atlantis; but we have already excluded Spence's hypothesis on geological grounds.

Although Spence disposes of the alternative theory, Atlantis-in-Carthage, in the same manner, Carthage (proposed by the eminent Homerist Victor Bérard) is harder to get rid of. It lies in the right direction from Greece, which Crete does not, and furthermore was not only an imperialistic sea-power, but also a city whose plan suggests that of the city of Atlas.

"The low, walled hill of the Byrsa or citadel on which stood the splendid temple of Aesculapius at Carthage was strengthened on the mainland side by three great ramparts which stretched across the breadth of the peninsula, and which were fortified at intervals by towers. Below the market-place and the Senate House a vast waterway, 1066 feet wide, had been constructed round a circular island on which stood the admiral's headquarters. The docks surrounding this waterway were roofed in by a great circular colonade supported by Ionic pillars, and were capable of accommodating war-vessels of the largest size." From this basin "a narrow channel ran southward into a mercantile harbour 1396 feet long. A great sea-wall masked this entrance and prevented attack by a hostile fleet. Marshes surrounded the landward side, and the water-supply was drawn from great cisterns on the neighboring hills, which seem also to have been used as baths."

Carthage was founded about 850 B.C. by the Tyrian princess Elissar ("Dido"), a daughter of King Mutton I, who fled from the tyranny of her brother Pygmalion. Though not the first Phoenician settlement in North Africa, it grew fast, and after Tyre fell to Nebuchadrezzar in 573 B.C. it became the ruler of the Western Mediterranean. To secure control of the western trade, especially with the tin region of southwestern Britain, Carthaginian warships stopped non-Carthaginian trading vessels and threw their crews overboard. This ironclad monopoly, not broken until the Punic Wars, accounts for the haziness of Greek knowledge of the Western Mediterranean in Plato's time.

Both Greeks and Carthaginians colonized Sicily and tried hard to throw each other out. The Carthaginian suffete Malchus almost conquered the whole island in 550 B.C., and wars continued, with time out for recovery and local revolutions, for three centuries. Hamilcar almost got the island again in 480 B.C. but the forces of Syracuse and Agrigentum beat him at Himera in a great battle almost as decisive in world history as the Greek victory over the Persians at Salamis in the same year. Plato had a chance to learn of the Carthaginian menace at close range during his visits to Syracuse, and the luxuriance and avarice of the Punic merchant aristocracy (or, to be fair, the luxuriance and avarice attributed to them by their Greek and Roman enemies) may have suggested the growth of those vices in Atlantis. On the other hand Carthage was a republic (not a kingdom like Atlantis) and in Plato's time, so far from disappearing, it was still growing in power.

That leaves Tartessos — the Biblical Tarshish, Jonah's destination — a flourishing old city-state in southwestern

Spain, near modern Cadiz, as a source for Atlantis.

The earliest reference to Tartessos is in *Isaiah,* where the prophet, sermonizing on the fall of Tyre, says: "Howl, ye ships of Tarshish; for it is laid waste. . . ." Arrian is probably wrong in saying that Tartessos was a Phoenician colony; on the other hand the affinities of its people are not known for sure. Cretan origin has been proposed, and the name of the place sounds like some of those applied to the Etruscans, whom the Greeks called *Tyrrhenoi* and the Egyptians *Tursha,* and who were said in Classical times to be from Lydia. The Etruscans, though, called themselves *Rasenna,* which sounds nothing like Tartessos.

Almost the only archeological relic of Tartessos is a ring inscribed with the following characters:

Outside: ΛＫＶ ＫＫ ＶＯＦＩΛＫＯ ＶＱＡ

Inside: ＶＯＶＫ ＶＯ ＫＫＶＯ ＶＫＯ

found on the site by Schulten in 1923. The alphabet seems to be related to those of the Greeks and Etruscans, and the inscription on the inner surface consists of a single four-letter word repeated three times — perhaps something like *psonr* or *khonr,* whatever that may mean. The repetition suggests a magical spell, something like "eeny, meeny, miny, mo." Pre-Roman Iberia used at least two alphabets, neither deciphered for sure yet.

Whatever its origin, Tartessos long flourished as a trading and mining city. The Phoenicians, arriving around 1000 B.C., found silver so common that in order to carry away as much as they could in return for olive-oil and other wares, they cast their anchor-stones of silver. During the tenth century B.C., when King Solomon and King Hiram of Tyre had their profitable partnership, their joint fleet used to make a round trip to Tartessos every three years, returning with "gold, and silver, ivory, and apes, and peacocks." And Ezekiel, lamenting the fall of Tyre, said:

"Tarshish was thy merchant by reason of the multitude of all kinds of riches; with silver, iron, tin, and lead they traded in thy fairs." The metals came from the mines of southern Spain, where a Rio Tinto mining town is still called Tharsis. The "apes" might be either from Africa, or from Gibraltar, where monkeys still live. The "ivory" might be from the Moroccan elephant, a smallish variety of the African elephant used in war by the Carthaginians and exterminated in Roman times, and the "peacocks," *thukkiyim,* might be either the now rare Congo peacock or an error for *sukkiyim,* "slaves." An allusion by Herodotos to "Tartessian weasels" suggests a fur-trade, and later Tartessos exported brasswork to Greece.

The Greeks made the acquaintance of Tartessos about 631 B.C., when a Samian ship under Kolaios, bound for Egypt, was blown far out of its course by an easterly gale and ended up at Tartessos — a record detour. The Samians made six talents from their voyage: an enormous sum for the time, equivalent to $75,000 or more in modern money. Next came men from Phokaia in Ionia, who also opened up the Adriatic and Tyrrhenian Seas to Greek commerce and founded Marseilles. For their Tartessian trade the Phokaians used, not the tubby merchantman of the time, but the swift narrow fifty-oared pentakonter, which, if it carried less cargo, had a better chance of getting away when a Carthaginian galley came crawling like a giant insect over the horizon with the intent of making the interlopers walk the plank.

The first Phokaian traders found Tartessos ruled by King Arganthonios ("Silver-locks") though you need not believe with Herodotos that he lived 120 years and reigned eighty of them. The king liked his guests so well (perhaps trade had been dull of late, Tyre having fallen and Carthage not yet risen to her later eminence) that he suggested that if the Persians pressed them too hard at home, they should all move to his dominions. When they declined, he gave them enough money to build a great wall about their city.

In 546 B.C., however, Cyrus of Persia sent his general Harpagus with orders to take Phokaia. The people, despairing of holding out even behind their fine wall, persuaded

Harpagus to withdraw for a while while they pondered
terms; then they crowded into their ships and rowed away
from their Asiatic homes. Hearing that Arganthonios's
long life had ended, they went to Corsica instead of to
Tartessos. Here they became embroiled with the Carthaginians and Etruscans, whose combined fleets they barely
beat in a battle in 536. The survivors picked up their
families in their twenty remaining ships and resettled on
the coast of Lucania in Italy.

In its heyday Tartessos was the leading city of southwestern Spain, then called Tartessis, whose people, the
Turdetani or Turduli, were rated the most civilized in
Spain. They had a caste system, and their own alphabet in
which were recorded their poems, laws, and history, which
told how the Ethiopians had once overrun North Africa
and how some of them remained in the Atlas Mountains.
These Turduli probably were among the people who lived
in Iberia before Kelts from Gaul overran it at the beginning
of recorded history. The unclassifiable Basque language
is a legacy from these pre-Indo-European Iberians, and if
we could read the pre-Roman Iberian inscriptions like
that on Schulten's ring, some of them might prove to be
written in languages related to Basque.

Tartessos stood at the mouth of the Tartessis or Baetis
River, now the Guadalquivir; a flat sandy region bordered
by a sea dangerous for strong tides and a heavy surf. The
region is sparsely populated today by a dark people taller
and broader-headed than most Spaniards. In ancient times
the river ended in a great bay which once reached inland
as far as Hispalis (Seville). Tartessos stood on a large island
that closed the mouth of the bay, so that the Guadalquivir
had two mouths. In historical times the bay had partly
silted up and grown islands, reminding us of the shoals
left by the sinking of Atlantis. Now the bay has become a
great malarial marsh, Las Marismas, and the northern arm
of the river is dry land.

As they became acquainted with Tartessos, the Greeks
introduced it into their legends. They had taken over from
the Phoenicians the whole cycle of the tales of Herakles,
the Tyrian Melkarth, probably a member of a widespread
Near-Eastern family of lion-slaying mythical heroes repre-

sented by the Sumerian Gilgamesh and the Hebrew Sampson. Herakles-Melkarth was persistently associated with the Far West; besides his journey thither to fetch the Apples of the Hesperides and that to capture the infernal watchdog Kerberos, he had also gone there to lift the cattle of Geryon, a three-bodied giant who lived on the island of Erytheia. Herakles stopped at Tartessos where he erected two pillars, sometimes identified with the hills of Gibraltar and Ceuta and sometimes with a pair of real pillars in the Temple of Herakles at Gades. Being overheated he threatened the sun with his bow, and the sun-god, admiring his nerve, lent him the golden goblet wherein he made his daily journey around the world. Herakles voyaged to Erytheia in this vessel, killed Geryon and his herdsman and dog, drove the cattle into the goblet, and sailed back to Tartessos, where he returned the mug to its owner. Therefore the Tartessians worshipped Herakles ever after.

Although Tartessos, according to Schulten, went back to the Neolithic Age, it acquired a rival in historic times. About 1100 B.C. the Phoenicians planted a colony on another island (since become a peninsula) twenty miles southwest of Tartessos, at the mouth of the Guadalete. They named their settlement *Ha-gadir*, "the hedge" or "the stockade," whence the Classical Gades and modern Cadiz. From the same root come the names of modern Agadir in Morocco, and Gadeiros, one of Poseidon's sons in Plato's story.

Tartessos and Gades coëxisted until the rise of the Carthaginian Empire. Between 533 and 500 B.C. the Carthaginians were active around the Pillars; they sent Hanno down the African coast, reduced Gades to subjection, and in 509 extorted a treaty from Rome confirming their western monopoly. At the same time they sent another admiral, Himilco, "to explore the outer coasts of Europe." Like his colleague Hanno, Himilco published a report on his return, which we know from a poetical synopsis by the late-Roman poet Rufus Festus Avienus. Himilco evidently set out to make his readers' flesh creep. He had gone, he said, to the Tin Islands, where the Tartessians used to trade, and the round trip took four months — not because of the distance, but because:

There no breeze exists to propel the ship forward;
Deadening is this sluggish sea's still vapor.
Many seaweeds grow in the troughs of the billows,
Slowing the ship like bushes, he says, thus showing
Here to no great depth the sea descends, and
Here the water barely covers the bottom.
Here the beasts of the sea move slowly wandering,
And among the sluggishly creeping vessels
Languidly swim the great monsters.

To make sure that prospective explorers got the idea, Himilco added that if one kept on one got into regions of impenetrable fog. This report was in line with the legend of the sticky water that Pytheas heard of in the North and that the ill-fated Sataspes reported in the tropics on his return to Persia from his failure to sail around Africa. In those days an explorer who got cold feet often excused himself later by saying his ship had stuck fast in some syrupy water.

Some geographers think that Himilco's weed-covered shallows were the Sargasso Sea, an elliptical patch between Florida and the west coast of Africa, about 1000 miles north-and-south and 2000 miles east-and-west, where floating sargassum weed is found most frequently. This weed, which grows along the Atlantic coasts of the Americas from Cape Cod to the Orinoco, breaks away from its attachment on the sea-shore and floats out to sea supported by its little gas-bladders. The broken pieces live for a long time in this condition, since they continue growing at one end while they decay at the other, and support a rich fauna of curious crabs and fish adapted to life in the weed. The stuff collects in the great eddy formed by the Gulf Stream and the North Equatorial Current, and drifts round and round for years until it finally sinks. A similar patch exists in the Indian Ocean.

Some odd ideas are current about the Sargasso Sea because in 1896 the novelist T. A. Janvier wrote a gripping novel, *In the Sargasso Sea,* in which he described the tract as an impenetrable tangle of weed holding fast the remains of ships of all ages from Spanish galleons down. Unfortunately this picturesque concept is not at all like the

real Sargasso Sea, in which the weed is so thinly scattered
that ships sail right through it without their passengers'
knowing they have done so. The Bermudas, for instance,
lie within the area of greatest density.

Babcock surmised that some Phoenician crew had
once penetrated the Sargasso Sea and, assuming that the
floating weed was growing from the bottom as seaweed
usually does, deduced that they were over a bank or shoal.
Hence Himilco's report and Plato's Atlantean shallows.
Not impossible, perhaps, but most unlikely in view of the
small sea-keeping capacity of Classical ships. A Mediterranean trireme of Himilco's time was after all less than
half the size of one of the big catamarans, misleadingly
called "canoes," in which the Polynesians ranged the Pacific. And even if the Phoenicians had known the Azores,
which is not certain, a Phoenician ship starting from these
islands would have had to sail nearly a thousand miles in
the teeth of the prevailing westerlies to reach the area of
dense weed — an impossible feat for an ancient square-rigger
with only one or two sails, no central rudder, and no compass to show the way when the sky was overcast.

More likely the whole shoals yarn was a Carthaginian
cock-and-bull story intended to frighten away competitors.
In any event it circulated widely in Greece, being adopted
without qualms by Plato and Aristotle and mentioned
by other writers.

And Tartessos? Nobody knows what happened to it,
but after Himilco's voyage nothing further is heard of the
city, and Himilco is suspected of having liquidated this
competitor in the course of his voyage. Later geographers
sometimes confused Tartessos with Gades, or with the little
towns of Calpe and Carteia near Gibraltar.

Atlantis is even more strikingly like Tartessos than it
is like Carthage or Crete: Tartessos, like Atlantis, lay in
the Far West, beyond the Pillars; it was enormously rich,
especially in minerals, and had wide commercial contacts
with the Mediterranean; it was associated with shoals; behind it lay a great plain bordered by mountains; and it
mysteriously disappeared. While the Tartessians are not
known to have performed a bull-ceremony, the region was
and still is a cattle country.

In the 1920's Professor Adolf Schulten of Erlangen, assisted by the archeologist Bonsor and the geologist Jessen, dug up the site of Tartessos. Besides the ring, Schulten found blocks of masonry which, he thought, showed the existence of two former cities, one dating from about 3000 and the other from about 1500 B.C. The high water-table prevented digging far down, and the investigators reluctantly concluded that the other remains of Tartessos had long since sunk deep in the mud of the Guadalquivir estuary.

Schulten also located the ruins of the Temple of Melkarth of Gades, on the little island of Santi Petri, with two wells mentioned by Polybios inside, recalling the springs in the Atlantean temple of Poseidon. Schulten's theory, popularized by Dr. Richard Hennig, was that all the necessary materials for Plato's story were right there in Spain.

About the same time Ellen M. Whishaw (widow of the archeologist Bernhard Whishaw, whom she succeeded as director of the Anglo-Spanish-American School of Archaeology) found relics in the same region that suggested to her a former great Hispano-African culture, the "Liby-Tartessian." She learned for instance that in the middle of the nineteenth century, "In a neolithic sepulchral cave, known as the Cave of the Bats, in the province of Granada, 12 skeletons were discovered, sitting in a circle round a central skeleton of a woman, dressed in a leather tunic. At the entrance of the cave where three more skeletons, one wearing a crown and dressed in a tunic of finely woven esparto grass. Beside the skeletons were hid bags containing carbonised food, and other bags filled with poppy heads, flowers and amulets; poppy heads were scattered all over the floor of the cave. Among a number of other objects were some little clay discs identified by archaeologists as necklace ornaments connected with the sun-cult, found in the land harbor of Niebla and in a building trench near Seville."

These skeletons were supposed to have been those of a royal family and their attendants who for some reason committed mass suicide by eating opium. Mrs. Whishaw cited other evidence for her Liby-Tartessian culture, such as a Neolithic cup found near Seville, decorated with a picture of a woman dressed like the one in the Cave of the

Bats fighting two Libyan warriors; prehistoric Iberian graves; and modern Spanish customs pointing to an ancient matriarchal society like that of the Berbers. She argued that Tartessos was, not Atlantis itself as Schulten asserted, but a colony of Atlantis — Plato's Atlantis, sunken island and all. But, as we have seen, geology (a science of which Mrs. Whishaw admitted she knew nothing) bars that view.

One set of comparisons remains to tie this mass of speculation into a neat bundle — or as neat as the straggly nature of the material allows. That is the comparison between Plato's Atlantis and the historical Tartessos on one hand, and Scheria, the land of the Phaiakes or Phaeacians in Homer's *Odyssey,* on the other.

Before his arrival at Scheria, Homer's hero Odysseus had set forth from the isle of Ogygia, "at the very center" (literally "boss" or "navel") "of the sea," where for eight homesick but not uncompensated years he had been kept by the nymph Kalypso, a daughter of Atlas. When the gods finally forced Kalypso to let him go, he built a raft and set sail for "fertile Scheria, the land of the Phaiakes, near kin of the gods," supposedly twenty days' sail away.

Here we have some peculiar coincidences. The Greek word for the raft on which Odysseus wafted was *schedia,* supposedly of Phoenician origin, meaning either a raft or a pontoon bridge. Besides, there were at least two Phoenician colonies and trading-centers named Schedia, one on the north coast of the island of Rhodes, the other on the coast of Egypt near Alexandria. And the Phoenician word for market is *schera.* . . . The study of Atlantis is full of such coincidences, which might be very useful if they all pointed in one direction instead of in many.

Following Kalypso's directions to sail "with the Bear on his left," Odysseus made most of his journey without incident. But on the eighteenth day, with his goal in sight, he was spied by Poseidon returning from a feast among the Ethiopians. The Earth-Shaker, having it in for Odysseus for having blinded his son Polyphemos, wrecked the raft. Odysseus would have drowned had not the nymph Leukothea taken pity on him and lent him her veil as a magical life-preserver.

By strenuous swimming through a savage surf the Wanderer made "the mouth of a fair-flowing river." He quickly prayed to the river-god, who "stayed the stream, held back the wave and made it smooth before him, and brought him safely into the river's mouth." When the exhausted Odysseus had crawled out of the water and gotten his wind back, he cast Leukothea's veil into the water, and "a great wave bore it down the stream" to the nymph.

Interpreters of Homer have sought geographical clues in these passages as to the whereabouts of Scheria. Although the country is "in the sea," implying an island, the river indicates a large land mass. The water's reversal at the river-mouth *might* be a description of oceanic tides, which would locate Scheria outside the Mediterranean whose tides are measured in inches.

Homer then tells a little about the Phaiakes: how their King Nausithoös (a son of Poseidon) had led them from "spacious Hypereia near the overbearing Kyklopes" to their present site, "far from wheat-eating men." This makes little sense, since the only Hypereia known to history is a fountain in the town of Pherai, in southwestern Thessaly, near Iolkos of Argonautic fame, while the usual euhemeristic interpretation of the *Odyssey* identified Polyphemos with Mount Etna and the land of the Kyklopes therefore with Sicily.

At any rate, Nausithoös now was deceased, and

> With god-given wisdom Alkinoös reigned
> in his stead.

The day after Odysseus's escape from the sea, Athena inspired Alkinoös's daughter, the white-armed Nausikaä, to collect her handmaidens and drive the royal laundry-wagon down to the river to do the palace wash. After washing they amused themselves with a ball-game until their chatter awakened Odysseus. The Wanderer, holding a bush in front of his nakedness with un-Hellenic modesty, appealed for help to Nausikaä, who alone of the girls had not run at the sight of him.

The charming little princess called back her girls, lent Odysseus clothes, and told him to follow them into town. She explained that he was among the godlike

Phaiakes, who "live farthest away on the loud-surging sea, and none else of mortals mingles with us." Their city had a lofty wall, a spacious harbor, a splendid temple of Poseidon, and a megalithic assembly-place. "For," she said, "the Phaiakes care not for bow or quiver, but for masts and oars and trim ships, with which they rejoice to cross the foaming sea."

Odysseus, much impressed by the bronze threshold of the palace of Alkinoös, its golden doors, silver doorposts, bronze walls, and golden statues of youths holding torches within, made his appeal to the king and was received with great hospitality. He learned that the Phaiakes' ships were "swift as the flight of a bird or even a thought"; that the Phaiakes were, like the Kyklopes, related to the gods; and that the farthest land they had visited was the island of Euboia off the east coast of Greece, whither they had once taken Rhadamanthos to visit his cousin Tityos. In Greek myth Rhadamanthos had been a brother of King Minos of Crete. Because of his integrity the gods had made him, after death, one of the judges of the dead and let him live in the Elysian Fields, which were, like Homer's Ogygia and the land of the Kimmerioi, somewhere in the Far West. The Phaiakes lived a luxurious life, fond of "the feast, the lyre, the dance, change of raiment, warm baths, and love."

Alkinoös, who seems to have been something of a jolly old souse, agreed to send Odysseus home. Next day he threw a fine party for the stranger, with athletic competitions, dancing, and the bard Demodokos singing lays of the lusts of the gods to his lyre. After the feast Odysseus was persuaded to tell his name and his adventures since leaving Troy. His tale, occupying Books IX to XII of the *Odyssey,* is the best-known part of the epic: how his fleet was first blown to Thrace where his men raided the Kikones but were beaten off; thence to the land of the Lotophagoi, the Lotus-eaters, in North Africa; thence to the land of the Kyklopes; to the floating island of Aiolos with its brazen wall; to the land of the cannibal Laistrygones; to Aiaia the island of Kirkë; to the land of the Kimmerioi on the banks of the river Okeanos, where Odysseus interviewed the ghosts of his comrades; past the Sirens' island and the Planktai or Clashing Rocks, through the strait of Skylla and Charybdis

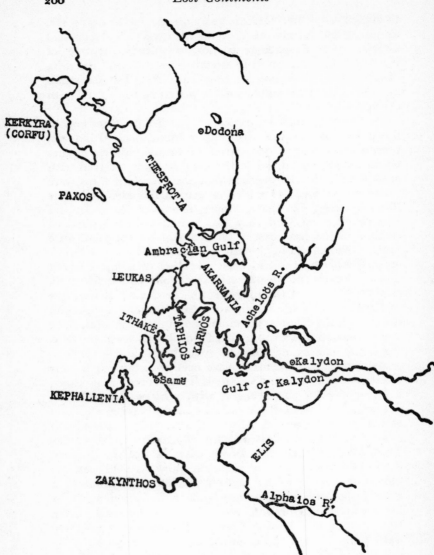

Fig. 15. THE KINGDOM OF ODYSSEUS — the west coast of Greece
and the islands off it mentioned by Homer.

to the island of Thrinakia, where his men's slaughter of the Cattle of the Sun completed the ruin of the expedition, now reduced to a single ship. Odysseus ended his account clinging to a piece of his foundered ship and drifting to Kalypso's isle.

Next day Alkinoös sent Odysseus off in one of the Phaiakes' magic ships with rich presents. After a night's run faster than a hawk could fly, they set Odysseus ashore on his native Ithaka, where he duly took vengeance on Penelopeia's suitors. But Poseidon, already jealous of the impunity with which the Phaeacian ships crossed his sea, and furious at Odysseus's safe homecoming, turned the Phaeacian ship to stone as it entered its home port. Moreover he announced his intention "to overshadow their city with a great mountain." To avert this last calamity the Phaiakes swore off their friendly custom of taking home any strangers who happened by their city.

Now, what is Scheria? It has been identified with almost as many places as Atlantis.

Apollodoros, Strabo, and other Classical writers thought that it was Corfu (also called Kerkyra or Corcyra) the erstwhile summer home of Kaiser Wilhelm II. Some modern Homerists stick to this traditional view; the Scot Shewan, for instance, thought it was Corfu with a Phoenician or Cretan population. Still, Corfu lacks tides and rivers and, so far from being "farthest away on the sounding sea," near the Elysian Fields, is actually in sight of Levkas, one of the group of islands that comprised Odysseus's domain. To bring Odysseus home from Corfu would hardly have needed a magical ship—unless, as Shewan thought, the Phaiakes were merely bragging when they told of their marine thaumaturgy.

Moreover Corfu has other claimants. Leaf thought it was Taphos, the realm of the King Mentes mentioned in the first book of the *Odyssey,* contrary to the usual belief that Taphos was the historical Taphios, a little island east of Levkas.

Others like Leutz-Spitta and Henning have proposed that Corfu was none other than Odysseus's Ithaka itself. Off the west coast of Greece, clustered around the entrance to the Gulf of Corinth, lies an archipelago of four large

islands and many small ones. Nowadays the four large
islands are called (reading from south to north) Zakynthos
or Zante, Kephallenia or Cephalonia, Ithakë or Thiaki
(the smallest of the four) and Levkas or Santa Maura. In
Classical times they were called Zakynthos, Kephallenia,
Ithakë, and Leukas or Leukadia respectively. (Remember
the goddess Leukothea?) Now, Homer's Odysseus repeat-
edly refers to his realm as comprised of "Zakynthos, Doulichion, Samë, and Ithaka," and it would seem no great trick
to fit these four names into the four islands in question.

It turns out, however, that Zakynthos is the only one
we can be sure of. Kephallenia might be either Doulichion
(whose description it answers) or Samë (since it has a town
of that name) and conversely Levkas might be either Samë
or Doulichion. Ithakë, despite its name, does not answer
very well to the description Odysseus gave Alkinoös of his
home island: "low-lying, farthest up in the sea towards the
gloom, while the others are away towards the dawn and the
sun." *Zophos*, "gloom" or "darkness," is here a poetical
term meaning, probably, "west," but possibly "north" or
even "northwest." And modern Ithakë lies right in the
midst of the group. Furthermore it is described as *chtha-
malē*, "low" or "flat," whereas Ithakë is a mass of rocky hills
rising to 2645 feet.

In the last century Draheim and Dörpfeld suggested
that Levkas was Homer's Ithaka. Others denied this, asserting that Levkas was not an island at all but a peninsula,
since it is separated from the mainland only by a shallow
ford. As if this fog of names were not opaque enough already, Homer also mentions the Kephallenians as subject
to Odysseus, though without saying where they live. Perhaps we had better not get ourselves involved in this ancient
argument, which has raged for 2400 years without settling
the question. Maybe Homer, describing the islands from
hearsay, got his geography a little mixed, confusing Ithakë,
Levkas, and Corfu with one another.

Walter Leaf gave up the hunt for Scheria with the
words: "No; if we take our measure from earthly maps,
Corfu is not Scherie. But Scherie has its place in the map
of poetry and fancy; and there I believe it can be identified. Is it not Homer's name for Plato's Atlantis? And if

we want some connexion with the real world, let us think of the ingenious and attractive idea which finds in Atlantis a recollection of the lost glories of the Minoan empire, and consider whether the Phaeacians who, in Nausikaa's words, 'care not for bow and arrow, but only for masts and oars and ships,' may not fairly remind us of the men of Knossos, who, secure in the rule of the sea, never cared to fortify their palace by the shore?" But Crete makes a pooor Scheria, since Scheria was well-fortified and Crete was celebrated for its archery. Moreover the *Odyssey* names Crete—in fact Odysseus mentions it to Alkinoös. (Likewise, although the Land of the Kyklopes was long held to be Sicily, the *Odyssey* also mentions Sicily under its old name of Sikania.)

Nor could Scheria be Carthage, since Carthage was only founded about the time the Homeric poems were taking shape. Some Homerists have even found Scheria in Palestine, Tunisia, Sicily, Gades, the Canary Islands, or the island of Socotra in the Arabian Sea. And some think it a fairyland, a pure creation of the imagination.

If Scheria be anything in the real world, which we shall probably never know, Tartessos would seem not unlikely: Tartessos the Silvery, as seen, not first-hand, but through a mist of rumors picked up from Phoenician seafarers who had been there by Greeks who had not. Tartessos lay in the right direction. Looking west from Greece it was "farthest away on the loud-surging sea"; it possessed the lavish metallic wealth attributed to Scheria—Polybios even said of a rich Iberian king that "he rivalled the luxury of the Phaiakes." It lay at the mouth of a large river, on a coast beaten by a powerful surf and strong tides. Finally Scheria's threatened doom of being encircled by a high mountain suggests the actual fate of Tartessos in being stranded in the midst of great mud-banks built up by the silting of the Baetis; and, more remotely, the submergence of Atlantis.

Not that Scheria *is* Tartessos, except in the sense I discussed at the beginning of the chapter. Like Atlantis, Scheria is a literary creation, used as the background for a piece of imaginative fiction. Therefore its author, no sober historian, need not stick to strict facts. His literary land may be based on a combination of several real places further touched up by his poetic imagination.

And Plato, even if he had not heard much about Tartessos, could well have drawn upon the *Odyssey's* account of Alkinoös's palace for his gleaming metal-sheathed Atlantean citadel.

The most we can say, then, is that both Homer and Plato possibly used traditions and rumors of Tartessos in building their literary creations of Scheria and Atlantis. In addition, Homer and Plato may have used traditions relating to Crete, and Plato may have used both Homer's Scheria and the real Carthage as sources of inspiration. The exact truth, like much else in such matters, is probbaly lost beyond recovery.

Such being the case, could there, nevertheless, have been great fallen civilizations whose remains have not yet been discovered? Perhaps.

Europe has been combed over by archeologists to the point where sensational finds of high cultures seem unlikely. Future discoveries will probably take the form of filling in the gaps, as by finding more information on the Tartessian and Minoan cultures.

In the New World the rise of the Mayas, Aztecs, and Andean peoples from barbarism has been fairly well charted. Their histories may be extended back a few centuries, and new sites and colonies disovered, but we are most unlikely to find a whole new civilization preceding those already known.

In Africa likewise we know the general outlines in space and time of the great Egyptiac civilization, and also of the high Negro civilization that flourished in the western Sudan for a thousand years until destroyed by White invaders (Moroccans and Spaniards) in the seventeenth century. Although pseudo-scientists have long speculated about the mysterious fortifications at Zimbabwe in South Africa, archeology seems to indicate that they were built by native Bantu-speaking people as late as 1300 A.D. Australia, and the Pacific Islands before the Polynesian invasion, seem never to have had civilized cultures.

Asia, however, has hardly been scratched archeologically. A great Indus Valley civilization of Sumerian times has come to light in the last few decades, and further discoveries are quite possible. Perhaps we shall some day find an early semi-civilized Central Asiatic culture whence the Sumerian and Chinese civilizations sprang. In fact, cryptic tales of buried cities in the Caucasus and Russian Turkestan, and of a great pyramid in western China, have drifted out of Asia in recent years. If the Russian archologists who have been digging up the wilder parts of the U. S. S. R. become less close-mouthed they may have exciting revelations to make. And even in a well-explored area like Iraq, archeologists have pushed the horizons of civilization back to 7000 years ago by the discovery of the oldest known farming villages at Qalat Jarmo and Hassuna.

Still, don't expect four-armed hermaphrodites with astral bodies and airplanes. The people who built up these early cultures were much like ourselves: the men tending herds and crops, building houses, and making utensils; the women cooking, sewing, and baby-tending; the kings waging war, dispensing justice, and indulging their lusts. There was no prehistoric machine-age. Of the hundreds of thousands of archeological relics found since men started to hunt them, not one bronze-age flashlight or fountain-pen has turned up, nor is any likely to.

CHAPTER IX

THE AUTHOR OF ATLANTIS

Mariners all, declare
    Where those lost islands lie,—
The Fortunate, the Fair,
    Under what shining sky,
Robed with what shining air?

Noyes

So far we have bent our attention upon the story of Atlantis itself, and sought through time and space for real events and things to match those of this colorful narrative. For all our efforts we have not achieved any very positive results beyond the statement that Plato might have obtained ideas from acounts or traditions of the real Tartessos or Carthage or Minoan Crete.

We have learned a good deal in a negative way—that is, we have eliminated a lot of possibilities as unlikely or impossible. We know now that Plato could not have been describing a real event, in any literal sense, because according to all the geological evidence his Atlantic continent never existed and no continent ever disappeared in the way he described. Furthermore, the arguments of Atlantists to prove the Atlantean origin of all civilization, from cultural similarities between various peoples, are quite useless for that purpose. These arguments, based upon mistaken ideas of archeology, anthropology, mythology, linguistics, and kindred sciences, are at worst ridiculous and at best can be used with equal ease to support entirely different theories, such as the diffusionist or the continental-drift hypotheses.

It is time, however, that we came to some more positive conclusions about the lost-continent question. Perhaps we can get a firmer grip on the Atlantis problem if we shift our focus from the modern printed page to the original *Timaios* and *Kritias* and to the man who wrote them. Plato was after all not a disembodied Voice of Eternity but a human being with a busy brain full of the knowledge and the error his time.

If you like you may imagine him sitting in his house in Athens, then a crooked little town of stinking muddy streets, from which the Akropolis, beautified by Perikles a century before, rose like a tiara on a garbage-heap.

"The solemn Plato," scratching away at his papyrus and musing on the might-have-beens of his long life, is a stocky, full-bearded figure, well-preserved for his seventy-odd years. It has been many decades since he dabbled in politics, wrote a lot of dubious love-poetry, wrestled at the Isthmian Games, and was decorated for his bravery in a battle with the Delians. His contemporaries have recorded his robust figure and soft voice; they sometimes made mild fun of his intense seriousness and his restless habit of pacing about as he talked.

Although Aristotle ascribed a godlike nobility of character to Plato, little is really known of his personality. His writings hint (to me at least) at a voluable, opinionated, didactic individual—imaginative and ascetic, full of charm, mystical intuitions, and headlong enthusiasm for world-reforming schemes. But whether he was actually like that we cannot be at all sure. Like many Classical Greeks he seems to have been mainly homosexual in his personal affections, though in addition to his lovers male he is said to have had a mistress and to have left a son behind him.

Now what was he writing? Not the story, which we know; but what kind of composition? A transcription of a real discussion, fiction, drama, or what?

For one thing, *Timaios* and *Kritias* are certainly not stenographic records of real conversations among Socrates and his friends, since the fictitious date of these dialogues is two-thirds of a century earlier when Plato was a child of about six and could neither have taken notes on them when they occurred nor have remembered them when he was old enough to write them out. They probably do not even record real speeches heard later in the Socratic circle, for in his young manhood Plato had more interest in his poetry and in a political career than in the philosophy of his middle-aged friend Socrates.

For that matter, there is no reason to regard any of Plato's dialogues as a stenographic record. Most of them, like *Timaios* and *Kritias,* are laid at too early a date for Plato to have heard them with understanding, and Plato sometimes juxtaposes people who in real life could not, for chronological reasons, have known one another.

It seems that the naïve modern reader is sometimes deceived as to the nature of the philosophical dialogue because this is now a rare form of literary expression. However, it was extremely common from Classical times down to recent centuries, for it let the author present several sides of controversial questions in a lively form without committing himself to any. Of the thousands of such dialogues written, none pretends to be an accurate transcript of a real discussion.

Furthermore putting imaginary speeches into the mouths of historical characters was also an accepted practice in Plato's day—even the conscientious Thucydides did it— and has not disappeared even yet. Writing such speeches was long a standard exercise in Classical schools of rhetoric, and in time some of these orations in the names of famous men came to pass as original and authentic works.

In fact, whatever Plato's virtues, literal accuracy was never one of them. He approved pious frauds, devoting a section of the *Republic* to a defense of the doctrine of the "noble lie" that rulers may tell their subjects to make them contented with their lot. He filled his dialogues with pseudo-myths like the story of Er the Pamphylian at the end of the *Republic* and imaginary speeches like that of Lysias in *Phaidros,* all of which, as far as we know, he made up himself. When he read one of his first dialogues, *Lysis,* in public, Socrates is said to have complained: "By Herakles, what a lot of lies this young man is telling about me!" And the sophist Gorgias was equally astounded at the words Plato put into his mouth. We have long known that there is little connection between the speeches of Plato's "Socrates" and the real Socrates. While the former may express some of the latter's ideas, Plato's "Socrates" is essentially a ventriloquist's dummy, and we cannot tell where the opinions of the real Socrates leave off and those of Plato begin.

Plato's elaborate build-up—the tale of Solon and the Egyptian priest, Socrates's assurance that the Atlantis story is "no invented fable but genuine history," and the talk of Kritias's old manuscript—are common literary devices. Not only is the "old manuscript" a well-worn literary artifice, used by story-tellers from ancient Egypt down to Poe and Lovecraft, but also Plato is not even consistent in using it: in *Timaios* Kritias says he lay awake all night trying to remember the story, whereas in *Kritias* he asserts that he has the notes that Solon took on his Egyptian tour at home. If he had this material in written form, he should not have had to lose sleep trying to recall the tale.

As Babcock said: "the Atlantis tale must be treated either as mainly historical, with presumably some distortions and exaggerations, or as fiction necessarily based in some measure (like all else of its kind) on living or antiquated facts." All indications are that the latter is the true explanation; Plato even hinted that such was the case by Kritias's remark about considering the prehistoric Athenians as identical with the citizens of *The Republic*.

Moreover Plato was not (as Atlantists assume) the kind of person interested in accurately passing on oral traditions —an illiterate primitive supernaturalist—but a sophisticated urban literateur, as well qualified as any man of his time to write imaginative fiction. The unity and lifelikeness of the tale, sometimes cited as evidence of its truth, are within the competence of any good story-teller—certainly one of Plato's powerful intellect.

It is absured to object, as does the mystical Merezhkovski, that it is "incredible" that one of Plato's uprightness should have lied about so important a matter as the Atlantis story. No matter how much fiction he used, Plato did not think of himself as lying. Instead he meant to express those "higher truths" that philosophers play with but that have nothing to do with the facts of science and history. Why not? He composed fictional allegories all the time in his writings, and so did his contemporaries. It was the custom, as it was also among Jewish philosophers.

As for the "living or antiquated facts" used by Plato, we cannot wholly reject the possibility that he derived something from a tale brought back from Egypt by Solon; but neither can we rely upon it. Plato is the only authority for Solon's unfinished epic poem; nobody else (except those who comment on Plato) mentions it. While that does not prove it never existed, Classical writers were much given to identifying their sources, so that, considering Solon's eminence, other Greek writers would probably have quoted from or referred to his Atlantean epic if it ever circulated at all. Moreover Plato had no very accurate knowledge of Solon's Egyptian visit, since in his account of it he put the wrong king on the Egyptian throne.

Furthermore we do have a slight gleam of information about the actual source of the *Timaios:* "Timon the Pyrrhonist (*circ.* B.C. 279) is the earliest authority for the statement that Plato founded the Timaeus upon a book which he bought. Later writers amplify the story: one asserts that Plato was himself a member of a Pythagorean brotherhood and was expelled; another tells us that the book was by Ocellus Lucanus; a third ascribes it to Timaeus Locrus; a fourth even goes so far as to mention the sum paid for it. The final version is given by Hermippus of Smyrna: it relates that the book was by Philolaus and was obtained through a kinsman of his. Hermippus does not say that it existed in his time; if it had, it would probably have been in the library of Alexandria, and we should have heard more of it."

E. K. Chambers, just quoted, goes on to say that a book purporting to be this very treatise by Philolaos, a leader of the Pythagoreans after Pythagoras, is quoted from by later writers, but Chambers considers this a late forgery. However that may be, while we need not take the elaborations of Hermippos and his colleagues seriously, the skeptical Timon wrote less than a century after Plato's death and may have known what he was talking about. Books were rare and expensive enough in those days to make the whereabouts of a single copy of interest to literary men.

Timon's mention of the *Timaios* without the *Kritias* suggests that the book in question dealt, not with Atlantis, but with the long philosophical dissertation by "Timaios" on Pythagorean philosophy. Still and all, Plato undoubtedly owned other books.

With the facts in mind, the Atlantis story makes an interesting comparison with Coleridge's poem *Kubla Khan:*

> In Xanadu did Kubla Khan
> A stately pleasure-dome decree:
> Where Alph, the sacred river, ran
> Through caverns measureless to man
> Down to a sunless sea.
> So twice five miles of fertile ground
> With walls and towers were girled around. . . .

This too is a vivid fragment describing luxurious surroundings and filled with a note of doom. As Lowes showed in his erudite *Road to Xanadu,* Coleridge got all his colorful images from old travel-books such as Purchas's *Pilgrimes* and *Pilgrimage* and William Bartram's *Travels.* Perhaps Plato obtained his ideas from similar sources. Bramwell, comparing Coleridge with Plato, suggests that if more of the books in the Alexandrian library had survived the ravages of war and religiosity, we might today be able to trace down Plato's actual sources in like detail.

Assuming that Plato did *not* get his Atlantean concepts from a lost treatise on Pythagoreanism that furnished material for the rest of *Timaios,* where then did he acquire them? Here we should perhaps ask about the ideas held by educated men of Plato's time concerning the world they lived in. Obviously Plato could not have used rumors of America as a source for Atlantis if no Greek had ever heard of the place.

Now, you cannot say off-hand that because an ancient author fails to mention something in his surviving works, he could have known nothing about it. Still, from the treatises of pseudo-Skylax, Strabo, and other Classical geographers, and from geographical references by other ancient writers, we can form a fair picture of the expansion of the knowledge of the Hellenes about the world beyond their rocky peninsula.

When any writer accurately describes an area, you may infer that he either has been there or has obtained his information from a reliable source. But if he makes flat misstatements about a place, as by saying that it is water when it is actually land, then that site is obviously beyond the bourne of the region of which he has sound knowledge. While he may or may not know about places he does not mention, *if* he knows a site he will probably know the main localities between there and his own home. Then, if he does not know a place, the chances are that neither will he know places on the far side of it from his home. "His home" we may in this case take to mean Athens and its surroundings, or more broadly Greece and the Aegean Sea.

Furthermore the age from Homer to Strabo was, in the Classical world, a time of more or less continuous advance in geographical knowledge. (There may however have been some small recession after Homer's time as a result of the rise of Carthage.)

Therefore if one of Plato's predecessors knows a place, the chances are that Plato either knew it too or could have found out about it if he wished. Conversely if neither Plato nor Aristotle knows a place, it is unlikely that their predecessors like Solon and Herodotos would know it either. Remember that there is no *sharp* boundary to a man's geographical knowledge, and also that educated men may have differed in their geographical beliefs; still there was a good over-all consensus of geographical opinion in Athens at any one time.

With these principles to guide us, let us see how knowledge of the world expanded from Homer to Plato. Anybody who discusses Greek history, science, or art starts with Homer as a matter of course.

But who was Homer, anyway?

The answer is far from simple. In Classical times people took it for granted that the *Iliad* and *Odyssey* had been composed by a blind Ionian poet named *Homēros* who wandered about the Aegean singing lays to the tune of his lyre. His birth was ascribed to a dozen or more places, and dated anywhere from 1159 to 685 B.C. In Hellenistic and Roman times several biographies of Homer circulated, all probably written in the time of Aristotle or later in response to the demand for non-existent information, and based mainly on conjecture from the poems themselves and on sheer romancing.

True, there was a small school of *chōrizontes* or "sepa-

ratists" like Xenon and Hellanikos, who asserted that the
two poems were by two different authors. But they had
little influence and were almost forgotten until quite modern times.

Then in 1795 Friedrich August Wolf of Halle and Ber-
lin startled the academic world by proclaiming that
"Homer" had been neither one man or two, but many.
"Homer," he said, was a collective name adopted by or
applied to a group of poets who composed a number of
heroic lays not combined into the present *Iliad* and *Odyssey*
until the time of Peisistratos, a dictator of Athens in the
sixth century B. C.

This radical opinion led to a tremendous war of words
among Greek scholars that has continued down to the
present with no decision in sight. Some adhere to the one-
Homer view; others the two-Homer or "separatist" opinion,
though they divide the two poems variously between the
authors, some for instance giving the *Catalogue of Ships*
in the second book of the *Iliad* to one author and all the
rest to the other. The Wolfians or multiple-Homer party
(who to me seem to have the better of the argument)
differ widely among themselves as to how and when the
various parts of the poems were combined into their present
form. And the arguments of the various groups of Homerists are contaminated with such rank subjectivism that an
outsider can hardly pursue them with profit.

Gilbert Murray, the most eminent of the Wolfians,
affirms that both poems were composed by a long line of
poets, one of the more gifted of whom may have been
named Homer. They are traditional books, dating back to
days when there was no reading public and writing was
confined to a few bards, each of whom had his own book:
a long roll of papyrus on which poems were scratched without table of contents, chapter-headings, punctuation, or
even divisions between words. Each bard added new mattter
when he could; he might for instance let a colleague copy
one of his lays in return for the same favor. Otherwise he
kept his manuscript as secret as he could. When in the
midst of a recitation he announced he would have to
consult the Muses, he ducked into the woods for a quick
look at his book to refresh his memory. Although poems

the length of the *Iliad* and *Odyssey* can be memorized by exceptional people, a written version is still a great convenience.

Furthermore the *Iliad* and *Odyssey* did not stand alone in early Greek literature. They formed part of the Trojan cycle of epics, which included half a dozen other poems like the *Sack of Ilion* and the *Homecomings*. In addition there were several other whole cycles such as the *Argonautika* and *Herakleia*. None except the *Iliad* and *Odyssey* has come down complete, and these probably survived the tooth of time by having been chosen for public recitation at the festival of the Panathenaia during the fifth century B.C. Many fragments of the other epics, which are attributed some to Homer and some to other more or less legendary bards like Stasinos, have survived in quotations, and we are familiar with their plots from the many later Greek plays, poems, and mythological treatises based upon them.

Concerning the content of the *Iliad* and *Odyssey*, opinions varied in ancient times from extreme skepticism to an idolatrous reverence for Homer's wisdom and truthfulness. Strabo boiled with rage whenever some skeptic like Eratosthenes of Kyrenë cast doubt on Homer's accuracy by saying that after all poets are paid to please, not to teach. However, while some Homerists tend to take the poet as literally as they can, modern criticism generally supports Eratosthenes. The poems contain many plainly fictional elements, such as the intervention of gods in human affairs, and private conversations that could never have been recorded.

As for the "substratum of truth" that devout Homerists suppose to underlie the fictions of the poems, there may well be scraps of history buried in these works. However, judging from similar poems dealing with periods for which we do know the history, such as the Charlemagne cycle of romances, the historical content of the Homeric epics must be so slight and so muddled that we cannot filter it out at this late date.

Some of Homer's characters were no doubt based upon real people: thus Atreus king of the Achaeans, father of Menelaos and Agamemnon, is probably the Atarissiyas king of the Akhiyawas mentioned in the Hittite royal archives dug up at Boğazköy in Turkey. Others again may be pure myths: Helen a fertility goddess, for example, since if she had experienced all the abductions attributed to her she would have been nearly ninety when Paris took her to Troy. The learned Murray thought that the fleet-footed Achilles among others might be a tribal god, or a personification of a tribe.

If we knew all the facts about the Homeric characters we might discover that most of the persons named in the poems combined (1) the name of a real person, (2) the deeds of some real persons and some folk-tale heroes, and (3) the attributes of real persons, fictional characters, and gods all rolled up together in varying proportions. The whole question of who Odysseus and his fellow-characters "really" were is so very obscure and disputed that we are lucky not to have to decide it once and for all in this book.

Now about the geography of Homer: The anti-Homer Classical writers like Kallimachos of Kyrenë tended to restrict Homer's knowledge to the eastern Mediterranean, while those of the pro-Homer party like Strabo and Plutarch affirmed that the poet had known the lands and seas all the way from the Atlantic to the Black Sea, and sought like Victor Bérard and others in modern times to identify all the places Odysseus visited in his travels with real places. Hence Homer's "Thrinakia" became Trinacria, a name for Sicily, and the strait of Skylla and Charybdis was identified both with the Strait of Messina between Italy and Sicily and with the Strait of Gibraltar. Samuel Butler even wrote a book in which, after vigorously attacking the Wolfians, he set out to prove that the *Odyssey* was written by a woman: a young lady of Drepanum (modern Trapani) who put herself into the poem as Nausikaä; and that all the places mentioned in the work are Sicilian. Butler's highly subjective argument, however, converted few people except the late George Bernard Shaw.

Of course *if* we knew that Homer (regardless of whether he was a man or a corporation) accurately knew the geography of the Western Mediterranean, we might be compelled to find such a relationship between fact and fiction, if only for our own peace of mind. In fact, though, we know nothing of the kind.

Irrespective of authorship, the poems (dealing with events supposed to have happened in the twelfth or eleventh century B.C.) probably took somewhat their present form between 900 and 600 B.C. Although they did not finally jell until after Plato's time, their author(s) and editors seem to have taken pains not to introduce "modern" inventions and institutions that would spoil the archaic flavor. Hence the heroes always use weapons of bronze, though Homer himself mentions iron in figures of speech.

On the other hand the Greeks did not begin to send out colonies to regions outside the Aegean until the eighth century B.C., and did not definitely penetrate the Western Mediterranean until the seventh and settle there permanently until the sixth. To be sure, Phoenicians and Cretans before them had navigated the western waters, but it does not follow that they had tried to impart their knowledge to the Greeks. If anything they tried to keep it secret. And even if a Greek bard picked up hints of western lands from some drunken Phoenician sailor, he still would not have possessed a clear picture of the region, especially since Greek cartography got started in earnest only when Anaximandros of Miletos attempted the first world map in the sixth century B.C.

In fact, Homer was not the least map-conscious; he even put the eastern Kimmerioi or Cimmerians in the West. He might well have heard rumors of Tartessos which he used in building his magical realm of Scheria; he may have transformed Madeira into Kalypso's isle of Ogygia; and perhaps tales of the eruption of Etna and Vesuvius became the bombardment of the ships of Odysseus with boulders by the Laistrygones. It has also been suggested with some plausibility that the skull of a fossil elephant (which seen from the front does look rather like that of a one-eyed human giant) became the Kyklops Polyphemos (who is also identified with a volcano) and that a large squid or octopus metamorphosed into the man-eating monster Skylla.

However, while for centuries Homerists have found the floating island of Aiolos in the Lipari Islands west of southern Italy, they ignored the fact that Odysseus, after leaving this island, sailed due east with the winds in a bag almost

to his native Ithaka—to accomplish which he would have to
sail right across southern Italy, mountains and all.

Judging from his use of names and description of
places, then, Homer knew the Aegean Sea and its shores
and islands from firm first-hand acquaintance. He had a
fairly reliable second-hand knowledge of the west coast of
Greece, where however he got the islands mixed up. That
is not surprising; after all Apollonios of Rhodes in his
Argonautika bungled the geography of that region just as
badly, though he lived long after Plato and had plenty of
information available.

Homer also knows of the nomadic Scythians and
Thracians north of Greece, whom he calls by such descrip-
tive names as *hippēmolgoi,* "mare-milkers." To the east,
although he has heard of the Mysians of the interior of
Anatolia, his knowledge seems not to reach as far as the
Hittite and Assyrian Empires, unless perchance the Keteioi
mentioned in the *Odyssey* (XI, l. 521) be the Khatti
or Hittites. Southward he has heard of Sidon of the
Phoenicians, of Egypt, and of the dark Aithiopes or Ethiopians ("burnt-faces") beyond Egypt. His Kikones of
Thrace and his Lotus-eaters of the Libyan coast are probably real people, and the seductive fruit eaten by the latter
is probably the jujube, still relished in those parts.

To the west he has heard vaguely of the tribes of
southern Italy and Sicily, the Sikeloi and Sikanioi, but they
are mere names to him. Hence when Odysseus leaves the
land of the Lotophagoi he sails off into Fairyland, finding
islands like Aiaia and Ogygia and monsters like Polyphemos
wherever it suits the story-teller to put them. Finally, Homer
has no true picture of the Atlantic Ocean, since for him the
world is still surrounded by a great *river,* the "flowing
river Ocean." When the Atlantic later became familiar to
the Greeks they called it the "Ocean" because it was located
where they had previously pictured this mythical river. Nobody ever explained why such a circular stream, holding its
tail in its mouth like the serpent Ouroboros of occult symbolism, should flow round and round as they evidently
thought it did.

Hesiod, another half-legendary poet of the eighth century B.C., heard of the Tyrrhenians or Etruscans of Italy, but

otherwise knew little more about distant places than Homer. However he described more imaginary regions like Erytheia and the Hesperides in the West, Hyperborea in the North, and the lands of the dog-headed men and gold-guarding griffins in Asia, all of which jostled real countries and peoples for places on the map for the next two thousand years.

By the sixth century, the age of Solon the statesman and Hekataios the first Greek historian, the Black and Caspian Seas were known, though opinions differed as to whether the latter were merely a bay of the encircling Ocean Stream. The large western islands—Sicily, Sardinia, and Corsica—were pretty well located, though the Balearics, whose men wore coats of grease in lieu of clothes and hired out as mercenary slingers, remained unknown for some time yet. The Strait of Gibraltar was known, and Tartessos beyond it, but beyond that—nothing.

Rumor had it that the Phoenicians sailed out through that strait to a group of islands, the Kassiterides, whence they returned with tin, but nobody knew which direction these islands lay in. The Tin Islands became such a fixture in the minds of Classical geographers that even after the real source of tin, Cornwall and the Scilly Isles, became known, the Kassiterides continued to lead a ghostly separate existence in the watery wilderness of the Atlantic.

The next century, that of Herodotos and Socrates, saw further advances in knowledge. Herodotos knew that the Ocean was more than a river, though he failed to learn, despite inquiry, whether it extended around to the north of Europe as the mythical Ocean River had been thought to do. He was sure, however, that it stretched around Africa to join the other great sea, the Erythraean or Arabian. No Greek is known to have explored the northern coasts of Europe until Pytheas, who came after Plato's time. The poet Pindar in this period described the Pillars of Herakles as "the furthest limit of voyaging." "All beyond that bourne cannot be approached," and "Beyond Gadeira toward the gloom we must not pass."

Plato was born two or three years before Herodotos died in 425 B.C. In his time geographic knowledge was still expanding; one of his contemporaries, Damastes, made the

Fig. 16. THE WORLD ACCORDING TO HEKATAIOS, or as conceived by educated Greeks between the time of Solon and that of Plato. (After Bunbury.)

first Greek mention of Rome. Another contemporary wrote a *periplous* or navigational guide under the name of Skylax of Karyanda, who many years before had been hired by Darius I of Persia to explore the Indus River to the ocean. Pseudo-Skylax mentioned places beyond the Pillars— Gadeira, Tartessos, Kernë—but in a muddled fashion that showed he got his information from Carthaginian sources, and either garbled it in the process or was deliberately misled by his informants. He mentions "much mud, and high tides, and open seas" in the Atlantic.

Thus, when Plato looked to the West, all was clear as far as Sicily. Beyond that, in Carthaginian-ruled regions, he saw as through a glass darkly. Aside from the Pillars of Herakles and the ocean beyond them he had merely some names and casual descriptions to go by; nothing one could draw a map from. Hence even if the shoals marking the site of Atlantis existed, Plato would probably have had no clear knowledge of them. All the more he could not have known of Britain, or Scandinavia, or the Americas and used them as the basis for Atlantis.

The same is true *a fortiori* of Solon, who had flourished two centuries previously. In Solon's day there was a vague sense that the Inner Sea narrowed to the Far West before opening out again into the Ocean River, but of details of that region the Greek mind was as bare as an egg is of hair. Solon could hardly have imagined the Atlantic Ocean as it really is, with or without bobbing continents.

Could he however have picked up such ideas in Egypt? As far as the evidence goes the Egyptians knew less than the Greeks of the world beyond their borders. To them the world was shaped like the inside of a shoe-box, the floor consisting mainly of Egypt with the Nile running length-wise down it. Around Egypt lay a meager border of seas and deserts, inhabited by barbarous tribes of no interest to the sons of the gods except when they got above themselves and invaded Egypt. No, the Egyptians were the last people on earth to consult on geography.

From Homer to Solon the Greek picture of the world comprised the more or less circular land-mass of Europe with its appendages of Asia and Africa, surrounded by the Ocean Stream, and beyond that remarkable river the all-surrounding Outer Continent. Between Solon and Plato, partly as a result of Greek expansion and partly from tales picked up from the ubiquitous Phoenicians, educated Greeks acquired a more realistic picture of the Atlantic. Herodotos first mentioned the Atlantic Ocean by its present name, though not for several centuries did this term altogether displace such terms as "Great Sea," "Outer Sea," and "Western Ocean."

This process of rectification was not yet complete in Plato's time. The old Ocean Stream had simply been widened until it was large enough to accommodate Plato's Atlantic continent between Europe and the outer or "true" continent, which not only Plato but later writers like Theopompos and Plutarch retained even after the geographers had discarded it. The old Ocean River of Homer and Hesiod was not wide enough for Atlantis. The idea of an Atlantic continent can hardly be older than the knowledge of an ocean to hold it; and since the knowledge of this ocean only began to penetrate Greek minds about the time of Herodotos, it is hard to see how the Atlantis concept can be older than about 500 B.C. at the earliest.

The legendary lands associated with Ocean River evolved parallel to it. In Homer, Hesiod, and Pindar all is exquisitely vague, though these places, it is implied, are either lands on the banks of the river or islands small enough to fit into it. Odysseus goes from Kirkë's island "to deep-flowing Okeanos, the outer bound of the earth, where lie the land and city of the Kimmerioi, veiled in fog and cloud." Here Odysseus summons up the ghosts of his friends by necromancy, and the shade of Achilles mournfully remarks that he "had rather be a poor man's serf than king over all the dead."

There too lie "the islands of the blessed along the shore of deep swirling Ocean," where dwell the "happy heroes for whom the grain-giving earth bears honey-sweet fruit flourishing thrice a year," "where the ocean-breezes blow . . . and flowers of gold are blazing," where Kronos rules, and fair-haired Rhadamanthos, the virtuous brother of Minos, judges the shades of the dead. In Greek theology, a soul that had worked out its karma by sufficiently virtuous lives was given a last incarnation as a statesman or

sage and then turned out to grass in the Isles of the Blest.
These concepts need not, however, have been based upon
any real knowledge of western geography, since many
people like the Samoans place the land of the dead in the
Far West, perhaps because of a subconsciously felt analogy
between a setting sun and a dying man.

When the Greeks borrowed the Herakles myth-cycle
from the Phoenicians, they took with it additional details
for their picture of Atlantic geography: the Pillars of
Herakles, Tartessos, Gades, and the island of Erytheia
where Geryon kept his kine. As accounts of far-western
lands percolated back to Greece, the Greeks naturally identified these lands with the mythical isles of their poets;
hence the name "Fortunate Isles" for the Canaries. Some
of the first accounts to reach Greece were far from the facts.
It was said, for instance, that the sea-going Semites had
found a great island with not only the delightful climate
and lush fertility that myth had attributed to the Isles of
the Blest, but navigable rivers as well. The Carthaginians
liked it so well that they planned to seek refuge there in
case of defeat in war, and cut the throats of any whom they
suspected of meaning to go there without permission.

The real Atlantic islands, the Canaries, Madeiras, and
Azores, answer the description as to woods and climate,
but none is big enough for navigable rivers. Whether some
account of Britain, or a section of the Atlantic coast of the
mainland, got mixed up with a report on Madeira we do
not know. Two writers after Plato tell the story: a member
of Aristotle's school who wrote under the master's name,
and Diodoros of Sicily. The first also told of a region of
islands and. weed-covered shallows swarming with tuna,
which sounds like the West African coast near Cape Blanco
and the Gulf of Arguin. No doubt the tale of the island
had circulated in Greece before it was written up in its
present form—perhaps as early as Plato—and might therefore have suggested Atlantis. Himilco's story of a vast weed-
covered shoal, perhaps based upon the mud-flats of the
Guadalquivir and perhaps on the West African coast explored by Hanno, reached Greece about the same time,
and was adopted by pseudo-Skylax, Plato, Aristotle, and
the pseudo-Aristotle who wrote of the delectable island.

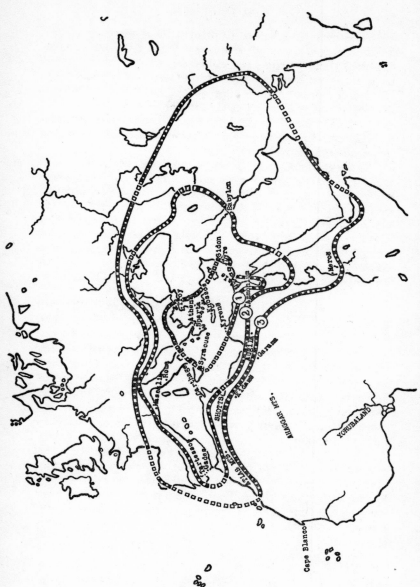

Fig. 17. THE GROWTH OF GREEK GEOGRAPHICAL KNOWLEDGE – approximate limits of accurate knowledge possessed by Greeks of the world in the times of (1) Homer, (2) Solon, and (3) Plato.

The encircling Outer Continent went through a similar development, though it never merged with the facts in Classical times because the Americas, which corresponded to it more or less, lay beyond the reach of Classical explorers. Originally beginning at the far bank of Ocean Stream, it was gradually pushed farther away as voyages into the Atlantic failed to find it, until it dissolved entirely, though one can detect faint ghostly traces of it in the Terra Australis of later times.

The Outer Continent seems to have been associated with the satyrs, those snub-nosed, horse-tailed supernaturals whose implacable lust is indicated by a baldly physical manifestation in Greek vase-paintings. Their patriarch was the drunken Silenos, a son of the Arcadian goat-god Pan and a nymph. King Midas of Lydia was supposed to have once caught Silenos by getting him drunk (a feat later transferred to the Neopythagorean mage Apollonios of Tyana) and kept him to hear him tell of the Outer Continent.

In later Greek times the satyrs were confused with Pan and given goat-legs in place of their original form. Pausanias, telling of a stone seat in Athens whereon Silenos was said to have sat, recounts that Euphemos the Carian told of having been blown on a voyage into the Outer Sea, where the ship made a landfall on some islands called Satyrides from the satyrs who lived there. The latter swarmed aboard and began raping the women without more ado, until the sailors pushed off, leaving one unlucky lady as an offering to their concupiscence.

So, you see, Greek geography expanded by the same process as that of the knowledge of any people during an age of exploration. A sedentary primitive tribe normally knows little about the country a few miles beyond its own territory. As far as it is concerned, its own land is surrounded by a belt of the unknown, and beyond that nobody knows—perhaps one comes to where the world-island ends or the sky-bowl comes down to meet the land.

The myth-makers of the tribe people the surrounding belt of *terra incognita* with the characters from their myths and legends: gods, monsters, the land of the dead, and so on. As knowledge increases the tribe hears tales, often distorted, about the real nature of this belt, and combines with these tales the existing myths to make a new mythos of monsters and supermen. Then as civilization arises and travel familiarizes the folk with the unknown lands, they realize that it does not contain the creatures they thought it did. However, since the monsters, giants, and the like have now become part of the nation's sacred literature, they must exist somewhere. Therefore they are pushed out farther into the new belt of unknown land surrounding the former one, and this process is repeated over and over. "Geographers," said Plutarch, "crowd into the edges of their maps parts of the world which they do not know about, adding notes in the margin to the effect that beyond this lies nothing but sandy deserts full of wild beasts, unapproachable bogs, Scythian ice, or a frozen sea."

Thus, Atlas was originally at home in Greece (a cave on the west coast of the Peloponnesos was pointed out as the setting of a myth about his seven daughters) and so, perhaps, were the Gorgons and the satyrs. When increasing knowledge expelled these lively figments from Greece, the whole lot were dumped in Africa, which made a congenial home for them since the Greeks knew only the fringe of the northern coast. The Berber tribes who controlled the interior, people notoriously disinclined to take any nonsense from strangers, told tall tales of dog-headed men and the like to the south, which stories the Greeks duly added to their stock of African legends.

Herodotos therefore described North Africa thus: As one went west from Upper Egypt one first came to the Oasis of Ammon (now the Oasis of Siwah). Thence one proceeded to Augila (modern Aujila) and thence to Garama (modern Jerma) the capital of the Garamantes, a Berber tribe whose country was later called Phazania (modern Fezzan). Farther yet one reached Mount Atlas, so tall that clouds always concealed its top, and around which lived the Atlantes who got their name from the mountain, ate no living creature, and never dreamed.

Maybe there really was a tribe in those parts whose name reminded some Greek of the word "Atlas" and thus helped fix the name to the region. Or perhaps the locus of Atlas and his associated myths gradually moved west,

like the Welsh Indians, as the country became known,
until it could go no farther:

> Close to the shore of the ocean, not far from the
> region of sunset,
> Farthest of all is the Aethiop land, where Atlas the
> mighty
> Turns on his shoulder the firmament studded with
> bright constellations . . .

Hence Plato, to relocate Atlas further west, had to invent
a continent in the ocean for him.

Later writers added to the picture: Diodoros told of
the Atlantioi (presumably the same as the Atlantes of
Herodotos), the Gorgons, and Queen Myrina of the African
Amazons. Pliny the Elder and the Spanish-Roman geographer Pomponius Mela in the first century A.D. furnished
dramatic details about the dreadful dwellers in the Saharan
waste: the Troglodytes lived in holes in the ground (as the
Matmata of Tunisia still do), ate snakes, and had no language but batlike squeaks; the Garamantes had no institution of marriage; the Aegipani ("Goat-Pans") were half
goat; the Gamphasantes went naked; the Himantopodes
("Strapfoots") had snakes for legs like the Titans whence
Atlas sprang; finally the Blemmyae were headless men
with faces in their chests. And at night Mount Atlas re-
sounded with the raucous revels of satyrs. Similar tales were
told of the remoter tribes of India, which also was credited
with satyrs, snake-footed men, and the mouthless Astomi
who subsisted by smelling flowers.

To confuse matters further Pliny described the Fortunate Isles, alluded to Plato's sunken Atlantis, and in
addition spoke of an existing island named "Atlantis" off
the Moroccan coast opposite Mount Atlas.

A series of military expeditions under the Roman
Empire into the African interior finally cleared up these
African mysteries. The first under Cornelius Balba in 20
B.C. captured Kidamë and Garama, and in the second century A.D. Generals Septimus Flaccus and Julius Maternus
penetrated south from Phazania to the Sudan. Needless to
say they met no Strapfoots or Goat-Pans. However all this
was long after Plato.

Thus Plato could evidently have acquired his geographical ideas, such as large islands in the Atlantic and impassible shoals that might be the remains of such islands, from beliefs current among his contemporaries. The same applies to his geological ideas of submergence of land by earthquakes, which the Mediterranean peoples were familiar with on a small scale. Severe quakes had shaken Greece in 426 and 373 B.C., and the former had already, in Plato's time, been described by Thucydides in language that strongly suggests the Atlantis story:

"At about the same time, while the earthquakes prevailed, the sea at Orobiai in Euboia receded from what was then the shoreline, and then coming on in a great wave overran a portion of the city. . . . In the neighborhood also of the island of Atalantë, which lies off the coast of Opuntian Locris, there was a similar inundation, which carried away a part of the Athenian fort there, and wrecked one of two ships which had been drawn up on the shore."

Now, if any passage in pre-Platonic literature, still extant, gave Plato his idea for the sinking of Atlantis, this is it. While we don't know for sure that Plato read Thucydides's history, the chances are that he did. His pupil Aristotle certainly did, as is shown by some of his remarks in his *Constitution of Athens*. Did Plato expand little Atalantë into great Atlantis? When we learn from Strabo that in consequence of this tremblor Atalantë "had been rent asunder" and "got a ship-canal through the rent" it certainly looks that way.

Plato's knowledge of history was of the same order as his knowledge of geography. Though an intelligent man who could handle abstract ideas with great adroitness, he had after all only limited materials to work on. Authentic Greek history starts between 700 and 650 B.C. with the institution of the archons or regents in Athens. Before that time we have only lists of dubiously real kings like Lykourgos in Sparta and Kodros in Athens, and the alleged dates of the founding of colonies. And this history becomes detailed only about 600 B.C. in Solon's time.

There is of course the voluminous literature of the Greek Heroic Age, headed by the Homeric poems and filled out with the narratives of the Siege of Thebes, the Labors of Herakles, and the Quest of the Golden Fleece, together with minor stories like those about Atalanta of Kalydon. All these events are supposed to have happened within a few generations between 1250 and 1100 B.C., after which the Heroic Age peters out, leaving Greek history an almost complete blank for more than four centuries. Judging by comparable bodies of literature, such as the sagas of the Germanic Heroic Age at the time of the Fall of Rome, the actual history embalmed in traditions of the Greek Heroic Age must be small.

Therefore Plato could not have secured an accurate history of Atlantis or of the prehistoric Athenian Empire, even had they existed, through Greek historical channels because these channels did not run back more than three centuries before his own time, let alone 9000 years. Accurate history, in fact, depends upon writing, which the Greeks had acquired only in the ninth or eighth century B.C. A preliterate people has no history in our sense of the term, but a body of myths, timeless and fantastic, which might embody an occasional fact like an insect in amber but which are practically useless for reconstructing the past of the tribe. The folk may also have the orally transmitted memory of striking events and outstanding personalities going back perhaps several centuries but not much farther.

When a folk learns to write, certain people fascinated by history try to find out and record what happened from the Creation to their own time. These tyronic historians assume that the myths and the memories of real events are the same sort of thing; myths, they suppose, describe the real events that happened just before the start of mundane history. Therefore they try to splice the two together by guesses and interpolations. Classical historians, trying to make a coherent tale out of their invincibly inconsistent mass of myths in order to complete their concept of world history, were driven to conceiving several Helens and several Zeuses. For example Cicero, one of these rationalizers, solemnly assures us that "The third Apollo is the son of the third Jupiter and Latona," and the scholiast on the *Timaios* tells of three Deluges.

Greek pseudo-history started with the Creation: "In

the beginning Ouranos (Sky) ruled the universe. By intercourse with Ge (Earth) he begat first those called the hun-
dred-handed—Briareos, Gyes, Kottos—unsurpassed in size
and strength, each having 100 hands and fifty heads. Then
Earth bore him Kyklopes. . . ."

After more begettings Ouranos's son Kronos became
God, under whom occurred the Golden Age described by
Hesiod, the time when everybody was happy and good.
The idea of progress, remember, is essentially modern; to
Greeks as to other pre-scientific peoples perfection was to
be sought in the past. The Golden Age ended when Kronos
was dethroned by his son Zeus, the greatest begetter of all.
Plato's Atlantis belongs to the following Bronze Age, closed
by Deukalion's Flood, which Plato says followed the Atlantean earthquarke. Then when the descendants of Deukalion and other survivors had repeopled the earth came
the Heroic Age of Herakles and Theseus, of the Argosy
and the Theban and Trojan Wars.

The same process of synthesis can be seen in Hebrew
literature, where authentic history starts about the time of
Samuel and King Saul, since only then did the Hebrews
begin keeping written records. Everything earlier—Adam
and Noah, Abraham and Moses—is more or less fictional,
and furthermore the fictions are of the most diverse origin,
some like the Flood being modified Euphratean myths
picked up during the Babylonish Captivity.

If Plato could not have gotten Atlantis from Greek
historical sources, might it have come down via Egypt as
Plato said it did? If less enterprising geographically than
the Greeks, the Egyptians had a good historical sense and
had kept records ever since the early dynasties. When Herodotos visited the country they told him that "Min was the
first king of Egypt," and gave details of the reigns of several
of the 330 kings who had ruled since. Some of the things
they told him were true and some not: they identified the
pyramid-builders of the Fourth Dynasty more or less correctly but placed them 2000 years too late.

Under the Ptolemies, Manetho, a priest of the Delta,
wrote a history of his people in Greek. The original has
perished, more's the pity, but later writers have preserved
extracts. As one might expect, Manetho started with

dynasties of gods and demigods who ruled Egypt after the Creation. Then: "In succession to the Spirits of the Dead, the Demigods, the first royal house numbered eight kings, the first of whom, Menes of This, reigned 62 years. He was carried off by a hippopotamus and perished." This Min or Menes was historical in the sense that he represents a merger of several kings who struggled over many years to extend their sway over the entire land: Ka-Ap, Narmer, and Aha-Mena, to name the most prominent.

Egypt therefore really did have history, however bare and inacurate, that went back 3000 years before Plato. But although that is a long time, it will not take us back to Atlantis by another 6000 years.

Furthermore Egyptian history, as you go backward, stops short with the First Dynasty. Manetho said that before that the gods ruled, but the real reason is that before that time the Egyptians had not kept written records— were in fact just emerging from neolithic primitivism. Most that we know of the histories of Egypt and Iraq before 3000 B.C. we have learned, not from records of the kind available to Plato, but from modern archeology. And, in general, the Greeks made no acheological investigations.

It seems, then, that even if Atlantis had existed when Plato said it did, he could hardly have learned about it because he lacked the necessary historical and geographical knowledge. Therefore he must have used the beliefs of his own time: the earthquake from one of the real tremblors that had shaken Greece, perhaps Thucydides's account of the quake of 426 B.C., in the sixth year of the Peloponnesian War. The idea of lands' rising out of or sinking into the sea was familiar to educated Greeks, and so were the fictional mud-banks of the Atlantic Ocean. While Plato could just possibly have obtained elements of his story from an account handed down by Solon about his Egyptian tour, this could hardly have been Plato's main source, since important elements of his story like the existence of the Atlantic were not current in the Eastern Mediterranean until well after Solon's time.

Finally, rumor had it that the Phoenicians had found one or more large islands with big rivers in the Atlantic. What more natural than to combine these concepts into a large Atlantic island sunk by an earthquake, leaving impassible shoals behind? Plato no doubt ended by thinking that he had not only composed an attractive story but had also, incidentally, hit upon the true explanation of the great mud-skerries. (Plutarch explained the suppositious shoals differently: as sediments deposited by the rivers of the Outer Continent.) It was not Plato's fault that these banks did not exist, and that large islands and continents do not behave that way, as scientific geology was 2000 years in the future. Neither would Plato have known that one great storm could not erode Greece from a lush peninsula to its present gaunt form, or that earthquakes don't swallow whole armies in their cracks.

Once we have impaled Plato's main concept upon the pin of reason, the other elements of his history fall into place as well.

For instance he could have procured the general idea of the defeat of barbarian invaders by the brave Athenians either from the Persian Wars, or from the Graeco-Carthaginian wars in Sicily, or from both, combined with the legend of the invasion from the Outer Continent which Theopompos repeated in another form. The Kleito romance was doubtless derived from the general stock of myths about liaisons between gods like Poseidon and mortals like Halia, Kleito's hot-and-cold water system from Greek beliefs about the Alpheios and other rivers they wrongly supposed to flow underground, and the time and place of Atlantis both from stories of the real Tartessos and from the standard myths of the Western Paradise (Elysion, Garden of the Hesperides) and the Golden and Bronze Ages.

Plato could have derived the city plan of Atlantis either from Babylon, changing the square plan of the latter (as described by Herodotos) to a circle, or from Carthage with its circular walls, or both. The metallic decorations of the Atlantean citadel are probably either from Homer's palace of Alkinoös or from travellers' tales of Tartessos. He could have obtained the inspiration for the harbor-works of Atlantis from Syracuse, the New York of the Greek world, and of Atlantean sea-power from legends of the sea-kings of Crete.

The theory of periodical catastrophes expounded by Plato's Egyptian priest is probably of Babylonian origin. Lastly, the religious ceremony could have come from the Orphic mystery-religion, with which Plato was probably familiar because he got many of his ideas from the Pythagorean school of philosophy and Pythagoras had been strongly influenced by Orphism.

The most reasonable way to regard Plato's story, then, is as an impressive if abortive attempt at a political, historical, and scientific romance—a pioneer science-fiction story—based upon materials of Plato's own times, and possibly also upon traditions of Crete and/or Tartessos, which has been kept alive partly by its literary merit, partly by Plato's philosophical reputation, and partly by its nostalgic emotional appeal. It fitted perfectly into the views of world geography, geology, and history current in Plato's Athens, even though it does not at all fit into our own ideas on these subjects. That is no reason for not enjoying it as a good story—we enjoy *Alice in Wonderland,* though it isn't history.

CHAPTER X

THE LAND OF HEART'S DESIRE

It is evening, Senlin says, and in the evening,
By a silent shore, by a far distant sea,
White unicorns come gravely down to the water.
In the lilac dusk they come, they are white and stately,
Stars hang over the purple waveless sea;
A sea on which no sail was ever lifted,
Where a human voice was never heard.*

Aiken

PLATO'S Atlantis story, then, belongs to that great class of imaginative narratives loosely called myths, legends, allegories, or fantasies. Such being the case, what are myths, and where do they come from, and what is their relation to reality?

The names "myth" and "legend" are usually applied to traditional stories of unknown authorship, which circulated among a people by word or mouth before they learned to write them down. Allegories and fantasies, on the other hand, are thought of as tales composed by known writers among people who can read and write, an allegory being a symbolic or metaphorical story and a fantasy one with supernatural elements like gods, ghosts, and magic. As far as myths are distinguished from legends, the former deal with gods and the latter with mortals.

These distinctions, however, are not hard-and-fast. We tend to take it for granted that a traditional story handed down among a preliterate people could not have been composed all at once as a modern writer concocts his tale, but must have been pieced together over a long time by relays of bards or medicine-men. This may not be true in all cases, however, and even a modern writer weaves together incidents from other stories and from real life to

---

* From *Senlin: A Biography*, by Conrad Aiken, in *Selected Poems* (Scribner's, 1929); copyrighted 1918-1929 by Conrad Aiken; by permission of the author.

make his narrative, so that all stories are in a sense products of more than one mind.

There are many kinds of myths: creation-myths; catastrophe-myths; culture-hero myths about demigods who, like Osiris or Moses, taught men their culture; sun-myths, storm-myths, migration-legends, and so on. Fifty and more years ago the pioneer mythologists tended to try to reduce all myths to a single type. Thus they would assert that all myths were fictionalized history, or moral fables, or explanations of natural phenomena; and that all mythical heroes were either real people, or phallic symbols, or sun-gods. Nowadays, however, it is generally realized that there are too many different sorts of myths to be all accounted for by one simple explanation.

Myths may be either homiletic (that is, trying to persuade the hearer to do or believe something) or lusory (intended mainly to entertain). Some simply tell a tale (apologal myths); some explain how things got the way they are (etiological myths); some explain what happens to you after death (eschatological myths).

While all myths and most legends contain supernatural elements, and while the myths of primitive peoples often show a rather childish irrationality, they still reflect the lives and customs of those who tell them. One does not, for instance, find King Arthur riding an elephant or the goddess Ishtar driving a dog-sled. Thus the Polynesian myths are concerned with the Polynesians' main amusements: war, water-sports, genealogy, and adultery, and the winner of a contest has the privilege of eating the loser. Myths express people's hopes and fears, embody their dreams and complexes, rationalize their customs and rituals, and comment upon their social organization, class conflicts, and personal frustrations.

Pseudo-scientists such as most Atlantists assure you that all myths are founded upon fact, and then exaggerate the realistic or historical elements in them in order to support their own theories of jack-in-the-box continents. But while, as Babcock said, all fiction (including myths) is founded on living or antiquated facts, it does not follow that one can reconstruct the fact from the fiction. For instance, Sinclair Lewis's novel *It Can't Happen Here* is based upon fact. Nevertheless a future historian who tried to reconstruct the history of the twentieth-century United States from that novel alone might end with the conclusion that Windrip the dictator was a real man whereas Franklin Roosevelt was a sun-god or a culture-hero like Prometheus.

Atlantists and diffusionists also like to deny primitive men the power of imagination, so that they can affirm that all myths contain a large element of literal truth, and that any two similar myths found in different parts of the world must have had a common origin and been spread by diffusion. However, as we have seen, the minds of preliterates are much like those of other people, and we have to decide the question of the diffusion of myths the same way we do the diffusion of other culture-traits.

For example the group of Old-World stories in which a child-hero—Sargon, Moses, Perseus, Romulus—is set adrift in a box or boat upon the water and later rescued, probably have a common origin, since they are found in adjacent areas and are not far separated in time. A common origin is also likely for the stories of the headless man in the bear's den told in Lapland and the Caucasus, because the combination of ideas is so unusual.

On the other hand it is more likely that the myths from ancient Egypt and modern New Zealand, in which the sky and the earth lie in sexual embrace until one of their children thrusts them apart, were invented separately, since they were told by peoples living on opposite sides of the earth and thousands of years apart. Any fiction-writer, who knows how hard it is to invent a really original plot, can appreciate why there is nothing incredible about different myth-makers' hitting upon similar stories independently.

Myths, you see, are not handed down to posterity for the simple love of historical fact—a rare quality even in our supposedly enlightened culture. People pass them on because they are entertaining and so provide a living for story-tellers, or because they have been found useful as magical spells in controlling supernatural beings, or answering children's questions, or serving as librettos for religious rites, or persuading the average tribesman to obey his chief, honor his priests, and observe the tribal tabus.

Consider the large class of catastrophe-myths in which the gods, offended at the impiety of men, destroy them by fire, flood, or plague. Usually one pair escapes to repopulate the world: Ziusudra and his wife, or Noah and his, or Deukalion and his, or the pair that survives each of the Aztec catastrophes. The Aztecs, who took a grim view of the universe, believed in a whole series of calamities: first jaguars ate everybody; next a hurricane destroyed the world; the third time a rain of fire fell; the fourth catastrophe was a flood; and the fifth time, yet to come, the world will be ended by an earthquake.

Now, catastrophes do happen, and the myth-makers no doubt use them in making up their catastrophe-myths, just as detective-stories are based upon the fact that real people are sometimes murdered under mysterious circumstances. Thus the Sumerian priests made their own great floods of the sixth and fifth millenia B.C. (which doubtless seemed world-wide to those caught in them) into the World Deluge of the fable of the pious Ziusudra who, warned of the flood by which the gods meant to destroy mankind, escaped from Shurippak with his family and livestock in a boat.

This tale, which appears in a later form in the *Gilgamesh* epic, is preserved somewhat garbled in the fragments of the writings of the Hellenized Babylonian priest Berossos, who taught the Babylonian pseudo-science of astrology on the island of Kos about 300 B.C., and finally it got into *Genesis* as the familiar Noah-story. This Berossos said that the world would be destroyed by fire every time the planets all came into conjunction in the sign Cancer, and by water when they gathered in Capricorn.

Plato could not of course have derived ideas directly from Berossos because he died half a century too early. However some of the Babylonian legends and theories that Berossos set forth must have reached Greece ahead of the influence of Berossos himself. Thus the Greek myth of Deukalion and Pyrrha, which goes back at least to Pindar and was well-known to Plato, shows meaningful parallels to the Babylonian Flood-story.

Furthermore, by calling the planet Mercury the "Star of Hermes" instead of by its older Greek name of *Stilbōn*, "twinkling star," Plato showed that the Babylonian custom of naming the heavenly bodies after the gods (a custom that gave rise to astrology in the first place) had already reached Greece in his time. And Plato's younger contemporary, the astronomer Eudoxos of Knidos, warned his pupils against the Babylonian superstition of astrology that was beginning to corrupt astronomy. Lastly Plato's old priest in *Timaios* expounds a doctrine of periodical catastrophes much like that of Berossos but without the astrology.

Evidently we have found, in Babylonia, another of Plato's sources — not of Atlantis itself, but of the general world-picture into which Plato fitted his lost-continent concept.

Thus while a flood-legend may be based upon a real flood, it is not so much a historical record of any special inundation as a fiction that has borrowed a plot-element from reality. As to why those who have charge of such things hand flood-legends down from generation to generation with such care, think how beautifully adapted they are to scare a congregation into putting up the money for a new roof to the temple when the old one has begun to leak!

One of the casualties of the rise of civilization is literal belief in the myths of the race. But if myths cannot be taken literally, how shall they be interpreted? The devout, alarmed by the discovery of obvious errors in their sacred scriptures, have often answered that they are allegories by which the gods reveal hidden truths.

In the fifth century B.C. Prodikos of Keos proposed another theory: that the gods were personifications of natural phenomena like the sun, the wind, and the sea. This hypothesis, which I suppose should be called prodicanism, has been used and abused in modern times, Fiske's *Myths and Myth-Makers* being an extreme example of equating gods and heroes with meteorological facts.

Still another rationalization of myths was proposed about 300 B.C. by the Greek-Siceliot philosopher Euemeros of Messana: that the gods were merely mortal men whose deeds and qualities posterity had exaggerated. This theory, called "euhemerism" after its founder, quickly became popular; euhemerists explained that Zeus had originally been a mere king of Crete, and the theory has come down to modern times. Modern euhemerists have not only made Jupiter smiting giants with thunderbolts into a human king putting down an uprising, but also insist that Achilles, Abraham, and other mythological characters must once have been flesh-and-blood people.

Eighteen years ago Lord Raglan in *The Hero* attacked the claims of legends and traditions to any historical foundation. He concluded that preliterate peoples never kept records of historical events, and that King Arthur, Cuchulainn (the Irish Achilles), Sir John Falstaff, Helen of Troy, and a host of other "quasi-historical characters" were never real people, but humanized gods. He thought that the stories wherein they appeared were literary versions of the ritual dramas which people once acted out as part of their religious ceremonials.

Raglan made many good points: Legendary heroes have no dates like real people. They talk in verse, stay the same age for decades on end, and perform magical feats. The legends report private conversations that could never have been recorded and show kings gadding about alone which real kings seldom do. The Siege of Troy is impossible, for one thing because of no half-barbarous army like that of the Achaeans could be kept together for one year, let alone ten. Robin Hood's story is practically all anachronisms: He was a longbow expert before the longbow was invented; led the oppressed Saxons though he and nearly all his followers bore Norman names; was Earl of Huntington when the title was actually held by the Scottish king's brother, and so on.

Raglan estimated that illiterate communities, which have enough to do without burdening their minds with historical facts, could remember important events at most about 150 years. This is perhaps too short; since the Hudson's Bay Eskimos are said to have remembered Frobisher's sixteenth-century expedition as late as the nineteenth century, the Greenland Eskimos the fifteenth-century Norse settlements, and the Solomon Islanders Mendaña's landing in 1567, a limit of 400 to 500 years would sound more plausible. Still, Raglan is probably right in his general thesis that such word-of-mouth records will not be carried down forever, but will be crowded out in a few centuries by later events.

However, like many men dominated by one idea, Raglan committed some staggering bloomers. He said, for example: "I can find no evidence that before [Herodotos's] time the idea of history ever occurred to anyone" — never, apparently, having heard of Hekataios of Miletos or the *Books of Kings*. Such extreme views are as wide of the mark as Spence's "Traditions, we know, survive for countless centuries . . .", or Donnelly's thesis that all myths are "a confused recollection of real historical facts."

To show that legends are never based upon history Raglan cites cases that actually prove just the opposite. Conceding that the Dietrich and the Etzel of medieval German legend are based respectively upon the real Ostrogothic king Theodoric the Great and Attila the Hun, he insists that "they are never represented in the stories as doing anything which they can be supposed to have done in real life."

Nevertheless Dietrich of Bern becomes king of Italy as did Theodoric; Wittich, who appears sometimes as Dietrich's follower and sometimes as his foe, corresponds to the Witigis who was king of the Ostrogoths after Theodoric's time, and his imprisonment parallels the beleaguerment of Witigis in Ravenna by Belisarius in 539. The story of Dietrich's youth at Etzel's court is probably due to confusion of Theodoric with his father Thiudemer, a vassal of Attila. Likewise the slaughter of Gunther and his followers at Etzel's court in the *Niebelungenlied* corresponds to the overthrow of the historical King Gundicar's Burgundian kingdom by Attila's Huns about 437.

True, the authors of such poems show no sense of time. In the old English poem *Widsith* a minstrel boasts of having visited the courts of Eormenric (Hermanaric), Guthhere (Gundicar), and Aelfwine (Alboin) who lived in the fourth, fifth, and sixth centuries respectively. They send Charlemagne on a crusade, surround Arthur with knights several centuries before knighthood was invented, and telescope several characters into one — an easy task since there were, for instance, four Gothic kings named Theodoric within half a century plus several Frankish kings of that name.

Other legends, outside medieval Europe, have also turned out to be based upon fact. Besides King Atreus and Noah's Flood which I have mentioned, the legendary Assyrian queen Semiramis is based upon the real princess Sammuramat, mother of King Adadnirari III and an important person even if she did not build Babylon or invade India as some Hellenic writers said she did. Golden Gyges, in Greek legend the owner of a ring of invisibility set with a stone from a dragon's eye, was King Gugu of Lydia whom the Gimirai or Cimmerians overcame and slew when he proved faithless to his alliance with Ashurbanipal of Assyria. And Mopsos the seer, who was said to have founded Mallos in Cilicia after the fall of Troy, turns out to have been a Hittite king named Mupsh. Finally we don't call Alexander the Great a myth because about 200 A.D. somebody borrowed the name of the philosopher Kallisthenes to write a mendacious *Life of Alexander* that took its hero to Rome, Carthage, and China where the real Alexander never went.

Raglan even thinks that primitives cannot make up their own myths, since: "The savage is interested in nothing which does not impinge on his senses, and never has a new idea even about the most familiar things." Many anthropologists who have lived among primitives tell a quite different story about "savages": that they are much like other people with the usual proportions of the brilliant and the stupid, and that the former do contribute new elements to the tribal culture. As an example of modern myth-making by primitives, when Russian scientists got around to investigating the Siberian meteor-fall of 1908, they found that after the catastrophe the local Tunguz built a new religion around the event, which in their mythology became the visit to earth of the fire-god Adgy.

Myths and legends, then, do often have a basis of fact. But the factual part of the myth may be so small and muddled that you cannot possibly reconstruct history from the legend. As the historian Grote said: "The lesson must be learnt, hard and painful though it may be, that no imaginable reach of the critical acumen will of itself enable us to discriminate fancy from reality, in the absence of a tolerable stock of evidence." We could not recover the history of Theodoric the Great if the Dietrich legends were our only source, since the legend, for instance, does not mention the Roman Empire, which is like a life of George Washington without the British Empire. If our civilization fell and its history were replaced by legends of the Dietrich-type, we might have the saga of President Abraham Jefferson Roosevelt who married Queen Victoria, invented the automobile, beat the Japanese at the Battle of New Orleans by killing their emperor Sitting Bull in single combat, and finally departed for the moon in a flying saucer, promising to return when his people needed him!

In Plato's time the scientific speculations of philosophers, the rationalizing theories of Prodikos, and the agnosticism of the skeptical sophists had all begun to break down belief in the myths of the Greeks. Plato himself, though a pious man, had refined the gods from Homer's hearty and lustful war-band of divine barbarians to a distant set of all-good, all-wise abstractions. In *The Republic* he makes Adeimantos complain that stories of the gods' thefts, rapes, and murders are nothing but a lot of lies invented by poets for their own benefit. Adeimantos and Socrates then agree that when they set up their ideal state their first step will be to banish poets and establish a strict censorship to keep such wicked tales from corrupting the young. As to whether the gods are really good, Plato with his usual subjectivity simply says they are and that is that.

However, if the stories of Kronos's castrating his father Ouranos and swallowing his own children were merely poetic fictions, why couldn't Plato turn this fact to account by making up his own myths that should teach moral lessons of unspotted propriety? Or use rumors of far lands and fantastic folk as a basis for pseudo-historical narratives illustrating his theories of the Good State? And that, I think, is precisely what he did.

Plato was not the only man of his time to entertain such projects. His younger contemporaries Antiphanes of

Berga the dramatist and Hekataios of Abdera the historian were writing similar fantasies. The latter (not to be confused with Hekataios of Miletos a century and a half earlier) collected the current stories about the Hyperboreans and wrote a treatise in which, giving his imagination free rein, he represented this people as living in innocent bliss on an island north of Europe.

Plato himself composed not only the Atlantis tale but also many other pseudo-myths. In the tenth book of *The Republic* he had Socrates tell a story of Er the son of Armenios, a Pamphylian whose soul (when its owner was knocked unconscious in battle) went to the other world but returned to its body just in time to save the latter from being burned. Er told how souls assembled on the Plain of Lethe or Forgetfulness, at one of the earth's poles, where some were given a rest and others punished for their crimes in life. Eventually they chose their next incarnations and returned to the world of the living. Er saw the Homeric hero Aias choosing the life of a lion because he hated everybody; Atalanta selecting that of a famous athlete so that she could indulge her bent in that direction; and Odysseus, having had enough excitement in his last life, picking that of an obscure private person leading a quiet existence.

Likewise in his other dialogues Plato used similar concepts. In *Phaidon* he speaks of "hollows of various forms and sizes . . . in all parts of the earth . . . into which the water and the mist and the air collect" and of islands floating in the sky; mortal men like us live in the hollows, but men of a superior kind and gods live in the high parts and the sky-islands. He also tells of Tartaros inside the earth, and of underground streams of hot and cold water, mud, and fire.

To continue, in *Gorgias* Plato mentions the Islands of the Blest. The *Politikos* describes the Golden Age of King Kronos, which lasted until Zeus relaxed his grip on the management of the universe, which thereupon went to ruin because of the growth of materialism. In the *Phaidros* Plato's Socrates tells a little fable about the Egyptian god Theuth and King Thamos, who told the god that he had been foolish to invent writing because people would come to trust in letters to the neglect of their memories. Theuth is the ibis-headed Tehuti, the Egyptian god of wisdom, better known as Thoth and later merged with the Greek Hermes and euhemerized to a wizard-king of Egypt. Thamos is no Egyptian king at all but the old Near-Eastern vegetation-god Tammuz or Dumuzi, the equivalent of the Greek Adonis. Since Plato's story is not confirmed from Egyptian sources, he probably made it up himself like his other allegories.

Finally in *Timaios* Plato set forth his version of the Pythagorean creation-myth, wherein the gods originally created mankind as a race of double hermaphrodites with four arms and four legs apiece. Since this arrangement proved inconvenient, the gods split these creatures into men and women. If all Atlantists would read the rest of Plato's writings as well as his Atlantis story, and realize what a fertile myth-maker he was, they might be less cocksure about his reliability as a historian.

Plato's Atlantis tale, like most allegories, implies more than it says. You can easily point out things in the real world that the story might symbolize, without being sure of whether Plato really had such a relationship in mind when he wrote. For example the main theme of the narrative, the Atlanto-Athenian War, is pretty obviously based upon the repulse of the Persians and the Carthaginians by the Greeks in the previous century.

Furthermore Plato's prehistoric Athens is no real state, but a fictional version of the ideal state he had set forth in *The Republic*: an Athens on which he had imposed the puritanical, militaristic, and grimly authoritarian institutions of Sparta. Thus he idealized the regimented anthill-state revived many centuries later by the Osmanli Turks and again in modern times by various totalitarian governments. Plato's contemporaries believed that this unlovely system had been set up in Sparta by King Lykourgos about the ninth century; the story is given in full in Plutarch's *Lives*. Although Lykourgos is probably more legend than flesh-and-blood man, Plato, having no reason to doubt his existence, put him into his intellectual melting-pot along with his other story-elements.

(Despite his authoritarianism, give Plato credit for

proposing a radical emancipation of women, whose lot in Platonic Athens was little better than it is in the more backward Muslim countries.)

In Plato's time these Spartan institutions were decaying, following the short Spartan rule over Greece between the battles of Aigospotamoi and Leuktra. However, Plato's theories allowed for this decay too. He thought that governments went through a natural course of degeneration from the perfect primitive Aristocracy through Timocracy, Oligarchy, Democracy, and Despotism, each less "just" than the one before it. Since this decline was a universal and inevitable process, perfect government was to be sought in the remote past, though present-day governments could no doubt be vastly improved if only philosophers like Plato were allowed to run them.

Plato considered the main cause of this moral decline to be avarice. Therefore commerce was base and bad while farming and soldiering were noble and good. Plato pushed to extremes the common theme in Classical philosophy of the iniquity of "vulgar" business. Having degenerated through love of gold, commercial and luxurious Carthage-Atlantis must succumb to frugal and bucolic Athens-Sparta; the wisdom and martial spirit of Athena must prevail over Poseidon's sea-borne traffic.

Plato may also have been jealous of the pretensions of the Egyptians to antiquity and occult wisdom. Ever since Solon's time well-to-do Greeks had been touring Egypt. If the priests who guided them around their temples pointed out hieroglyphics which they said gave all the details of Egyptian history from the Creation down, the tourists, unable to read Egyptian, had to take their word for it. By putting his tale back 9000 years Plato tried to prick these pretensions by showing that Greece, too, had received a civilization directly from the gods at the time of the Creation. And by putting his story back thousands of years before the earliest Greek and Egyptian history he protected himself from criticism, since nobody could well contradict him; to the Athenian mind such a period took one into times of limitless remoteness and might just as well have been a hundred thousand years.

However, Plato seems to have run into trouble with

his plot. Here the strange ending of the *Kritias* may mean something. Perhaps old Plato lost interest; or more likely he found that he had written himself into a corner whence there was no exit. He had started out to show how his ideal "Republic" would work in practice, and incidentally to justify the ways of the gods to men and to clear up such questions as fate versus free-will and the origin of human evils.

Now, to make a sensible story he had to get rid both of Atlantis and his prehistoric Athens, since neither was known to history. For the first, the submergence by earthquake would do, since it fitted what was believed about the powers of earthquakes and the existence of Atlantic shoals. For the second he caused the earthquake also to devastate proto-Athens to provide a gap in history between the fictitious Athens and the real one. Such a hitch is always necessary in imaginative stories to account for the world as it is. Thus Setnau Khaemuast is prevailed upon to return to *Book of Thoth* with its invincible spells to Neferkaptah's tomb; Roger Bacon's brazen head is destroyed through the stupidity of his servant Miles; and Arthur's sword Excalibur is thrown back into the lake. Aristotle knew this principle perfectly well as he showed by his gibe about the wall of the Achaeans in the *Iliad*.

Athens represented Plato's ideal state; Atlantis, a state that started with the same virtues and advantages but that decayed through the dilution of its people's divine blood and their love of wealth. So far so good. The decline of Atlantis fitted his theory of political evolution, too.

But here comes a snag. If this decline is a natural and inevitable process, as he said in *The Republic*: "Since everything that has a beginning also comes to an end, even a constitution like yours will not last forever, but must suffer dissolution," why did the process not affect Athens at the same time as Atlantis? Moreover the gods had set up both states. If the gods were, as Plato assumed, good, omnipotent, and omniscient, why then did Poseidon's experiment miscarry?

Furthermore the dilution of the divine heredity that brought Atlantis to its downfall is a natural process, and can you justly punish people for doing what circumstances beyond their control compel them to do? No doubt Plato meant Zeus to explain the rationale of his actions in the speech he was about to make to the gods — but when it came to writing the speech Plato's inspiration failed. Nor is that all. If the Atlanteans are destroyed by the earthquake, where is the credit due the virtuous Athenians for defeating them? And is it divine justice for the gods to destroy virtuous Athens along with decadent Atlantis? And if Zeus, like a good father, meant merely to reform his erring children by suitable punishment, why did he wipe them utterly out?

The truth is that Plato had hung himself on the same philosophical hook upon which generations of theologians have squirmed ever since. If, as Plato assumed, God or the gods are all-good, all-powerful, and all-knowing, how came evil into God's good world? For if He made everything, being Himself perfectly good, all He made must have been perfectly good also, and so must the things made by or descended from these things. Furthermore, being omniscient, God would foresee exactly how everything He made would turn out, and if anything contained the germs of badness He would have known it and taken steps to correct the condition.

Jewish, Christian, and Muslim theologians have tried to answer the question in various ways, as by saying that evil is illusory and non-existent, or that God gave man the power of choice between good and evil so that he could earn his salvation. But all these explanations are merely smoke-screens to hide the logical dilemma. This dilemma has been solved to date only by the Zoroastrians, who assume two rival gods of equal powers, one good and the other evil.

In addition, many a fiction-writer, striving to compose an entertaining story and at the same time to show the triumph of good over evil, has ended up by creating a fascinating villain and a stuffed shirt of a hero. Plato fell into just this trap. Athens, his hero, turns out a bleak, dull sort of place while his villain Atlantis has fascinated men for centuries with its glamor. Even Plato himself succumbed to the seductions of Atlantis, for he gave it nearly three times the space in *Kritias* that he did Athens.

By the time he got to Zeus's speech, then, Plato probably realized that the *Kritias* was not going to turn out as he intended at all. A thrilling story, perhaps, but morally confused. Maybe he put it aside with a slight shudder of repugnance; he, the world's moral beacon, to descend to the level of a lower-class travelling minstrel? In any case, in his last dialogue *The Laws* he turned his back upon the charm and imagination that had featured his previous writings and devoted himself to straight exhortation which few read nowadays.

Plato would no doubt be astonished to learn that later generations paid almost as much attention to his Atlantis story, which he himself did not think enough of to finish, as to all the rest of his works put together. He would be piqued to find that these people concentrated on Atlantis, the villain of his piece, and resolutely ignored his stuffy hero Athens. He would moreover be exasperated to discover that they neglected the moral and philosophical aspects of the tale, to him its main excuse for existence, in favor of its geological, anthropological, and historical sides, to him material matters of no importance.

Plato's was not the first utopian romance written, for some of his contemporaries were working along similar lines. Aristophanes had written a play poking fun at would-be world-reformers like Phaleas of Chalkedon who wanted all property divided equally. Similar compositions occur in the literatures of other civilizations like the Chinese, and some of the messianic prophecies of the Hebrew prophets such as Isaiah, several centuries before Plato, contain utopian sketches of an ideal Israel where "the lion shall eat straw like an ox."

Plato's story is merely the first Classical work of the kind to come down to us whole. His successors wrote many such stories: Zeno of Kition, the founder of the Stoics, tried his hand at this kind of composition; so did Amometos who described the idyllic lives of a Himalayan tribe. Euemeros, who traced the gods back to men, wrote a full-fledged utopia about an archipelago in the Arabian Sea, Panchaia, of vast wealth and fertility. He described its capital city Panara on the island Hiera, its temple of

Zeus Triphylios with its white marble statues, its golden column on which were written the histories of Ouranos, Kronos, and Zeus, its sacred spring, and its castes of priests, soldiers, and farmers. Resemblances to Atlantis need not mean an actual connection between the two, but a tendency of Greeks to think in similar terms when they described their ideals of beauty, order, and opulence.

Following Euemeros, Iamboulos (likewise preserved in Diodoros) went him one better in the third century B.C. with another account of an ideal commonwealth on a group of islands in the Indian Ocean. Possibly both descriptions were inspired by tales of Ceylon, which Classical geographers knew vaguely as Taprobanë, and whose size, wealth, and virtuousness they much exaggerated.

Iamboulos wrote that when he was a commercial traveller in his youth he was captured by Ethiopians who, as part of a religious observance, set him and his companion adrift in the Arabian Sea in a small boat with provisions. After four months they reached seven wonderful islands where the people had flexible bones and forked tongues with which they could carry on two conversations at once. Their Sun State found complete communistic equality practical because they were all exactly alike, so that everybody took turns at all the jobs in the commonwealth. When finally expelled for misbehavior, Iamboulos and his companion sailed to India, whence the narrator found his way back to Hellas.

Curiously, no cult has ever arisen to insist that Iamboulos's Heliopolis was a real place, although the author claimed to have seen it with his own eyes. Perhaps the flexible bones and forked tongues were too much for the speculators, though they are really no more incredible than Plato's prehistoric Athenian empire or Helena Blavatsky's Lemurian hermaphrodites.

These stories had a considerable effect upon the people of their time. Take Naomi Mitchison's *The Corn King and the Spring Queen,* one of the best historical novels ever written in spite of the fact that Mrs. Mitchison's ancient Greeks have the disconcerting habit of bursting into Marxian jargon. This novel represents one of the utopian stories as helping to inspire the communist king

Kleomenes III and his associates to the drastic changes they inflicted upon Sparta. The idea is by no means incredible, since in 132 B.C. Aristonikos led a proletarian revolution in Pergamon, where the last king had bequeathed his realm to Rome, and seriously tried to set up a Sun State like that of Iamboulos in fact. An Italian Stoic, Blossius of Cumae who had been a supporter of Tiberius Gracchus, abetted him. Aristonikos destroyed a Roman consular army and its consul before he was captured and died in prison, whereupon Blossius killed himself.

Then in the third century A.D. the Egyptian Neoplatonist Plotinos, having set himself up as a philosopher in Rome, tried to persuade his friend the Emperor Gallienus to rebuild the Campanian ruin called the City of Philosophers, rename it Platonopolis, and set up an ideal community to be run by Plotinos according to Plato's prescriptions. The emperor declined, however, either because others jealously dissuaded him or because he thought the idea foolish.

With the rise of Christianity men's minds turned away from hopes of improving their earthly state. Utopias gave way to treatises on the joys of the future life, like Augustine's *City of God*. Then in 1516 utopianism returned with Sir Thomas More's *Utopia*, wherein a Portuguese sailor, Raphael Hythloday, tells the narrator how he sailed with Vespucci and visited Utopia, an island shaped like a thick crescent 200 miles across at its thickest point. There are fifty-four cities, the largest being Amaurot. A hierarchy of elected magistrates and officials appointed by them governs the people, who lead simple lives of agriculture and home industry.

Utopians may not travel without special permission, and criminals are enslaved for punishment. The country avoids war when possible, but if compelled to fight tries to achieve victory by guile and assassination, having a poor opinion of the conventional military virtues. Young couples wishing to marry are shown each other naked to forestall unpleasant surprises later. The Utopians (very moral and philosophical, but not scientific) call their supreme god Mithras and have a number of religious sects. Their language resembles Persian, and their culture suggests Greek colonization. This description takes up only half the work, the rest being devoted to criticism of the Europe of More's time.

Recently Arthur E. Morgan wrote a book in which he tried to prove that Hythloday was a real man who, going on one of the secret expeditions the Portuguese were sending out about Columbus's time, had reached Peru some decades before Pizzaro's conquest in the 1530's. More's Utopia, said Morgan, is a description of the Inca Empire: an interesting idea, but far from proved. For one thing More's democratic state is as unlike the Inca despotism in some ways as it is like it in others; for another, equally good sources for More can be found in Classical literature, for instance Plutarch's *Lykourgos* and Tacitus's *Germania*.

A century later Sir Francis Bacon, the "glittering serpent," undertook a similar project. His *New Atlantis* tells how on a voyage from Peru to China the narrator's ship was blown off course to an undiscovered South Sea land whose turbaned people had a perfect democratic limited monarchy. These folk call their country "Bensalem" and tell how they came thither from Plato's Atlantis, which was in America. The original Atlantis, an empire reaching from Mexico to Peru, was devastated by a great but temporary flood. Bacon's utopia is gaudy but solemn, and much more scientific than More's, with submarines, airplanes, microphones, air-conditioning, and a great research foundation. The Bensalemites are Christians, having received the gospel miraculously from St. Bartholemew. Like Plato's Atlantis story the work is unfinished. Bacon meant to write a second part, about the laws of the perfect commonwealth, but never got around to it.

About the same time a Dominican friar from Calabria, Tommaso Campanella, wrote a similar work while spending twenty-eight years in jail for a plot to free Naples from the Spaniards. Derived from Plato and other Classical writers, the work is called *The City of the Sun* like Iamboulos's capital, and is described as a "Dialogue between a Grandmaster of the Knights Hospitaller and a Genoese Sea-Captain, his Guest."

The captain tells how "In the course of my journeying I came to Taprobanë, and was compelled to go ashore at a place, where through fear of the inhabitants I remained in a wood. When I stepped out of this I found myself on a large plain immediately under the equator. . . . I came upon a large crowd of men and armed women, many of whom did not understand our language, and they conducted me forthwith to the City of the Sun." This city stands on a hill surrounded by seven great walls or rings, one for each planet. The people wear togas over union suits, white by day and red by night. Like Bacon's New Atlanteans they are very scientific; like Plato's Republicans they practice communism of women and children.

And another contemporary of Bacon, the German Lutheran divine Johann Valentin Andreä, wrote a similar work in 1619: *Christanopolis,* an ideal city combining More's austerity, Bacon's science, and Andreä's own oppressive piety. Andreä was probably the founder of Rosicrucianism, for he is thought to be the author of a series of anonymous manifestoes that appeared at that time calling upon the wise and learned to form a secret society, to be called the Rose-Cross Order, for the reform of Europe. Although Andreä failed with this project, occultists and charlatans have exploited these pamphlets ever since by claiming to be the true and original Rosicrucians.

All these utopias give free rein to their authors' prejudices. Thus while the bachelors, Plato and Campanella, preach communism of women, the married men like More and Andreä are strong for family life. Those who (like Bacon) enjoy pomp and circumstance indulge their utopians in such shows, while those who do not would inflict their own drab naturalism on everybody. It is like the story of the agitator who was haranguing the crowd to the effect that, come the Revolution, the bosses would have to work and sweat while they, the workers, would eat strawberries. *Voice from audience*: "But mister, I don't *like* strawberries!" *Orator*: "Come the Revolution, you eat strawberries and like 'em!"

In fairness to Plato, be it noted that in *The Laws* he abandoned pure communism as unsuited to this imperfect world. Instead he advocated a "second-best state" much like the aristocratically-ruled Greek city-states of his own time, with normal family relationships. He even found the Spartan constitution too militaristic, since it neglected to develop the peaceful virtues along with the martial.

In the seventeenth century James Harrington carried on the utopian tradition with his *Oceana,* a work very popular for some decades, but, being like Plato's *Laws* straight exhortation without fictional decoration, seldom heard of today. Harrington was a republican who advocated a system of checks and balances, and separation of the legislative, executive, and judicial functions of government. Since some of his ideas like the Electoral College (which has not however worked out as intended) were incorporated in the constitutions of the United States and some of its component states, utopias evidently do influence worldly affairs. One of Harrington's less practical ideas was to re-settle the world's Jews in Ireland ("Panopea" in *Oceana*) on the ground that the wretched Irish were dying out any-way and England ("Oceana") might as well have an industrious and adaptable people next door to tax.

While Plato and More relied heavily on a suppositious native virtue to make their ideal states work, the later utopians, beginning with Bacon, put more and more faith in natural science as the answer to the human question. This trend grew in the works of the early nineteenth-century socialists (mostly Frenchmen like Saint-Simon and Cabet) until it reached its climax half a century ago in the utopian novels of Edward Bellamy and H. G. Wells, wherein the combination of science and socialism was held out as the hope of mankind for heaven on earth.

More recently Aldous Huxley in *Brave New World* showed how dismal even a scientific socialism might be if pushed to extremes. Fortunately for man, he seldom pushes any program to its logical conclusion, or he would probably have long since become extinct. Utopias have been few in recent decades. Perhaps the utopians have been oppressed by a feeling that everything, including science and socialism, has been tried, and nothing seems to work so well as its prophets promised.

One recent novel, A. T. Wright's *Islandia,* represents the back-to-nature school of utopianism, which brings us back to Atlantis again. For Wright's reaction against the usages of civilization (call it primitivism, archaism, or naturalism) is an example of one of the sentiments that keeps the Atlantist cult afloat.

This feeling goes back to ancient stories of lost Edens and Golden Ages, but did not take virulent form until Jean Jacques Rousseau in the late eighteenth century spread the idea of the Noble Savage — never having met any savages himself. Rousseau and his contemporaries Morelly and Babeuf gave utopianism a belief in a natural goodness of man, along with an extreme equalitarianism and a dogmatic environmentalism. These ideas passed into early socialism, and thus are responsible in the long run for such phenomena as the war of the Michurin-Lysenko cult on scientific genetics in the U.S.S.R.

Wordsworth, Coleridge, Byron, and other romantic writers of the early nineteenth century popularized the back-to-nature ideal, and Byron's friend John Galt wrote a Roussellian tragedy called *The Apostate, or Atlantis Destroyed,* published in 1814. In this play the wicked Antonio, a shipwrecked mariner, converts King Yamos of Atlantis to Christianity and the Western arts and sciences while striving to seduce the queen and forcing the high-priest of the old Atlantean religion to flee into the forest with his few followers. Antonio attains his objective with the queen and persuades the Atlanteans to build a big wicked city on the European model. Finally Antonio is unmasked, and the priest incites the Atlanteans to burn the city that is threatening the Arcadian bliss of Atlantean life.

Half a century later John Ruskin used the Atlantis theme to lecture an audience of astonished businessmen of Bradford, England, on the evils of the Industrial Revolution. If they continued to worship gold, deface the countryside, and oppress the workers, he warned, they could learn from Plato's narrative what sort of fate to expect.

The Atlantis theme has indeed been put to many uses. The early Christian Father Arnobius employed it as an example of the catastrophes that afflicted the world before Christianity, to refute the arguments of pagans who (like Edward Gibbon long afterwards) thought that this religion had brought disaster upon civilization by undermining the Roman Empire. Kosmas dragged Atlantis into his flat-earthian argument. Bacon used it as an example of the scientific society to which he looked eagerly forward, whereas Galt cited it as an instance of the pre-scientific society to which he looked regretfully back. Benoît exploited Atlantis to illustrate the theme of the world's being well lost for love—a common motif in French novels of a certain type. And Theosophists and other occultists have fitted Atlantis into their own bizarre cosmogonies.

In recent years Lewis Spence has taken Atlantis so seriously that he warns us that, if we don't reform, God will sink us as He did Atlantis. So does Dmitri Merezhkovski, the emigré Russian novelist. Merezhkovski is an example of that fascinating but nearly extinct breed: the pre-Revolutionary Russian mystical-intellectual who broods in a state of feverish despondency on God, sex, death, and other large subjects. He interprets the Atlantis story both as a historical record and as an allegory of things to come. The West, he says, has released the power of the Machine, which if not controlled will destroy us in our pride as the insolence and avarice of Atlantis caused its downfall. In view of late scientific developments, maybe Merezhkovski has a sound idea here despite his rant.

Chapter XI

EVENING ISLES FANTASTICAL

Into the west, unknown of man,
Ships have sailed since the world began.
Read, if you dare, what Skelos wrote,
With dead hands fumbling his silken coat;
And follow the ships through the wind-blown wrack —
Follow the ships that come not back.*

Howard

W E have seen the part that the lost-continent concept has played in the sober sciences: archeology, anthropology, and paleogeography, and we have followed this glittering little ghost through the mazes of history, philosophy, and occultism.

But how about the use of the concept in the field in which (I think I have shown) Plato himself meant to use it: the field of fiction, of fantasy, of romance? For here is ground on which we can meet in reconciliation with the most ecstatic Atlantist. Here the question is not: Is it objectively true? but: Has the writer so craftily contrived his illusion that it seems true while we read it?

The poets, while they have made some use of Atlantis, have not exploited the theme so extensively as one would expect. One would have thought, for instance, that it would make a natural subject for the early nineteenth-century English romanticists. After Blake, however, they practically ignored it. True, Shelley gave "The Mouth of a great River in the island Atlantis" for one of the scenes of his long philosophical poem *Prometheus Unbound,* but made no

---

* From *The Pool of the Black One,* in *The Sword of Conan* by Robert E. Howard (Gnome Press, 1952); copr. 1933 by Popular Fiction Pub. Co. for *Weird Tales;* by permission of Otis Kline Associates, exclusive literary agents for the estate of Robert E. Howard.

use of the idea except the name.

Since then almost the only use of Atlantis in English-language verse has been the recent poems by Aiken and Masefield, quoted at the heads of Chapters I and IV: charming pieces, but far from these poets' best-known. Aiken tells how modern voyagers, sailing over the site of Atlantis, hear an elfin singing and the sound of ethereal bells from the haunted sea below. Masefield describes his pale Atlanteans, knowing all wisdom, moving silently through their bright cities, their thoughts flitting about like golden birds. . . .

But to the novelists Atlantis has been a gift of the gods, and recently the lost continent has become, like the other planets and the remote future, a standing setting for stories of the science-fiction *genre*. The fictional revival of Atlantis started with the publication in 1869 of Jules Verne's best-known novel, *Vingt Milles Lieues Sous les Mers,* translated four years later as *Twenty Thousand Leagues Under the Sea* and kept in print more or less continuously ever since.

In this tale the narrator and Captain Nemo leave the submarine *Nautilus* in diving-suits to walk on the bottom of the Atlantic, and presently come to where the lava flowing from a submarine volcano lights up the sea-bottom for miles around. By this light:

There indeed under my eyes, ruined, destroyed, lay a town, its roof open to the sky, its temples fallen, its arches dislocated, its columns lying on the ground, from which one could still recognize the massive character of Tuscan architecture. Further on, some remains of a gigantic aqueduct; here the high base of an acropolis, with the floating outline of a Parthenon; there traces of a quay, as if an ancient port had formerly abutted on the borders of the ocean, and disappeared with its merchant vessels and its war-galleys. Further on again, long lines of sunken walls and broad, deserted streets — a perfect Pompeii escaped beneath the waters. Such was the sight that Captain Nemo brought before my eyes!

Where was I? Where was I? I must know at any cost. I tried to speak, but Captain Nemo stopped me with a gesture, and picking up a piece of chalk stone, advanced to a rock of black basalt, and traced the one word:

ATLANTIS

While Verne's characters simply return to their submarine and go on to other adventures, Verne's successors soon found ways to bring, not merely the ruins of Atlantis, but also live Atlanteans on the stage. More than fifty novels on the lost-continent theme have been published in book form, plus a vast number of magazine stories. The motif came into common use after publication of the treatises of Donnelly and Le Plongeon in the 1880's, the greatest outburst of such stories occurring around the turn of the century, when the interest generated by these pseudo-scientific works was at its height. At least sixteen such novels appeared in the decade 1896-1905, some of them now very rare books.

The stories have used Atlantis in every conceivable manner. While in some cases the hero finds Atlantis in ruins, in others he finds it still a going concern. Whereas in some he goes back to the Atlantis of ancient times by time-travel, dream, or "racial memory," in others the story is simply laid in ancient Atlantis without a modern tie-in. Sometimes a continent sinks; sometimes instead it rises from the deeps. The lost continent may occupy the center of the stage, or it may play but a peripheral rôle, as in Gerhart Hauptmann's celebrated *Atlantis*: despite its title, not a lost-continent story at all but a modern novel about a German physician's visit to America, in which Atlantis is casually alluded to once or twice. And though most of the stories deal with Plato's Atlantis, a few like G. Firth Scott's *The Last Lemurian* concern other continents.

One of the earliest products of the great outburst of literary Atlantism was Oliver's *Dweller on Two Planets*, which though a singularly bad novel is still in print, and which is significant as the first of a long line of occult Atlantis novels, most of them equally unreadable.

But curiously enough, one of the best Atlantis novels was also one of the early ones, and a story with occult overtones: Cutcliffe Hyne's *The Lost Continent* (1900). The "frame" of the story is the finding by two Englishmen of a bundle of waxen sheets inscribed with Egyptian hieroglyphics in a cave in the Canaries. The manuscript proves to be the autobiography of Deucalion, a warrior-priest and politician of old Atlantis.

As the story opens, Deucalion is recalled home from his job as governor of Yucatán. He learns that the throne of Atlantis has been usurped by the base-born adventuress Phorenice, who meets him on his arrival: a small luscious redhead riding a tame mammoth. As a result of her reckless and tyrannical rule, a revolt has flared up and the rebels are besieging the capital city.

At a banquet old Zaemon, the head of Deucalion's priesthood who comes and goes as he pleases by means of his occult powers, appears to warn Phorenice to mend her ways. She defies him, and later lets Deucalion know that she intends him for her consort. Deucalion falls in love with Zaemon's daughter Naïs, who is with the rebels, but Zaemon orders Deucalion to marry Phorenice for the good of Atlantis.

Phorenice's soldiers capture Naïs, and Phorenice, suspecting her love for Deucalion, orders Deucalion to bury her alive in the base of a throne to be built in a public square. Deucalion smuggles a pill to provide indefinite sleep to Naïs in her cell, and after the burial Phorenice declares Deucalion her husband. Deucalion leads a sally against the rebels and while chasing the rebel army learns that Phorenice has heard of his last tryst with Naïs and wants his head.

Accordingly he flees to a wild volcanic part of Atlantis and for nine years lives the simple life, dodging dinosaurs. When he returns to civilization at Zaemon's behest he finds that Phorenice, having trampled down all other opposition, is now besieging the sacred mountain of Deucalion's priesthood. Although Deucalion rescues and revives Naïs and takes charge of the defense, the priests are overborne by numbers. Zaemon and the other senior priests sink the continent by magic, destroying all life except Deucalion and Naïs who are saved in an ark.

Despite the author's weakness for plesiosaurs and other anachronistic Mesozoic reptiles (who disappeared at least

60,000,000 years before the rise of man) the novel is a competent piece of story-telling: fast, well-constructed, colorful,
with the leading characters well-drawn and occasional
flashes of rather grim humor. Moreover the story is not
burdened with the sentimentality and didacticism that oppress so many novels of this group; thus the author does
not try to project our modern Western moral code on
people of an ancient and supposedly different culture.

David M. Parry's well-known *Scarlet Empire* (1906)
shows a quite different treatment of the Atlantis theme.
The author's main purpose is not to entertain but to
ridicule socialism and the labor movement. Walker, a
young socialist who has despaired of reforming the world,
jumps into the ocean at Coney Island to end it all. He comes
to in Atlantis, which is at the bottom of the sea under an
immense crystalline dome. Having rescued him the Atlanteans (who conveniently speak English) welcome Walker
and make him a citizen.

Atlantis, however, is a socialist land where equalitarianism has been carried to the point of making everybody
wear the same garb (a red robe — hence the title), eat the
same meals, and talk no more than a prescribed number
of words a day. Jobs are assigned by a wheel of fortune,
and marriages are arranged on the eugenic basis of pro-
ducing as uniform a population as possible. People have
numbers instead of names, and a quarter of the population
are inspectors who spy upon the rest. Those who display
individualism are locked up as "atavars." Naturally the
hero falls in love with an atavar and out of love with
socialism.

The main amusement of these wretches is the Festival
of Kuglum, in which wrongdoers and incurable atavars
are thrust through a lock in the dome into the water outside
to be devoured by the kraken, a kind of super-octopus.
The hero and his girl nearly undergo that fate, but escape
in a salvaged submarine with two other Atlanteans who
secretly sympathize with them. When the submarine comes
opposite the theater where the Festival is taking place, the
kraken grapples with it. Walker fires a torpedo which
misses the monster and blows a hole in the dome, and
that is the end of Atlantis.

As a story it is so-so except near the end, where it gets away from political theory and builds up to a fast and furious climax. Characters are simple black-and-white and the political satire is too one-sided to be interesting for long. Still, the novel is one of the earliest fictional warnings against the dangers of statism, a theme that Aldous Huxley and George Orwell have since put to such excellent use in their anti-utopian novels.

Two years after *The Scarlet Empire,* Richard Hatfield took the opposite point of view in his *Geyserland,* which represented the Atlanteans as living a blissful life of pure ideal communism at the North Pole. The book is based upon a theory propounded in 1852 by Alphonse-Joseph Adhémar, that from time to time ice piles up at the poles until the earth becomes unstable and does a flip so that the new poles are where the equator used to be. The last of these cataclysms, of course, formed the basis of Atlantis, Noah's Flood, and so on. Not only was the theory wrong, but also the author had no notion of how to tell a story. The result is a stupendous mass of politico-economic argument thinly disguised as fiction and quite unreadable.

World War I saw the disappearance of the lost-continent theme from fiction, but this motif revived in 1920 with the publication of Benoît's popular *L'Atlantide.* Since then such books have continued to appear at a modest rate: one every year or two. They fall into several well-defined classes according to their basic plots; some for instance (like Hyne's *Lost Continent*) being romances laid in a lost continent without modern characters. Thus several tales in Clark Ashton Smith's *Out of Space and Time* and *Lost Worlds* are laid in Hyperborea, Atlantis, or the future continent Zothique. These are volumes of stories (mostly reprinted from *Weird Tales*) carrying on the peculiar type of horror-fantasy tradition made popular by H. P. Lovecraft, who contributed a few short stories of his own to the genre. While some critics object to Smith's heavily euphuistic style, based ultimately on Poe and Dunsany, many connoisseurs of fantasy find the stories delightful.

The late Robert E. Howard presented a similar picture of human prehistory in his Kull and Conan stories, though he relied more upon swashbuckling swordplay and less upon sinister sorceries for his excitement. The three Kull stories deal with Atlantis, which Howard threw back into the late Pleistocene, while the twenty Conan stories are laid in Howard's imaginary "Hyborian Age" between the sinking of Atlantis and the dawn of recorded history.

Conan the Cimmerian is surely the toughest hero of imaginative fiction. When his enemies crucify him, and a vulture flies down for the purpose of pecking his eyes out, Conan bites the bird's head off. The mighty adventurer overcomes fearsome foes, both natural and supernatural, and wades through rivers of gore to seize the throne of Aquilonia — from which his enemies promptly plot to oust him.

These stories suffer from the crudity and carelessness of mass-production pulp-fiction writing, but still possess extraordinary zest and vitality. So vividly did Howard visualize his setting, people, and events that, however implausible his Hyborian Age may seem when coldly analyzed, it comes to gorgeous and furious life on his pages. Conan was a wishful idealization of his unhappy creator, a young Texan, who suffered from delusions of persecution and, at the age of thirty, killed himself in an excess of emotion over his mother's death.

One of the better-known books in this group of "straight" Atlantean novels is Lillian Elizabeth Roy's *The Prince of Atlantis* (1929) which, like Oliver's novel, presents a vaguely Theosophical view of the lost Atlantic continent. The story (unbearably dull and amateurish) tells how Atlantis, by lowering its immigration barriers, let in a lot of wicked pagan idolaters who corrupted the official religion, the Law of One, whose devotees are called Templars. The Master Qoka warns the remaining Templars to flee the continent when it is due to be dunked for its sins. There is also a tedious intrigue involving King Atlas, his mistress Lias, his son Atlan (the hero), the sorcerer Ritaro, and others. The author even uses the ancient gag of swapping a royal baby for a commoner.

Within the last decade two very similar occult-Atlantis novels have come out. Marjorie Livingston's *Island*

Sonata, while much more literate than the Roy book, is almost as dull.

Much more successful is Phyllis Cradock's *Gateway to Remembrance,* which, like the two previously described novels, tells how Atlantis fell when her people deserted the austere True Faith (a glorified Theosophical Spiritualism) for materialistic beliefs and hedonistic practices. The heroine Clio, daughter of the last Atlantean king and betrothed of Porlas (one of the seven elected Adjudicators who now rule the land) falls in love with Divros, a leader of the atheistic Emancipates. She runs off to live in sin with Divros while her fiancé is busy stamping out a depraved and orgiastic heretical cult that has sprung off from the state religion, and of course disaster follows. Though the story is slow at the start, very feminine in outlook, and overwritten with beautiful prose, and though its solemn occult premises may repel the non-believer, nevertheless I cannot help admitting that I found it a skillfully wrought and absorbing narrative.

If these occult-Atlantist novels have any moral, it would seem to be that religious liberty is evil and that the ideal state is a priestly dictatorship with a yogi or Theosophist at the head of it.

Then there are the stories wherein, as in the novels of Parry and Benoît, a modern character finds Atlantis still flourishing at the bottom of the ocean or elsewhere. Conan Doyle, for instance, exploited this idea in *The Maracot Deep,* originally run as a serial in *The Saturday Evening Post* in 1928 and subsequently published as a book with several of his shorter science-fiction stories to pad it out. This story presents the Spiritualist version of Atlantis.

The narrator, Headley, sets out on a bathysphere expedition with Professor Maracot and the American mechanic Scanlan. When they are suspended over an abyss in the Atlantic, the Maracot Deep, a gigantic crustacean snips the cable and down they go to the bottom. Atlanteans in diving-suits rescue them and take them to their submarine city.

It transpires that when the time neared for Atlantis to be sunk for its wickedness, the virtuous leader Warda

built a waterproof building in which he collected his followers. Now there are thousands of them, dwelling in
their suboceanic warren and getting their living by hunting
and farming on the ocean-bottom. After minor adventures
the surface-men fall afoul of a demon, Baal-Seepa, who
intends to destroy Atlantis. Maracot, with the help of the
spirit of Warda, exorcizes the evil spirit, and all ends
happily.

The story is very poor Doyle, with a thin plot, flagrant
scientific blunders, a big dose of Spiritualist propaganda,
and no characters of the stature of the immortal Holmes
or Challenger. Still, being told by a master story-teller, the
tale is better than some in the field. For instance, probably one of the worst novels ever written is Alaric J. Rob-
erts's *New Trade Winds for the Seven Seas* (1942) in which
the heroes find not only Lemurians living on Pacific Islands
but also a flourishing Atlantis and Lemuria in hollows
under the beds of the oceans.

Among the better novels of this group are *The Sunken
World* by Stanton A. Coblentz and *They Found Atlantis* by
Dennis Wheatley. The Coblentz book originally appeared
in the old *Amazing Stories Quarterly* in 1928, when the
magazine was under Gernsback's editorship. Later it was
reprinted in a magazine and finally republished as a book
in 1948, slightly reëdited to bring it up to date.

This story uses the undersea-crystal-dome concept of
The Scarlet Empire. An American submarine, crippled by
a German submarine, sinks to the bottom of the Atlantic
and is sucked into the inlet of the Salt River of Atlantis.
The narrator, Ensign Harkness, finds he can talk to the
Atlanteans because he was a teacher of Greek in civilian
life and the Atlanteans speak an archaic Greek dialect.
The Atlanteans are handsome long-lived vegetarianists and
eugenists whose society is a utopian communism without
money, officials chosen by competitive examination, and
careers assigned by a committee of personnel experts.

Harkness learns that the ancient Atlanteans, disgusted
with the way other nations were developing, built the dome
over their island and then caused the island to sink by
exploding atomic bombs under the ocean bed. He is assigned the task of writing a history of the upper world,

which when published so horrifies the Atlanteans that a movement for reëstablishing contact with the upper world is nipped in the bud. Harkness marries an Atlantean woman, Aelios, and prepares to spend the rest of his life there.

Then, however, it transpires that when the submarine was drawn into the inlet it bumped against the dome and cracked the glass. Although the crack has been patched, the patch gives way and water pours into Atlantis faster than the atomic pumps can force it out again. The Atlaneans send Harkness and his bride to the surface in one of their own submarines to ask help of the upper nations, but by the time a rescue party reaches Atlantis the dome is entirely filled.

While not unreadable, the story somehow lacks vitality. The writing is mediocre and the author knows little of Navy life. Moreover it suffers from a fault that appears in many utopian satires on modern Western civilization, such as Wheatley's *They Found Atlantis*, C. S. Lewis's *Out of the Silent Planet*, S. Fowler Wright's *The World Below*, and H. G. Wells's *Men Like Gods*.

The author, in his eagerness to make a strong case against modern life, sets up a straw-man: a picture of the modern world with its evils exaggerated and its positive achievements ignored. Then to this he opposes an ideal society in which all these faults have been arbitrarily banished. The result is a setup in which poor Western Civilization never has a chance. Now, there is no evidence that such an ideal society as Coblenz's Atlantis has ever existed, and whether such an ideal state could exist, or whether a species other than man might do better at culture-building, are unanswerable questions.

Wheatley's *They Found Atlantis* (1936) uses the bathysphere, the submarine-Atlantis, and the utopian-communism themes. A Dr. Tisch persuades the American heiress Camilla Hart to finance his Atlantean expedition. She brings along a group of her friends, including three suitors, but when they get to sea a gang of criminals takes over the ship in an attempt to extort Camilla's fortune from her. The gangsters, however, let their prisoners proceed with their underwater exploration.

Then a warship comes to the rescue just as the whole party is down on one of their dives, and in the resulting fracas the bathysphere cable is broken. They sink to the bottom and are sucked into the Atlanteans' waterworks. Here they fight a horde of sub-men who live in tunnels and feed on the fish drawn into the pumping system. Escaping these, they win their way to Atlantis proper, an island in a lake in a huge illuminated cave.

There are just twelve Atlanteans, six men and six women, who lead a lotus-eater's life, making love, working communally, and sleeping for days at a time during which they send their souls out to explore the upper world. They are descended from the few Atlanteans who survived the submergence by being in watertight underground chambers, and the sub-men are an artificial race created by Atlantean sorcerers. (These wicked Atlanteans appear lurking in Antarctica in another of Mr. Wheatley's adventure-novels, *The Man Who Missed the War.* The Antarctic continent is also the residence of Atlantean survivors in Charles B. Stilson's *Polaris and the Goddess Glorian.*) The Atlanteans welcome the explorers and all goes well until one of Camilla's suitors, a worthless movie-actor, unintentionally kills Tisch in a brawl. The peaceful Atlanteans are so appalled by this crime that they expel the party. After more escapes from the sub-men the travellers reach the surface on the island of Pico in the Azores, but are never able to find the tunnel entrance again.

The book shows many features of popular English novel-writing of the *Sheik* era. Thus Camilla and her cousin Sally Hart have "ripe-corn colored hair," Sally's friend Captain McKay is a typical bluff sea-dog, and the gangtsers speak a grotesque attempt at American underworld slang. The book contains many long passages of Atlantist argument paraphrased from Donnelly or Spence, and equally long rhapsodies on the beauties of the deep paraphrased from Beebe. Still, while not outstandingly good, the novel does hold the attention during passages of fast action.

In another sizeable group of stories a modern hero visits ancient Atlantis or its like by some pseudo-scientific gimmick; thus for instance in Nelson Bond's *Exiles of Time* a whole group of modern characters is transported by time-machine to Mu just as that continent is about to be destroyed by a comet.

The best story of this group is probably J. Leslie Mitchell's *Three Go Back* (1932) in which the author, who also wrote the scholarly *Conquest of the Maya,* presented the diffusionist and pacifist angles of the subject. He made two assumptions, both widely held when he wrote, though later events have largely discredited them. One was that the future of transoceanic air transport belonged to the large rigid dirigible airship. The other was that all wars are caused by wicked munitions-makers, "merchants of death" who foment quarrels among the nations from base motives of gain.

In the story, a transatlantic airship runs into a sort of time-warp that puts it back in the Pleistocene Period, though nobody realizes the fact until the ship runs into a mountain of Atlantis at night. The only survivors of the crash are the English woman writer Clair Stranlay, the disagreeable but forceful American professional pacifist Sinclair, and the suave and elderly British munitions-maker Mullaghan. As Sinclair has just been expelled from Germany for advocating the immediate lynching of all munitions-manufacturers, the party of survivors is less congenial than it might be.

They wander about in their nightclothes, seeing a mammoth and other Pleistocene animals, until they are found by a party of Basque-speaking Crô-Magnon hunters. These improbable cavemen are noble savages on the Roussellian model: jolly children of nature without clothes, fears, inhibitions, superstitions, aggressiveness, possessions, government, war, or any other of the supposed vices of civilization. These deplorable things only came into being when the Egyptians invented them, along with civilization, and spread them over the earth.

To the horror of the other two men, Clair (a fairly uninhibited young lady herself) mates with the hunter Aerte because he reminds her of a former lover killed in World War I. Mullaghan, sickening from the unaccustomed diet, dies after repenting his wickedness. The submergence of part of the continent by an earthquake forces the tribe to migrate south, in the course of which journey they are attacked by Neanderthal men. The Neanderthalers are "awful" — dirty, ferocious, and cannibalistic, just like armaments manufacturers.

Now the conflict between Crô-Magnon and Neanderthal is a well-worn theme of science-fiction, and since the poor Neanderthalers are no longer here to defend themselves, modern writers tend to give them the short end of the stick. Actually, seeing that they disappeared while the Crô-Magnons survived, it looks as though if anybody was aggressive and cannibalistic it was our own ancestors.

At any rate, in fighting a rear-guard action to enable their tribe to escape through a mountain pass, Clair, Sinclair, and Aerte are overwhelmed by numbers and are just being killed when Clair and Sinclair find themselves unhurt back in the twentieth century on what turns out to be one of the Azores. At the end they are about to marry and to carry on Sinclair's crusade against the warmongers. Despite certain faults, both scientific and literary, the story is written with a slick professional touch and holds the attention throughout.

In addition to the socialist, pacifist, and other aspects of the lost-continent theme, Francis Ashton used Hörbiger's Cosmic Ice Theory as the basis for another pair of lost-continent novels belonging to this class. In *The Breaking of the Seals* (1946) the young Englishman Melville is persuaded by the psychologist Kurdt to undertake an experiment. After psychoanalysis and yogic exercises, Kurdt puts Melville into a trance wherein he relives the life of Maht, a young warrior of the city of Mahbahste on a continent somewhere in the Atlantic. The Mahbahsteans worship Bahste, the satellite of the earth preceding the present moon, which has come so close that it takes up much of the sky and goes around so fast that it eclipses the sun thrice a day. The plots, escapes, and battles of Maht's active life are interrupted when Bahste begins to break up.

First the ice-coat of the satellite falls as a terrific hailstorm, followed by a deluge of mud. Then the solid core of Bahste crumbles and the earth is subjected to a rain of huge meteors, eruptions, and earthquakes. Finally the waters of the equatorial tidal bulge flow poleward, completely submerging Maht's continent. Maht and his love, the Princess Runille, escape in an ark built by the patriarch Nodah.

The same author's *Alas, That Great City* deals with the next phase of Hörbiger's cosmogony, when the present moon was captured by the earth. Jonathan Grant agrees to sail an unknown man, Allanson, to a spot in the Atlantic on a certain date in his small sailing-yacht. He is disconcerted when his passenger turns out to be a girl, Joyce Allanson, who explains that her father can't come. For the first hundred pages the story deals amusingly with the poor girl's efforts to learn enough seamanship to please her skipper, for Grant, feeling he has been tricked, takes a dour and hypercritical view of the expedition. Then at the designated spot a huge wave dismasts and nearly sinks them. While waiting for his head to emerge from the water Grant, like Melville, relives the life of a remote forebear — Larentzal, a young-man-about-Atlantis.

Larentzal and his sweetheart Cleoli (who resembles Joyce) are involved in a plot against the wicked Queen Nethali, who rules while her husband King Ramanzal is away fighting. The plot fails when Larentzal, instead of assassinating Nethali as he is supposed to do, lets himself be seduced by her and made into a male concubine. Cleoli, arrested, is saved from being publicly beaten to death only by becoming a priestess of Ra-Zatthwal, the sun-god. When Ramenzal returns from his war, Larentzal is saved from the king's vengeance by an earthquake caused by the approach of our present moon. As the continent sinks, Larentzal and Cleoli (to whom every misfortune has happened, including the loss of a leg) escape in a boat, but are overwhelmed by an earthquake wave and presumably drowned. Then we are back in modern times with Grant, half-drowned himself, pumping the water out of Joyce (whom he has naturally come to love) and learning that he has proved her father's theories of "collective unconscious" and "universal mind."

While not really bad, these two stories are not altogether successful either, though it is hard to tell just why. They suffer from sentimentality and bad editing, their climaxes are drawn out to inordinate length, and the second novel is less interesting than the first.

Quite a number of lost-continent stories have brought a single survivor, or a small group of them (as in Wheatley's story) down to the present. Thus G. C. Foster's *The Lost Garden* (1930) is an amiable spoof about a small group of Atlanteans made quasi-immortal by On-Ra, the High Priest of Poseidon. He and they survive the submergence and live adventurous lives through all ages down to the present; On-Ra for instance is an Anglican bishop at the end of the story. The most amusing passage is that in which the ancient Atlanteans fall to arguing whether the legendary Lemuria ever existed, using the arguments now employed for and against Atlantis.

Again, Owen Rutter's *The Monster of Mu* (1932) sends a party of treasure-hunters to an island near Easter Island in the Pacific, where the white priests of Mu reign over the last remnant of their lost continent. The work on the island is done by little brown men. All these Muvians are immortal, but since women are absolutely forbidden on the island the priests decide that the heroine must be tied to the rocks to be sacrificed like Andromeda to the monster, a reptilian creature something like a blind plesiosaur. After much lively action the island sinks beneath the sea, the hero and heroine alone escaping. While straight adventure-fiction of no great stature, the story is, like the Foster novel, good for an evening's light entertainment.

Another novel of this group appeared in Germany in the 1920's: Otto Schulz's *Tlavatli*, wherein a German named Justus, studying occultism, is directed by his Mahatma in India to take his yacht to a designated place in the ocean. There from the submerged ruins of a city Justus dredges up and revives the Princess of Atlantis, with whom he promptly has a love-affair. Subsequently he fetches up the Chief Sorcerer of Atlantis, who originally put the princess into her long sleep when she refused to marry him. When the sorcerer awakens he invokes a demon in the form of a monstrous toad, but then the Mahatma appears and in the ensuing occult fracas demon, sorcerer, and princess all perish.

Of all the stories bringing a handful of Atlanteans or

the equivalent down to the present, however, the outstanding one is *Out of the Silence* (1927) by Erle Cox, an Australian writer. An Australian farmer, Alan Dundas, while
digging on his grape-farm, uncovers the top of a huge sphere
of some hard substance. After much experimentation he gets
a door open and finds inside a series of museum halls
exhibiting the science of a vanished civilization. In the last
chamber a tall beautiful woman lies sleeping, and there
are directions for awakening her. Dundas swears his friend
Dr. Barry to secrecy and brings him in to revive the
woman.

She is Earani, entombed here twenty-seven million years
before when her civilization was about to be wiped out by
some catastrophe. The scientists of that day, to preserve
their knowledge, made up three of these time-capsules, two
containing a man and one a woman. By her science Earani
learns that one of the men, Andax, is still alive in his
sphere in the Himalayas.

When the language difficulty has been overcome, Earani
reveals her plans for reviving Andax and reforming the
world according to their own drastic ideas. They will start
by exterminating over half the world's population — all the
"colored" peoples, whom she (and apparently the author)
consider biologically inferior to the Whites. Although Dundas becomes her compliant slave for love of her, Barry
breaks his oath and goes to Melbourne to warn the Prime
Minister of this menace. They have agreed that assassination is their only hope and the Prime Minister is looking
over his revolver when Earani appears, disarms him by
her mental powers, and warns both that any further inter-
ference will mean instant death.

It looks as though nothing can stop Earani when she
is unexpectedly stabbed in a jealous rage by Dundas's
former sweatheart. Dundas carries Earani's body back into
the sphere and pushes the lever that blows up the capsule
and himself along with it.

It is a well-written, gripping tale that holds the attention partly because it presents, not merely the conventional
conflict between good and evil, but the more interesting
struggle between two different and incompatible ideas of
good.

In addition there is a group of novels that deal with an imagined modern catastrophe of the sort that supposedly sank Atlantis. Thus the novels *Deluge* (1928) and its sequel *Dawn* (1929) by S. Fowler Wright (a skilled novelist of great power but another Roussellian primitivist) tell of a catastrophe that submerges much of the world's present land surface, including most of the British Isles. The hero, an English lawyer, takes command of the few British survivors (mostly coal-miners) and gradually brings order out of chaos.

Karl zu Eulenburg's *Die Brunnen der Grossen Tiefe* (1926) takes another tack: a passenger-liner is stranded on Atlantis when that lost land rises to the surface. Some of the people go exploring and find an Atlantean museum full of mounted dinosaurs and other extinct animals. When a lady playfully spins a wheel the beasts come to life —for, it seems, they were not stuffed, only paralyzed—and the people flee. After much excitement the ship's passengers are rescued and the continent sinks again.

The most impressive novels of this group, however, are probably two stories by Francis H. Sibson that, like Eulenburg's work, use the sinking-continent theme in reverse. In *The Survivors* (1932) two ships are stranded on an island about the size of Greenland which suddenly appears where the Sargasso Sea is now. The passenger liner *General Longstreet* is completely wrecked, the only survivors being the second engineer and an American heiress. The other ship, the British cruiser *Maple Leaf*, comes through with a hole in her bottom but most of her complement alive. The Navy people at once begin making plans to patch and refloat their ship, and the story deals thenceforth with the strenuous rescue of the two from the *Longstreet* by a party from the *Maple Leaf*.

The sequel, *The Stolen Continent* (1934) is more complicated.

These stories were written, you will remember, when American gangsters were at the peak of their notoriety. The author assumes that gangs have taken over all the United States except the armed services, which are waging a guerilla war of extermination against them. Facing defeat, the gangsters build a secret air-base in the interior of New

Canada (as the new land is called) from which they plan to bomb the Atlantic shipping-lanes to extort tribute like the Barbary pirates of old. To divert attention from themselves they have set up a front-organization, the United States League of Defense, which foments a quarrel between Great Britain and the United States over the ownership of New Canada. They also kidnap several people, some from the previous book, and take them to the secret base. The kidnapees, however, blow up the base and escape just in time to avert an Anglo-American war.

Both stories move right along. The author can describe upheavals of nature in terms to glue the reader to his chair, and in narrating men's struggle with natural forces he is superb. Moreover he seems to know life in the British Navy from first-hand experience.

Besides the stories based upon the writings of Plato, Donnelly, Churchward, and other "orthodox" Atlanto-Lemurists, a few mention neither Atlantis nor the submergence of continents, but nevertheless deal with ultra-ancient civilizations supposed to have flourished many millenia before those of which we have historical records. Cox's novel, previously described, might be classed with these, and the concept plays a part in H. Rider Haggard's *She* (Ayesha, Earani, and Antinéa being sisters under the prose) and in Fowler Wright's *Island of Captain Sparrow* and *The Vengeance of Gwa*. The same idea forms the basis of Pierrepont B. Noyes's *Pallid Giant* (1927; republished in 1946 as *Gentlemen: You Are Mad!*) This last novel comprises a warning against the dangers of armament-races. A group of modern men find records in France of a pre-Crô-Magnon civilization that came to an end when scientists discovered a death-ray and the warring powers of that time all but wiped out the human race with it.

Nor does this survey by any means exhaust the uses to which the lost-continent theme has been put in fiction. Thus in Victor Rousseau's *Eye of Balamok* a miner prospecting for gold in Australia finds the descendants of the Atlanteans living on the inside of the supposedly hollow earth — the locale of a whole family of science-fiction stories, such as Burroughs's "Pellucidar" tales. Abraham Merritt's

Creep, Shadow deals with the legend of Ys and its pro-
fligate princess Dahut, while A. E. Van Vogt's *Book of
Ptath* is laid in the remote future on the continent of
"Gonwonlane" — Suess's Gondwanaland resurrected. . . .

In fact all that is needed to make the assortment
complete is a lost-continent story based on the premise that
no lost continents ever existed.

You will note that a large proportion of the foregoing novels are of British origin. Now, while the British
writers in question show, I think, a higher average level of
technical skill in writing than their American colleagues,
the American reader may be rather taken aback by certain
features that seem to run through British lost-continent
stories with surprising consistency.

For one thing, these writers make much use of national stereotypes: thus Germans are almost always pro-
fessors, while Americans are millionaires, heiresses, or
gangsters. American characters speak a weird mixture of
the slang and colloquialisms of all parts of the United
States, all periods of its history, and all classes of its people.
British novelists seem convinced that all Americans begin
every third sentence with "I reckon." There are touches
of racism and anti-Semitism. And some of the authors show
a pronounced aristocratic bias: They sneer loudly at the
"vulgarity" of the "American plutocracy," and their work-
ing-class characters are nearly always either knaves or fools,
utterly incapable of managing their affairs in a civilized
manner until an upper-class hero steps in to take charge.

In addition to all the foregoing out-and-out novels,
which do not pretend to be anything but fiction, there is
a group of books about Atlantis that would be classed as
fiction by most criteria, but whose authors aver them to
be factual accounts obtained by occult means. Among these
are the books of Leslie and Phelon, mentioned in the third
chapter, and, more recently, Clara von Ravn's *Selestor's
Men of Atlantis* and Daphne Vigers's *Atlantis Rising*.
Whereas Mrs. von Ravn's book purports to have been
dictated by an Egyptian "spirit control" with the unlikely
name of "Selestor," Miss Vigers claims to have visited Atlantis personally in her astral body. In neither case, however, did these occult operations produce anything worth

reading.

Thus the Atlantis theme has been tied to communism, socialism, anti-socialism, Spiritualism, Theosophy, racism, pacifism, romanticism, diffusionism, Roussellian primitivism, and Hörbigerism. When these tales are taken as a whole, I think the best entertainment is still provided by those which, like Hyne's *Lost Continent,* have no ideological axes to grind but simply tell a lively story in competent, professional style.

Since the rise of the science-fiction pulp-magazines in the last quarter-century, lost-continent stories published in these magazines have multiplied beyond count. (In fact I have written a few myself.) They come good, bad, and indifferent, like other stories, and a complete study of them would take a book in itself. Most of them are of a fairly elementary action-and-romance type, usually with a more or less ingenious pseudo-scientific assumption to carry the plot. The strangest (and perhaps the worst) of these Atlantean stories to appear so far are the tales comprising the Shaver Hoax, which ruffled the pulp-publishing business for three years.

The Shaver Hoax began with a story, *I Remember Lemuria,* by Richard S. Shaver in the March, 1945, issue of *Amazing Stories.* This was the first of a long series of Shaver stories that appeared in the magazine until March, 1948. *Amazing,* the bulkiest of the science-fiction pulps, has a large Midwestern circulation and at that time aimed at a lower mental level than did most of its competitors.

The plot of these stories was as follows: The earth was once, in the days of Atlantis and Lemuria, inhabited by divine Titans and Atlans who left this planet for other parts of the universe when the sun began sending out harmful rays. Of those that stayed behind, some became the human race and others the *deros,* a species of fiends who live in a network of caves inside the earth and cause all human crimes and ills by telepathically planting evil ideas in the minds of men. The stories are crude to the point of childishness (God and Satan dash about the universe in space-ships) and so badly written as to be almost incoherent in places.

Nevertheless the Shaver stories had a somewhat spectacular history. The author, like Plato himself, used the familiar "frame" of pretending that the stories were literally true. He had, he averred, obtained them through his "racial memory." The then editor of the magazine, Raymond A. Palmer, coöperated by pretending that he, too was convinced of the tales' truth — unless as some think he really *did* believe them. Visitors to office of *Amazing* were amazed to find Palmer cowering behind his typewriter and listening for deros. In this way he built up a large lunatic-fringe circulation, while the science-fiction fan-clubs and conventions denounced him for bringing discredit upon their favorite *genre* of literature.

In conclusion, then, Plato wrote a fascinating fiction that has had a large and continuing influence on Western literature and thought, but that has nothing to do with geology, anthropology, or history, about which Plato knew very little. While there may have been lost continents, they could have had nothing to do with Plato's tale because of the time-factor. Geological changes take place over millions of years, and do not happen as Plato described. Moreover the oral traditions of primitive men could not preserve the memory of such changes long enough to matter.

There have also been forgotten civilizations. Those of Crete and Tartessos, especially the latter, may have influenced Plato, but more ancient or more distant cultures like those of the Sumerians and the Mayas could not have because Plato had no way of finding out about them. He got the ideas for his Atlantis story from the knowledge and beliefs of his own time—from the legends of Scheria, Atlas, Poseidon, the Garden of the Hesperides, and the Golden Age; from rumors of wild Atlantes in Africa and fertile islands and impassible shoals in the Atlantic; from the real cities of Babylon, Syracuse, and Carthage; and from true accounts of earthquake damage in the Mediterranean, especially the flooding of Atalantë. Farther afield for his source material he neither could have gone nor needed to go.

Moreover, his purpose in writing the story was neither historical nor scientific, but moral and philosophical. He

meant to show how his ideal "Republic" would work in
practice, and probably left the task unfinished because of
plot-trouble. He had the materials for such a story and
the talent and imagination to write it. He wrote a number
of such allegories, and it would probably have never occurred to him that later generations would take Atlantis
as literally true to the point of erecting a cult about it.
Even if it had, he was not too scrupulous in such matters;
to him the idea was all, the material fact unimportant.

I hope I have shown that the arguments of most members of the Atlantist cult are not to be taken very seriously.
For the most part they repeat the long-since-refuted assertions of Donnelly, Le Plongeon, and other earlier Atlantists,
and the wilder speculations of eighteenth and nineteenth-
century historians, anthropologists, and geologists, ignoring
the immense progress that has been made in these fields of
science during the past half-century. From a few resemblances between places, cultures, and languages in various
parts of the world they instantly deduce connections, as do
the diffusionists, the Ten Tribists, and those who locate
Homer's Scheria all over the map. Obviously no system of
reasoning that gives nine-and-sixty different results when
applied to one set of facts can be worth much. To discover
true connections among scattered human culture-traits you
must do more than consider resemblances: you must take
into account all the other possibilities, and also consider differences as well as similarities.

Several sources supply the Atlantis theme with its
singular vitality. The search for lost continents enables
the bumptious amateur in the sciences to play at being a
historian, an archeologist, or a paleogeographer. The bor-
derland between the known and the unknown is the most
fascinating field of human knowledge to study, and statements about times and places for which no real history
exists are hard to disprove. Atlantis provides mystery and
romance for those who don't find ordinary history exciting
enough, and can be readily turned to account to point a
moral lesson — in fact, any of many different and contradictory moral lessons.

But most of all it strikes a responsive chord by its
sense of the melancholy loss of a beautiful thing, a happy

perfection once possessed by mankind. Thus it appeals to
that hope that most of us carry around in our unconscious,
a hope so often raised and as often disappointed, for assurance that somewhere, some time, there can exist a land
of peace and plenty, of beauty and justice, where we, poor
creatures that we are, could be happy. In this sense Atlantis
—whether we call it Panchaia, the Kingdom of God,
Oceana, the Classless Society, or Utopia — will always be
with us.

APPENDIX A

Classical texts having to do with theories of Atlantis.

(See also the text, p. 145, for a quotation from Seneca, *Medea;* p. 177, for a quotation from Homer, *Odyssey;* and p. 194 for a quotation from Avienus, *Ora Maritima.* The following translations are all more or less revised and abridged.)

Homer, *The Odyssey.*

I, ll. 44-54: And the goddess, bright-eyed Athena, answered: "Our father, son of Kronos, most exalted of rulers . . . my heart is torn for wise Odysseus, wretched man, who so long parted from his friends suffers on a seagirt isle at the very navel of the sea. On this forested isle dwells a goddess, daughter of crafty Atlas who knows the depth of every sea and keeps the tall pillars that sunder earth and heaven."

XI, ll. 13-19: "And then [Odysseus's ship] reached the boundary of the world, deep-flowing Okeanos. There lie the land and city of the Kimmerioi, veiled in fog and cloud. Never does the sun shine upon them with his rays, either in his climb into the starry heavens or in his return to earth, but deadly night covers these miserable men."

Hanno, *Periplous* (Cory translation).

In the neighborhood of the mountains lived the Troglodytes, men of various appearances, whom the Lixitae described as swifter in running than horses. Having procured interpreters from them, we coasted along a desert country, toward the south, for two days. Thence we proceeded towards the east in the course of a day. Here we found a recess of a certain bay a small island, five stadia around, where we settled a colony, and called it Kernë. We judged from our voyage that this place lay in a direct line with Carthage, for the length of our voyage from Carthage to the Pillars equalled that from the Pillars to Kernë.

Pindar, *Odes.* (Sandys translation.)

Nemean Ode III, ll. 19-26: . . . no further is it easy for him to sail across the trackless sea beyond the Pillars of Herakles, which that hero and god set up as far-famed witnesses of the furthest limit of voyaging. He quelled the monstrous beasts amid the seas, and tracked to the very end the streams of the shallows, there where he reached the bourne that sped him home again; and he made known the limits of the land.

Herodotos, *The Histories.* (Rawlinson translation.)

I, 163: Now the Phokaians were the first of the Greeks to perform long voyages, and it was they who made the Greeks acquainted with the Adriatic and Tyrrhenia, with Iberia, and the city of Tartessos. The vessel they used in their voyages was not the round-built merchant-ship, but the long fifty-oared galley. On their arrival at Tartessos, the king of the country, named Arganthonios, took a liking to them. This monarch reigned over the Tartessians for eighty years, and lived to be 120 years old. He regarded the Phokaians with so much favor as, at first, to beg them to quit Ionia and settle in whatever part of his country they liked. . . .

IV, 152: . . . A Samian vessel, under the command of a man named Kolaios . . . quitted the island; and, anxious to reach Egypt, made sail in that direction, but were carried out of their course by a gale from the east. The storm not abating, they were driven past the Pillars of Herakles, and at last, by some special guiding providences, reached Tartessos. This trading-town was in those days a virgin port, unfrequented by merchants. Hence the Samians made by the return-voyage a profit greater than any Greeks before their day, excepting Sostratos, son of Laodamas, with whom no one else can compare. From the tenth part of their gains, amounting to six talents, the Samians made a brazen vessel, in shape like an Argive wine-bowl, adorned with the heads of griffins standing out in high relief.

IV, 184: At the distance of ten days' journey from the Garamantians there is again another salt-hill and spring of water; around which dwell a people, called the Atarantes, who alone of all known nations are destitute of names. The title of Atarantes is borne by the whole race in common, but the men have no particular names of their own. The Atarantes, when the sun rises high in the heavens, curse him, and load him with reproaches, because (they say) he burns and wastes both their country and themselves. Once more at the distance of ten days' journey there is a salt-hill, a spring, and an inhabited tract. Near the salt is a mountain called Atlas, very taper and round; so lofty, moreover, that the top (it is said) cannot be seen, the clouds never quitting it either summer or winter. The natives call this mountain the Pillar of Heaven, and they themselves take their name from it, being called Atlantes. They are reported not to eat any living thing, and never to have any dreams.

IV, 188: The inhabitants of the parts about Lake Tritonis worship in addition Triton, Poseidon, and Athena, the last especially.

Thucydides, *The Peloponnesian War* (Jowett translation).

III, 89: In the ensuing summer the Peloponnesians and their allies . . . intended to invade Attica, but were deterred from proceeding by numerous earthquakes, and no invasion took place in this year. About the time when these earthquakes prevailed, the sea at Orobiai in Euboia, retiring from what was then the line of the coast and rising in a great wave, overflowed a part of the city; and although it subsided in some places, yet in others the inundation was permanent, and that which was formerly land is now sea. All the people who could not escape to high ground perished. A similar inundation occurred in the neighborhood of Atalantë, an island on the coast of the Opuntian Locri, which carried away a part of an Athenian fort, and dashed in pieces one of two ships which were drawn up on the beach. At Perarethos also the sea retired, but no inundation followed; an earthquake, however, over-threw a part of the wall, the Prytaneus, and a few houses.

Plato, *Timaios* (Jowett translation).

20D: Then listen, Socrates, to a tale which though strange is certainly true, as Solon, the wisest of the Seven, declared. He was a relative of my great-grandfather, Dropides, as he himself says in several of his poems; and Dropides told Kritias, my grandfather, who remembered and told us: That there were of old great and marvelous actions of this city, which have passed into oblivion through time and the destruction of the human race, and one in particular, which was the greatest of them all, the recital of which will be a suitable testimony of our gratitude to you, and also a hymn of praise true and worthy of the goddess, which may be sung by us at the festival in her honor . . . Kritias said: "At the head of the Egyptian Delta, where the river Nile divides, there is a certain district which is called the Saïtic, and the great city of the district is also called Saïs, and is the city from which King Amasis was sprung. And the citizens have a deity who is their foundress; she is called Neïth, and in Greek, they say, is Athena. Now the citizens of this city are great lovers of the Athenians, and say that they are in some way related to them.

Thither came Solon, who was received by them with great honor; and he asked the priests, and made the discovery that neither he nor any other Hellene knew anything worth mentioning about the times of old. On one occasion, when he was drawing them on to speak of antiquity, he began to tell about the most ancient things in our part of the world — about Phoroneus, who was called 'the first,' and about Niobe; and after the Deluge, to tell of the lives of Deukalion and Pyrrha; and he traced the genealogy of their descendants, and tried to reckon how many years old were the events of which he was speaking, and to give the dates. Thereupon one of the priests, of a very great age, said: 'O Solon, Solon, you Hellenes are but children, and there is never an old Hellene.' Solon, hearing this, said: 'What do you mean?'

" 'You are all young in mind,' he replied; 'there is no old opinion handed down among you by ancient tradition; nor any science hoary with age. The reason of this is as follows: There have been, and will be again, many destructions of mankind arising out of many causes; the greatest have been brought about by the agencies of fire and water, and lesser ones by countless other causes. . . .'

23B: " 'As for those genealogies which you have recounted to us, Solon, they are no better than the tales of children; for in the first place you remember but one deluge, whereas there were many of them; and in the next place, you do not know that there dwelt in your land the fairest and noblest race of men that ever lived, of whom you and your whole city are but a seed or remnant. And this was unknown to you because for many generations the survivors of that destruction died and made no sign. For there was a time, Solon, before the great deluge of all, when the city that is now Athens, was first in war and was preëminent for the excellence of her laws, and is said to have performed the noblest deeds and to have had the fairest constitution of any of which tradition tells, under the face of heaven.' Solon marvelled at this, and earnestly requested the priest to inform him exactly and in order about these former citizens.

" 'You are welcome to hear about them, Solon,' said the priest, 'both for your own sake and for that of the city, and above all for the sake of the goddess who is the common patron and protector and educator of both our cities. She founded your city a thousand years before ours, receiving from Ge and Hephaistos the seed of your race, and then she founded ours, the constitution of which is set down in our sacred registers as 8000 years old. As touching the citizens of 9000 years ago, I will briefly inform you of their laws and of the noblest of their actions. . . .'

24D: " 'Many great and wonderful deeds are recorded of your state in our histories. But one of them exceeds all the rest in greatness and valor. For these historians tell of a mighty power which was aggressing wantonly against the whole of Europe and Asia, and to which your city put an end. This power came forth out of the Atlantic Sea, for in those days the sea was navigable; and there was an island situated in front of the straits which you call the Pillars of Herakles; the island was larger than Libya and Asia together, and was the way to other islands, and from the islands you might pass to the whole of the opposite continent which surrounds the true ocean; for this within the Straits is only a harbor, having a narrow entrance, but that other is the real sea, and the surrounding land may be most truly called a continent.

" 'Now in this island of Atlantis there was a great

and wonderful empire which had rule over the whole island and several others, as well as over parts of the continent, and besides these they subjected parts of Libya within the Straits as far as Egypt, and of Europe as far as Tyrrhenia. This vast power, thus gathered into one, tried to subdue at one blow our country and yours and the whole of the land which was within the Straits; and then, Solon, your country shone forth, in the excellence of her virtue and strength among all mankind, for she was the first in courage and military skill and was the leader of the Hellenes. And when the rest fell off from her, being compelled to stand alone after having undergone the very extremity of danger, she defeated and triumphed over the invaders, and preserved from slavery those who were not yet subjected, and liberated all the others who dwell within the limits of Herakles. But afterwards there occurred violent earthquakes and floods, and in a single day and night of rain all your warlike men in a body sank into the earth, and the island of Atlantis in like manner disappeared, and was sunk beneath the sea. Wherefore the sea in those parts is impassible and impenetrable, because there is so much shallow mud in the way, caused by the subsidence of the island.' "

25E: I have told you shortly, Socrates, the tradition which the aged Kritias heard from Solon. And when you were speaking yesterday about your city and citizens, this very tale I am telling you came into my mind, and I could not help remarking how, by some strange coincidence, you agreed in almost every particular with the account of Solon. . . .

26A: Therefore, as Hermokrates has told you, on my way home yesterday I imparted my recollection to my friends in order to refresh my memory, and during the night I thought about the words and have nearly recovered them all. . . .

26C: And now, Socrates, I am ready to tell you the whole tale of which this is the introduction. I will give you not only the general heads, but the details exactly as I heard them. And as to the city and citizens, which you yesterday described to us in fiction, let us transfer them to the world of reality; this shall be our city, and we will suppose that the citizens whom you imagined were our venerable ancestors, of whom the priest told. . . .

Plato, *Kritias* (Jowett translation).

108E: Let me begin by observing first of all that 9000 was the sum of years which had elapsed since the war that was said to have occurred between all those who dwelt outside the Pillars of Herakles and those who dwelt within them; this war I am now about to describe. Of the combatants on the one side, the city of Athens was reported to have been the ruler and to have directed the contest; the combatants on the other side were led by the kings of the Island of Atlantis, which as I was saying once had an extent greater than that of Libya and Asia, and when afterwards sunk by an earthquake became an impassible barrier of mud to voyagers sailing from home hence to the ocean. The progress of the history will unfold the various tribes of barbarians and Hellenes which then existed, as they successively appear on the scene; but I must begin by describing first of all the Athenians as they were in that day, and their enemies who fought with them; and I shall have to tell of the power and form of government of both of them. Let us give the precedence to Athens:

In former ages, the gods had the whole earth distributed among them by allotment. . . .

109C: Hephaistos and Athena, siblings sprung from the same father, having a common nature and being also united in the love of philosophy and of art, both obtained as their allotted region this land, which was naturally adapted for wisdom and virtue; and there they implanted the brave children of the soil, and put into their minds the order of government; their names are preserved, but their actions have disappeared by reason the destruction of those who had the tradition, and the lapse of ages. . . .

110C: Now the country was inhabited in those days by various classes of citizens; artisans, tillers of the soil, and a warrior class originally set apart by divine men; these dwelt by themselves, and had all things suitable for nurture and education; neither had any of them anything of their own, but they regarded all things as common property; nor did they require to receive of the other citizens anything more than their necessary food. . . . The land was the best in the world, and for this reason was able in those days to support a vast army, raised from the surrounding people. . . .

111A: Now this country is only a promontory extending far into the sea away from the rest of the continent, and the surrounding basin of the sea is everywhere deep in the neighborhood of the shore. Many great deluges have taken place during the 9000 years, for that is the number of years that have elapsed since the time of which I am speaking; and in all the ages and changes of things, there has never been any settlement of the earth flowing down from the mountains as in other places, which is worth speaking of; it has always been carried round in a circle and disappeared in the depths below.

Hence, in comparison of what then was, there are remaining in small islets only the bones of the wasted body, as they may be called; all the richer and softer parts of the soil having fallen away, and the mere skeleton of the country being left. But in former days, and in the primitive state of the country, what are now mountains were only regarded as hills; and the plains, as they are now termed, of Phelleus were full of rich earth, and there was an abundance of wood in the mountains. . . .

111E: Such was the natural state of the country, which was cultivated, as we may well believe, by true farmers, and had a soil the best in the world, and abundance of water, and in the heaven above an excellently tempered climate. Now the city in those days was arranged thus: in the first place the Akropolis was not as now. For the fact is that a single night of excessive rain washed away the earth and laid bare the rock; at the same time there were earthquakes, and then occurred the third extraordinary inundation, which immediately preceded the great deluge of Deukalion. . . . Outside the Akropolis and on the sides of the hill there dwelt artisans, and such of the farmers as were tilling the ground near; at the summit the warrior class dwelt by themselves around the temples of Athena and Hephaistos, living as in the garden of one house, and surrounded by one enclosure.

On the north side they had common houses, and had

prepared for themselves winter places for common meals, and had all the buildings which they needed for the public use, and also temples, but unadorned with gold and silver, for these were not in use among them; they took a middle course between meanness and extravagance, and built moderate houses in which they and their children's children grew old, and handed them down to others who were like themselves, always the same. . . . And they took care to preserve the same number of men and women for military service, which was to continue through all time, and still is, namely, about 20,000. . . .

113A: Yet, before proceeding further in the narrative, I ought to warn you, that you must not be surprised if you should hear Hellenic names given to foreigners. I will tell you the reason of this: Solon, who was intending to use the tale for his poem, made an investigation into the meaning of the names, and found that the early Egyptians in writing them down had translated them into their own language, and he recovered the meaning of the several names and retranslated them, and copied them out again in our language. My grandfather had the original in writing, which is still in my possession, and was carefully studied by me when I was a child. . . .

113C: And Poseidon, receiving for his lot the island of Atlantis, begat children by a mortal woman, and settled them in a part of the island which I will describe. On the side toward the sea and in the center of the whole island there was a plain said to have been the fairest of all plains and very fertile. Near the plain again, and also in the center of the island at a distance of about fifty stadia,* there was a mountain not very high on any side. There dwelt one of the earth-born natives of that country, named Euenor, and he had a wife named Leukippë, and they had an only daughter called Kleito. The maiden was growing up to womanhood, when her father and mother died; Poseidon fell in love with her and had intercourse with her, and breaking the ground inclosed the hill in which she dwelt all round, making alternate zones of sea and land larger and smaller encircling one another, two of land and three

---

* One stadium (Greek, *stadion*) = 600 Greek feet (607 to 631 U. S. feet).

of water, which he turned as with a lathe out of the center of the island, equidistant in every way so that no man could get to the island, for ships and voyages were not yet heard of. He, being a god, easily effected special arrangements for the center island, bringing two streams of water under the earth, which he caused to ascend as springs, one of warm water and the other of cold and making every variety of food to spring up abundantly in the earth.

He also begat and reared five pairs of twin sons, dividing the island of Atlantis into ten portions; he gave to the first-born of the eldest pair his mother's dwelling and the surrounding allotment, which was the largest and best, and made him king over the rest; the others he made princes and gave them rule over many men and a large territory. And he named them all; the eldest, who was the king, he named Atlas, and from him the whole island and the ocean received the name of Atlantic. To his younger twin brother, who obtained as his lot the extremity of the island towards the Pillars of Herakles, as far as the country which is still called the region of Gadeira in that part of the world, he gave the name which in the Hellenic language is Eumelos, in the language of the country which is named after him, Gadeiros. Of the second pair of twins he called one Ampheres and the other Euaimon. Of the third pair of twins he gave the name Mneseus to the elder and Autochthon to the one who had followed him. Of the fourth pair of twins he called the elder Elasippos and the younger Mestor. And of the fifth pair he gave the elder the name of Azaës and to the younger that of Diaprepes. All these and their descendants were the inhabitants and rulers of divers islands in the open sea; and also, as has been already said, they held sway in the other direction over the country within the Pillars as far as Egypt and Tyrrhenia.

Now Atlas had a numerous and honorable family, and his eldest branch always retained the kingdom, which the eldest son handed on to his eldest for many generations; and they had such an amount of wealth as was never before possessed by kings and potentates and is not likely ever to be again, and they were furnished with everything that they could have both in the city and country. For because of the greatness of their empire many things were brought to them from foreign countries, and the island itself provided much of what was required by them for the uses of life. In the first place they dug out of the earth whatever was to be found there, mineral as well as metal, and that which is only a name and was then something more than a name, orichalc, was dug out of the earth in many parts of the island, and except gold was then the most precious of metals. There was an abundance of wood for carpenter's work and sufficient maintenance for tame and wild animals.

Moreover there were a great number of elephants in the island, and there was provision for animals of every kind, both for those that live in lakes and marshes and rivers and also for those that live in mountains and on plains, and therefore for the animal that is the largest and most voracious of them. Also whatever fragrant things there are in the earth, whether roots, or herbage, or woods, or distilling drops of flowers and fruits, grew and thrived in that land; and again, the cultivated fruit of the earth, both the dry edible fruit and other species of food that we call by the general names of vegetables. . . .

115C: And they arranged the whole country thus: First they bridged the zones of sea which surrounded the ancient metropolis, and made a passage into and out of the royal palace; and then they began to build the palace in the habitation of the god and their ancestors. This they continued to ornament in successive generations, every king surpassing the one who came before him to the utmost of his power, until they made the building a marvel to behold for size and beauty. And beginning from the sea they dug a canal of three plethra* in width and 100 feet in depth, and fifty stadia in length, which they carried through to the outermost zone, making a passage from the sea up to this, which became a harbor, and leaving an opening sufficient to enable the largest vessels to sail through. Moreover they divided the zones of land which parted the zones of sea, constructing bridges of such a width as would leave a passage for single trireme to pass out of one into the other, and roofed them over; and there was a way underneath for the ships, for the banks of the zones were raised considerably above water. Now the largest of the zones into

---

* One plethrum (Greek, *plethron*) = 100 Greek feet.

which a passage was cut from the sea was three stadia in breadth, but the next two, as well the zone of water as of land, were two stadia, and the one which surrounded the central island was a stadium only in width.

The island in which the palace was situated had a diameter of five stadia. This and the zones and the bridge, which was a plethrum in width, they surrounded by a stone wall, on either side placing towers, and gates on the bridges where the sea passed in. The stone which was used in the work they quarried from underneath the central island and from underneath the outer and inner circles. One kind of stone was white, another black, and a third red, and as they quarried they at the same time hollowed out docks double within, having roofs formed of the native rock. Some of their buildings were simple, but in others they put together different stones which they intermingled for the sake of ornament, to be a natural source of delight. The entire circuit of the wall, which went around the outermost one, they covered with a coating of brass, and the third, which encompassed the citadel, flashed with the red light of orichalc.

The palaces in the interior of the citadel were constructed in this wise: In the center was a holy temple dedicated to Kleito and Poseidon, which remained inaccessible, and was surrounded by an enclosure of gold; this was the spot in which they originally begat the race of the ten princes, and thither they annually brought the fruits of the earth in their season from all the ten portions, and performed sacrifies to each of them. Here too was Poseidon's own temple of a stadium in length, and half a stadium in width, and of a proportionate height, having a sort of barbaric splendor. All the outside of the temple with the exception of the pinnacles they covered with silver, and the pinnacles with gold. In the interior of the temple the roof was of ivory adorned everywhere with gold and silver and orichalc; all the other parts of the walls and pillars and floor they lined with orichalc. In the temple they placed statues of gold. . . .

117A: In the next place they used fountains both of cold and hot springs; these were very abundant, and both kinds wonderfully adapted to use by reason of the sweetness and excellence of their waters. They constructed buildings about them and planted suitable trees; there were the king's baths, and separated baths for private persons, also separate baths for women and others again for horses and cattle, and to each of them they gave as much adornment as was suitable for them. The water which ran off they carried, some to the grove of Poseidon, where were growing all manner of trees of wonderful height and beauty, owing to the excellence of the soil; the remainder was conveyed by aqueducts which passed over the bridged to the outer circles; and there were many temples built and dedicated to many gods; also gardens and places of exercise, some for men and some set apart for horses in both of the two islands formed by the zones, and in the center of the larger of the two there was a race-course of a stadium in width and in length allowed to extend all around the island for horses to race in. Also there were guard-houses at intervals for the bodyguard, the more trusted of whom had their duties appointed to them in the lesser zone, which was nearer the Akropolis, while the most trusted of all had houses given them within the citadel and about the persons of the kings. The docks were full of triremes and naval stories, and all things were quite ready for use. . . . Crossing the outer harbors, which were three in number, you would come to a wall which began at the sea and went all round: this was everywhere distant fifty stadia from the largest zone and harbor and enclosed the whole, meeting at the mouth of the channel towards the sea. . . .

118A: The whole country was described as being very lofty and precipitous on the side of the sea, but the country immediately about the surrounding city was a level plain, itself surrounded by mountains which descended towards the sea; it was smooth and even, but of an oblong shape, extending in one direction 3000 stadia, and going up the country from the sea through the center of the island, 2000 stadia; this whole region of the island faces towards the south, and is sheltered from the north. The surrounding mountains are celebrated for their number and size and beauty. . . .

118C: I will now describe the plain, which had been

cultivated during many ages by many generations of kings. It was rectangular and for the most part straight and oblong, and what it wanted of the straight line followed the line of a surrounding ditch. The depth, width, and length of this ditch were incredible. . . . It was excavated to the depth of 100 feet, and its breadth was a stadium everywhere; it was carried round the whole of the plain, and was 10,000 stadia in length. It received the streams which came down from the mountains and, winding round the plain and touching with city at various points, was there led off into the sea. From above likewise, straight canals 100 feet wide were cut in the plain, and again led off into the ditch towards the sea: these canals were at intervals of 100 stadia, and by them they brought down the wood from the mountains to the city and conveyed the fruits of the earth in ships, cutting transverse passages from one canal into another, and to the city. . . .

118E: As to the population, each of the lots in the plain had an appointed chief of men who were fit for military service, and the size of the lot was to be a square ten stadia each way, and the total number of all the lots was 60,000. And of the inhabitants of the mountains and of the rest of the country there was a vast multitude having leaders to whom they were assigned according to their dwellings and villages. The leader was required to furnish for the war the sixth portion of a war-chariot, so as to make up a total of 10,000 chariots; also two horses and riders upon them, and a light chariot without a seat accompanied by a fighting man on foot carrying a small shield and having a charioteer mounted to guide the horses; also he was bound to furnish two heavy-armed, two archers, two slingers, three stone-shooters, and three javelin-men who were skirmishers, and four sailors to make up the complement of 1200 ships. . . .

119C: Each of the ten kings had his own division and in his own city had the absolute control of the citizens and in many cases of the laws, punishing and slaying whomsoever he would. Now the relations of their governments to one another was well regulated by the injunctions of Poseidon as the law had handed them down. These were inscribed by the first men on a column of orichalc, which was situated in the middle of the island at the temple of Poseidon, whither the people were gathered together every fifth and sixth years alternately, thus giving equal honor to the odd and to the even number. And when they were gathered together they consulted about public affairs, and inquired if any one had transgressed in anything and passed judgment on him accordingly, and before they passed judgment they gave their pledges to one another thus: There were bulls who had the range of the temple of Poseidon, and the ten who were left alone in the temple, after they had offered prayers to the gods that they might take the sacrifices that were acceptable to them, hunted the bulls without weapons but with staves and nooses; and the bull they caught they led up to the column; the victim was then struck over the head by them and slain over the sacred inscription. . . .

120A: When therefore after offering sacrifice according to their customs they had burned the limbs of the bull, they mingled a cup and cast in a clot of blood for each of them; the rest of the victim they took to the fire after having made a purification of the column all round. Then they drew from the cup in golden vessels and, pouring a libation on the fire, they swore that they would judge according to the laws on the column and would punish any one who had transgressed, and that for the future they would not if they could help transgress any of the inscriptions, and would not command or obey any ruler who commanded them to act otherwise than according to their father's laws. This was the prayer which each of them offered up for himself and for his family, at the same time drinking and dedicating the vessel in the temple of the god, and after spending some necessary time at supper, when darkness came on and the fire about the sacrifice was cool, all of them put on most beautiful dark-blue robes and, sitting on the ground at night near the embers of the sacrifices on which they had sworn and extinguishing all the fire about the temple, they received and gave judgment, if any of them had any accusation to bring against any one; and when they had given judgment, at daybreak they wrote down their sentences on a gold tablet, and deposited them as memorials with their robes. . . .

120D: Such was the vast power which the god settled in that place, and this he later directed against our land on the following pretext, as traditions tell: For many generations, as long as the divine nature lasted in them, they were obedient to the laws and well-affected towards their divine kindred; for they possessed true and in every way great spirits, practising gentleness and wisdom in the various chances of life and in their intercourse with one another. They despised everything but virtue, not caring for their present state of life and thinking lightly of the possession of gold and other property, which seemed only a burden to them; neither were they intoxicated by luxury, nor did wealth deprive them of their self-control, but they were sober and saw clearly that all these goods are increased by virtuous friendship with one another, and that by excessive zeal for them and honor of them the good of them is lost. . . . But when this divine portion began to fade away in them and became diluted too often with too much of the mortal admixture, and the human nature got the upper hand, then they, being unable to bear their fortune, became unseemly, and to him who had an eye to see they began to appear base, and had lost the fairest of their precious gifts. . . . Zeus, the god of gods, who rules by law and is able to see into such things, seeing that an honorable race was in a most wretched state and wanting to punish them that they might be chastened and improve, collected all the gods into his most holy abode, which being placed in the center of the universe sees all things that partake of generation. And when he had called them together he spake as follows:

Aristotle, *Meteorologica* (Ross translation).
II, i, 354a: Outside the Pillars of Herakles the sea is shallow owing to the mud, but calm, for it lies in a hollow.

"Aristotle," *De Mundo* (Ross translation. Probably not by Aristotle but by a member of his school).
iii, 392b: The earth is diversified by countless kinds of verdure and lofty mountains and densely wooded copses and cities, which that intelligent animal man has founded, and islands sit in the sea and continents, ignoring the fact

that the whole of it forms a single island round which the sea that is called Atlantic flows. But it is probable that there are many other continents separated from ours by the sea that we must cross to reach them, some larger and others smaller, but all, save our own, invisible to us. For as our islands are in relation to our seas, so is the inhabited world in relation to the Atlantic, and so are many other continents in relation to the whole sea; for they are as it were islands surrounded by the sea.

"Aristotle," *On Marvelous Things Heard* (Hett translation).

84: In the sea outside the Pillars of Herakles they say that a desert island was found by the Carthaginians having wood of all kinds and navigable rivers, remarkable for all other kinds of fruits, and a few days' voyage away; as the Carthaginians frequented it often owing to its property, and some even lived there, the chief of the Carthaginians announced that they would punish with death any who proposed to sail there, and they massacred all the inhabitants, that they might not tell the story, and that a crowd might not resort to the island and get possession of it and take away the prosperity of the Carthaginians.

135: It is said that the first Phoenicians who sailed to Tartessos took away so much silver as cargo, carrying there olive-oil and other pretty wares, that no one could keep or receive the silver, but that on sailing away from the district they had to make all their other vessels of silver, and even all their anchors.

136: They say that Phoenicians who live in what is called Gades, on sailing outside the Pillars of Herakles with an east wind for four days, came to some desert islands full of brushes and seaweed, which were not submerged when the tide ebbed but were covered when the tide was full, upon which were found a quantity of tuna of incredible size and weight when brought to shore; pickling these and putting them into jars they brought them to Carthage. These alone the Carthaginians do not export, but owing to their value as food they consume them themselves.

"Skylax", *Periplous* (Warmington translation).

112: After the Pillars of Herakles, as you sail into the

outer spaces, having Libya on your left, there is a great gulf extending as far as Cape Hermaion. . . . From Point Hermaion onwards stretch great reefs, in fact from Libya towards Europe, not rising above the surface of the sea, though in some places breakers flood in on them. This barrier of rocks stretches to another headland opposite in Europe, the name of this headland being the Sacred Promontory. . . . The stretches of sea beyond Kernë are not further navigable because of the shallows of the sea and because of the mud and seaweed; the seaweed is one hand's breadth in width and at the tip is sharp enough to prick.

Apollodoros, *The Library* (Frazer translation).

II, v. 10: As a tenth labor [Herakles] was ordered to fetch the kine of Geryon from Erytheia. Now Erytheia is an island near the ocean; it is now called Gadeira. . . . So journeying through Europe to fetch the kine of Geryon he destroyed many wild beasts and set foot in Libya, and proceeding to Tartessos he erected as tokens of his journey two pillars over against each other at the boundaries of Europe and Libya. But being heated by the sun on his journey, he bent his bow at the god, who in admiration of his hardihood gave him a golden goblet in which he crossed the ocean. And having reached Erytheia he lodged on Mount Abas. However the dog, perceiving him, rushed at him; but he smote it with his club, and when the herdsman Eurytion came to the help of the dog, Herakles killed him also. But Menoites, who was there pasturing the kine of Hades, reported to Geryon what had occurred, and he, coming up with Herakles beside the river Anthemos, as he was driving away the kine, joined battle with him and was shot dead. And Herakles, embarking the kine in the goblet and sailing across to Tartessos, gave back the goblet to the sun.

III, x, 1: Atlas and Pleionë, daughter of Okeanos, had seven daughters called the Pleiades, born to them at Kyllenë in Arkadia, to wit: Alkyonë, Meropë, Kelaino, Elektra, Steropë, Taÿgetë, and Maia. . . . And Poseidon had intercourse with two of them, first with Kelaino, by whom he had Lykos, whom Poseidon made to dwell in the Islands of the Blest, and second with Alkyonë. . . .

Diodoros of Sicily, *The Library of History* (Oldfather translation).

III, 53: We are told, namely, that there was once in the western parts of Libya, on the bounds of the inhabited world, a race which was ruled by women and followed a manner of life unlike that which prevails among us. For it was the custom among them that women should practise the arts of war. . . .

As mythology relates, their home was on an island which, because it was in the west, was called Hespera, and it lay in the marsh Tritonis. This marsh was near the ocean which surrounds the earth and received its name from a certain river Triton which emptied into it; and this marsh was also near Ethiopia and that mountain by the shore of the ocean which is the highest of those in the vicinity and impinges upon the ocean and is called by the Greeks Atlas. The island mentioned above was of great size and full of fruit-bearing trees of every kind. . . . They subdued many of the neighboring Libyans and nomad tribes, and founded within the marsh Tritonis a great city which they named Cherronesos ("Peninsula") after its shape.

III, 54: Setting out from the city of Cherronesos, the account continues, the Amazons embarked upon great ventures, a longing having come over them to invade many parts of the inhabited world. The first people against whom they advanced, according to the tale, were the Atlantioi, the most civilized men among the inhabitants of those regions, who dealt in a prosperous country and possessed great cities; it was among them, we are told, that mythology places the birth of the gods, in the regions which lie along the shore of the ocean, in this respect agreeing with those among the Greeks who relate legends, and about this we shall speak in detail a little later.

Now the queen of the Amazons, Myrina, collected, it is said, an army of 30,000 foot-soldiers and 3000 cavalry. . . . Upon entering the land of Atlantioi they defeated in a pitched battle the inhabitants of the city of Kernë. The Atlantioi, struck with terror, surrendered the cities on terms of capitulation and announced that they would do whatever should be commanded them, and that the queen Myrina, bearing herself honorably towards the Atlantioi, both established friendship with them and founded a city to bear her name in place of [Kernë] which had been razed; and in it she settled both the captives and any native who so desired. . . . And since the natives were often being warred upon by the Gorgons, as they were named, a folk which resided upon their borders, and in general had that people lying in wait to injure them, Myrina, they say, was asked by the Atlantioi to invade the land of the aforementioned Gorgons. But when the Gorgons drew up their forces to resist them a mighty battle took place in which the Amazons, gaining the upper hand, slew great numbers of their opponents and took no fewer than 3000 prisoners; and since the rest had fled for refuge into a certain wooded region, Myrina undertook to set fire to the timber, being eager to destroy the race utterly, but when she found that she was unable to succeed in her attempt she retired to the borders of her own country. . . .

III, 55: But the Gorgons, grown strong again in later days, were subdued a second time by Perseus, the son of Zeus, when Medusa was queen over them; and in the end both they and the race of the Amazons were entirely destroyed by Herakles, when he visited the regions to the west and set up his pillars in Libya. . . . The story is also told that the marsh Tritonis disappeared from sight in the course of an earthquake, when those parts of it which lay towards the ocean were torn asunder. . . .

III, 56: But since we have made mention of the Atlantioi, we believe that it will not be inappropriate in this place to recount what their myths relate about the genesis of the gods. . . .

III. 60: After the death of Hyperion, the myth relates, the kingdom was divided among the sons of Ouranos, the most renowned of whom were Atlas and Kronos. Of these sons Atlas received as his part the regions on the coast of the ocean, and he not only gave the name of Atlantioi to his peoples but likewise called the greatest mountain in the land Atlas. They also said that he perfected the science of astrology and was the first to publish to mankind the doctrine of the sphere; and it was for this reason that the idea was held that the entire heavens were supported upon the shoulders of Atlas. . . .

Atlas, the myth goes on to relate, also had seven daughters, who as a group are called Atlantides after their father, but their individual names were Maia, Elektra, Taygetë, Steropë, Meropë, Halkyone, and the last Kelaino. These daughters lay with the most renowned heroes and gods and thus became the first ancestors of the larger part of the race of human beings. . . . These daughters were also distinguished for their chastity and after their death attained to immortal honor among men, by whom they were both enthroned in the heavens and endowed with the appelation of Pleiades. . . .

V, 19: But now that we have discussed what relates to the islands which lie within the Pillars of Herakles, we shall give an account of those which are in the ocean. For there lies out in the deep off Libya an island of considerable size, and situated as it is in the ocean it is distant from Libya a voyage of a number of days to the west. Its land is fruitful, much of it being mountainous and not a little being a level plain of surpassing beauty. Through it flow navigable rivers which are used for irrigation, and the island contains many parks planted with trees of every variety and gardens in great multitudes which are traversed by streams of sweet water; on it also are private villas of costly construction, and throughout the gardens banqueting houses have been constructed in a setting of flowers, and in them the inhabitants pass their time during the summer season. . . . There is also excellent hunting of every manner of beast and wild animal. . . .

And, speaking generally, the climate of this island is so altogether mild that it produces in abundance the fruits of the trees and the other seasonal fruits of the year, so that it would appear that the island, because of its exceptional felicity, were a dwelling-place of gods and not of men.

V, 20: The Phoenicians, then, while exploring the coast outside the Pillars for the reasons we have stated and while sailing along the shore of Libya, were driven by strong winds a great distance out in the ocean. And after being storm-tossed for many days they were carried ashore on the island we mentioned above, and when they had observed its felicity and nature they caused it to become known to all men. Consequently the Tyrrhenians, at the time when they were masters of the sea, purposed to dispatch a colony to it; but the Carthaginians prevented their doing so, partly out of concern lest so many inhabitants of Carthage should remove there because of the excellence of the island, and partly in order to make ready a place in which to seek refuge against an incalculable turn of fortune, in case some total disaster should overtake Carthage.

Strabo, *Geography* (Jones-Sterrett translation).

I, ii, 26: Ephoros says the Tartessians report that Ethiopians overran Libya as far as Dyris, and that some of them stayed in Dyris, while others occupied a great part of the seaboard. . . .

I, iii, 20: Demetrios of Kallatis, in his account of all the earthquakes that have ever occurred throughout all Greece, says that the greater part of the Lichades Islands and of Kenaion was engulfed. . . . And they say, also, of the Atalantë near Euboia that its middle portions, because they had been rent asunder, got a ship-canal through the rent, and that some of the plains were overflowed even as far as twenty stadia, and that a trireme was lifted out of the docks and cast over a wall.

II, iii, 6: On the other hand, [Poseidonios] correctly sets down in his work the fact that the earth sometimes rises and undergoes settling processes, and undergoes changes that result from earthquakes and the other similar agencies, all of which I too have enumerated above. And on this point he does well to cite the statement of Plato that it is possible that the story about the island of Atlantis is not a fiction. Concerning Atlantis Plato relates that Solon, after having made inquiry of the Egyptian priests, reported that Atlantis did once exist, but disappeared—an island no smaller in size than a continent; and Poseidonios thinks that it is better to put the matter that way than to say of Atlantis: "Its inventor caused it to disappear, just as did the Poet the wall of the Achaeans." And Poseidonios also conjectures that the migration of the Cimbrians and their kinsfolk from their native country occurred as a result of an inundation of the sea that came all of a sudden.

III, i, 6: They call the country [of southwestern Spain] Baetica after the river, and also Turdetania after the inhabitants; they call the inhabitants both Turdetanians and Turdulians. . . . The Turdetanians are ranked as the wisest of the Iberians; and they make use of an alphabet, and possess records of their ancient history, poems, and laws written in verse that are 6000 years old, as they assert. And also the other Iberians use an alphabet, though not letters of one and the same character, for their speech is not one and the same, either.

III, ii, 11: Not very far from Castalo is also the mountain in which the Baetis is said to rise; it is called "Silver Mountain" on account of the silver-mines that are in it. . . . The ancients seem to have called the Baetis River "Tartessos"; and to have called Gades and the adjoining islands "Erytheia"; and this is supposed to have been the reason why Stesichoros spoke as he did about the neat-herd of Geryon, namely, that he was born "about opposite famous Erytheia, beside the unlimited, silver-rooted springs of the river Tartessos, in a cavern of a cliff." Since the river had two mouths, a city was planted on the intervening territory in former times, it is said, — a city which was called "Tartessos," after the name of the river; and the country, which is now occupied by the Turdulians, was called "Tartessis." Further, Eratosthenes says that the country adjoining Calpe is called "Tartessis," and the Erytheia is called "Blest Isle.". . .

III, ii, 14: The wealth of Iberia is further evidenced by the following facts: the Carthaginians who, along with Barcas, made a campaign against Iberia found the people in Turdetania, as the historians tells us, using silver feeding-troughs and wine-jars. And one might assume that it was either from their great prosperity that the people there got the additional name of "Makraiones" ("Long-livers") and particularly the chieftains; and that this is why Anakreon said as follows: "I, for my part, should neither wish the horn of Amaltheia, nor yet to be king of Tartessos for one hundred and fifty years"; and why Herodotos recorded even the name of the king, whom he called Arganthonios. . . . Some, however, call Tartessos the Carteia of today.

XIII, i. 36: However, the Naval Station, still now so called, is so near the present Ilion that one might reasonably wonder at the witlessness of the Greeks and the faint-heartedness of the Trojans . . . for Homer says that the wall had only recently been built (or else it was not built at all, but fabricated and then abolished by the poet, as Aristotle says). . . .

Philo Judaeus, *On the Incorruptibility of the World* (Yonge translation).

xxvi: Consider how many districts on the mainland, not only such as were near the coast, but even such as were completely inland, have been swallowed up by the waters; and consider how great a proportion of land has become sea and is now sailed over by innumerable ships. Who is ignorant of that most sacred Sicilian strait, which in old times joined Sicily to the continent of Italy? and where vast seas on each side being excited by violent storms met together, coming from opposite directions, the land between them was overwhelmed and broken away; . . . in consequence of which Sicily, which had previously formed a part of the mainland, was now compelled to be an island.

And it is said that many other cities also have disappeared, having been swallowed up by the sea which overwhelmed them; since they speak of three in Peloponnesos—

> Aigira and fair Boura's walls,
> And Helika's lofty halls,
> And many a once renowned town
> With wreck and seaweed overgrown,

as having been formerly prosperous, but now overwhelmed by the violent influx of the sea. And the island of Atalantes which was greater than Africa and Asia, as Plato said in the *Timaios*, in one day and night was overwhelmed beneath the sea in consequence of an extraordinary earthquake and inundation and suddenly disappeared, becoming sea, not indeed navigable, but full of gulfs and eddies.

Gaius Plinius Secundus, *Natural History* (Rackham & Bostock-Riley translations).

II, xcii: Cases of land entirely swept away by the sea are, first of all (if we accept Plato's story), the vast area covered by the Atlantic, and next, in the inland sea also, the areas that we see submerged at the present day, Acarnania covered by the Ambracian Gulf. . . .

IV, xxxvi: Opposite to Celtiberia are a number of islands, by the Greeks called Cassiterides, in consequence of their abounding in tin: and, facing the Promontory of Arrotrebae, are the six Islands of the Gods, which some persons have called the Fortunate Islands. At the very commencement of Baetica, and 25 miles from the mouth of the straits of Gades, is the Island of Gadis, 12 miles long the three broad, as Polybius states in his writings. . . . On the side which looks toward Spain, at about a hundred paces distance, is another long island, three miles wide, on which the original city of Gades stood. . . . Timaeus says, that the larger island used to be called Cotinusa, from its olives; the Romans call it Tartessos; the Carthaginians Gadir, that word in the Punic language signifying a hedge. It was called Erythia because the Tyrians, the original ancestors of the Carthaginians, were said to have come from the Erythaean, or Red Sea. In this island Geryon is by some thought to have dwelt, whose herds were carried off by Hercules.

VI, xxxi: Polybius says that Cerne is situated at the extremity of Mauritania, over against Mount Atlas, and at a distance of eight stadia from the land; while Cornelius Nepos states that it lies very nearly in the same meridian as Carthage, at a distance from the mainland of ten miles, and that it is not more than two miles in circumference. It is said also that there is another island situated over against Mount Atlas, being itself known by the name of Atlantis. Five days' sail beyond it are deserts, as far as the Aethiopian Hesperiae and the promontory, which we have mentioned as being called Hesperu Ceras, a point at which the face of the land takes a turn towards the west and the Atlantic Sea. Facing this promontory are also said to be islands called the Gorgades, the former abodes of the Gorgons, two days' sail from the mainland, according to Xenophon of Lampsacus. . . . Beyond these even, are said to be the two islands of the Hesperides. . . .

V, vii: In the middle of the desert some place the Atlantes, and next to them the half-animal Goat-Pans and the Blemmyae and Gamphasantes and Satyrs and Strapfoots.

The Atlantes have fallen below the level of human civilization, if we can believe what is said; for they do not address one another by names, and when they behold the rising and setting sun they utter awful curses against it as the cause of disaster to themselves and their fields, and when they are asleep they do not have dreams like the rest of mankind. The Troglodytes hollow out caverns which are their dwellings; they live on the flesh of snakes, and they have no voice, but only make squeaking noises, being entirely devoid of intercourse by speech. The Garamantes do not practice marriage but live with their women promiscuously. The Augilae worship only the powers of the lower world. The Gamphasantes go naked, do not engage in battle, and hold no intercourse with any foreigner. The Blemmyae are reported to have no heads, their mouth and eyes being attached to their chests. The Satyrs have nothing of ordinary humanity about them except human shape. The form of the Goat-Pans is that which is commonly shown in pictures of them. The Himantopodes are people with feet like leather thongs, whose nature it is to crawl instead of walking.

Pomponius Mela, *Description of the World.*

III, x: At the promontory whereof we speak begins that coast which, turned westward, is bathed by the Atlantic Sea. The first parts are inhabited by Ethiopians; those of the middle are uninhabitable, since they are either burning, or covered with sand, or infested by snakes. Facing the sections burnt by the sun are placed the isles which it is said were inhabited by the Hesperides. In the midst of the sandy region is Mt. Atlas, rearing its enormous mass, steep and inaccessible by reason of the sharp-pointed rocks that surround it on all sides; the higher it gets, the more it diminishes in size; its summit is higher than the eye can reach: it loses itself in the clouds; also it is fabled not only to touch with its top the sky and the stars, but even to support them.

Opposite are the Fortunate Isles, where the soil of its

own accord produces an abundant quantity of fruits which re-seed and succeed themselves incessantly, so that the natives pass their days without anxiety and more happily than those who live in magnificent cities. This place is remarkable for the presence of two springs characterized by a singular property: the water of one makes those who drink it laugh themselves to death, while that of the other cures all ills.

Plutarch, *Parallel Lives* (Dryden translation).

Sertorius, vii: [Sertorius] escaped with difficulty, and after the wind ceased, ran for certain desert islands scattered in those seas, affording no water, and after passing a night there, making out to sea again, he went through the straits of Gades, and sailing outward, keeping the Spanish shore on his right hand, landed a little above the mouth of the river Baetis, where it falls into the Atlantic Sea, and gives the name to that part of Spain. Here he met with seamen recently arrived from the Atlantic Islands, two in number, divided from one another only by a narrow channel, and distant from the coast of Africa 10,000 furlongs. These are called the Isles of the Blest; rain falls there seldom, and in moderate showers, but for the most part they have gentle breezes, bringing along with them soft dews, which render the soil not only rich for plowing and planting, but so abundantly fruitful that it produces spontaneously an abundance of delicate fruits, sufficient to feed the inhabitants, who may here enjoy all things without trouble or labor. The seasons of the year are temperate, and the transitions from one to another so moderate . . . that the firm belief prevails, even among the barbarians, that this is the seat of the blessed, and that these are the Elysian Fields celebrated by Homer.

Solon, xxxi: [Solon's] first voyage was for Egypt, and he lived, as he himself says:

"Near Nilus' mouth, by fair Canopus' shore," and spent some time in study with Psenophis of Heliopolis, and Sonchis the Saite, the most learned of all the priests; from whom, as Plato says, getting knowledge of the Atlantic story, he put it into a poem, and proposed to bring it to the knowledge of the Greeks. . . . Now Solon, having begun the great work in verse, the history or fable of the Atlantic Island, which he had learned from wise men in Saïs, and thought convenient for the Athenians to know, abandoned it; not, as Plato says, by reason of want of time, but because of his age, and being discouraged at the greatness of the task . . . Plato, willing to improve the story of the Atlantic Island, as if it were some fair estate that wanted an heir and came with some title to him, formed, indeed, stately entrances, noble enclosures, large courts, such as never yet introduced any story; but, beginning it late, ended his life before his work; and the reader's regret for the unfinished part is the greater, as the satisfaction he takes in that which is complete is extraordinary. For as the city of Athens left only the temple of Olympian Zeus unfinished, so Plato, amongst all his excellent works, left only this piece about the Atlantic Island imperfect.

Plutarch, *On the Apparent Face in the Moon's Orb* (King translation).

xxvi: "The great continent by which the great sea is surrounded on all sides, they say, lies less distant from the others, but about 5000 stadia from Ogygia, for one sailing in a rowing-galley; for the sea is difficult of passage and muddy through the great number of currents, and these currents issue out of the great land, and shoals are formed by them, and the sea becomes clogged and full of earth, by which it had the appearance of being solid. That sea-coast of the mainland Greeks are settled on, around a bay not much smaller than the Maiotis, the entrance of which lies almost in a straight line opposite the entrance to the Caspian Sea. Those Greeks call and consider themselves *continental* people, but *islanders* all such as inhabit this land of ours, inasmuch as it is surrounded on all sides by the sea. . . . But when the star of Kronos . . . comes into the sign of the Bull every 30 years, they having got ready a long while beforehand all the things required . . . they send out people appointed by lot in the same number of ships, furnished with provsions and stores necessary for persons intending to cross so vast a sea by dint of rowing, as well as to live a long time in a foreign land. . . . Those who escape from the sea, first of all, touch at the foremost isles, which are inhabited by Greeks also, and see the sun setting for less than one hour for thirty days in succession. . . . For wonderful are both the island and the mildness of the climate; whilst the deity himself has been an obstacle to some when contemplating departure. . . . For Kronos himself is imprisoned in a vast cavern, sleeping upon a rock overlaid with gold; for his sleep has been contrived by Zeus for his chaining. . . ."

Plutarch, *Of Isis and Osiris* (Babitt translation).

x: Eudoxos, they say, received instruction from Chonouphis of Memphis, Solon from Sonchis of Saïs, and Pythagoras from Oinouphis of Heliopolis.

Arrian, *The Anabasis of Alexander* (Chinnock translation).

II, xvi: So I think that the Herakles honored in Tartessos by the Iberians, where are certain pillars named after Herakles, is the Tyrian Herakles; for Tartessos was a colony of the Phoenicians, and the temple to the Herakles there was built and the sacrifices offered after the usage of the Phoenicians.

Pausanias, *Description of Greece* (Jones translation).

I, xxiii, 5: Euphemos the Carian said that on a voyage to Italy he was driven out of his course by winds and was carried into the outer sea, beyond the course of seamen. He affirmed that there were many uninhabited islands, while in others lived wild men. . . . The islands were called Satyrides by the sailors, and the inhabitants were red-haired, and had upon their flanks tails not much smaller than those of horses. As soon as they caught sight of their visitors, they ran down to the ship without uttering a cry and assaulted the women in the ship. At last the sailors in fear cast a foreign woman on to the island. Her the Satyrs outraged not only in the usual way, but also in a most shocking manner.

Athenaeus, *The Deipnosophists* (Gulick translation).

XIV, 640d: Plato in his account of Atlantis calls dessert *metadorpia* in these words: [Here follows a quotation from *Critias*, 115B.]

Appian, *Roman History: The Wars in Spain* (White translation).

VI, i: In like manner the Greeks visited Tartessos and
its king Arganthonios, and some of them settled in Spain;
for the kingdom of Arganthonios was in Spain. It is my
opinion that Tartessos was the city on the sea which is
now called Karpessos.

Tertullian, *On the Ascetics' Mantle* (Thelwall translation).

ii: Even now [the earth's] shape undergoes local muta-
tions, when some spot is damaged; when among her islands
Delos is no more, Samos a heap of sand, and the Sibyl thus
proved no liar; when in the Atlantic the isle equal in size
to Libya or Asia is sought in vain; when formerly the side
of Italy, severed in the center by the shivering shock of the
Adriatic and the Tyrrhenian seas, leaves Sicily as its
relics. . . .

Aelian, *Various Anecdotes* (Stanley translation).

III, xviii: Theopompus relates that . . . Silenus told
Midas that Europe, Asia, and Africa were islands surrounded by the ocean; that there was but one continent,
which was beyond this world, and that in size it was infinite; that in it were bred, besides other very great creatures, men twice as big as those here, and they lived double
our age; that many great cities are there, and peculiar
manners of life; . . . that there are two cities far greater
than the rest, nothing like each other; one named Machimus ("Warlike") and the other Eusebes ("Pious"); that
the Pious people live in peace, abounding in wealth, and
reap the fruits of the earth without plows or oxen, having
no need of tillage or sowing. . . . The inhabitants of the
city Machimus are very warlike, continually armed and
fighting, and this one city predominates over many. The
inhabitants are not fewer than 2,000,000. They die sometimes of sickness, but this happens very rarely, for most
commonly they are killed in the wars by stone or wood,
since they are invulnerable to steel. . . . He said that they
once planned a voyage to our isles, and sailed the ocean,
being 10,000,000 in number, till they came to the Hyperboreans; but understanding that they were the happiest

men amongst us, they despised us as persons who lead a
mean and inglorious life, and therefore thought it not worth
their while to go further. He added what is yet more wonderful, that there are men living among them called
Meropes, who inhabit many great cities; and that at the
farthest end of their country is a place named Anostus
("No-Return") resembling a gulf. It is neither very light
nor very dark, the air being a kind of dark red. There are
two rivers there, one of pleasure, the other of grief, and
along each river grow trees the size of a plane-tree. Those
which grow up by the River of Grief bear fruit of such
nature that if anybody eats of them, he spends all the rest
of his life in tears and grief until he dies. The other trees,
growing by the River of Pleasure, produce fruit of the
opposite kind, for whoever tries it is eased from all his
former desires. If he loved anything he quite forgets it;
and in a short time he becomes younger, and lives over
again his former years . . . becoming first a young man,
then a child, and lastly an infant, and so dies. This any
man may believe if he thinks the Chian worthy of credit;
to me he seems an egregious romancer in this as in other
things.

Aelian, *On the Nature of Animals* (Bramwell translation).
      XV, ii: Dwellers by the ocean tell the story that the
ancient kings of Atlantis who traced their descent from
Poseidon wore head-bands of the skin of male sea-rams, as
a sign of authority. The queens likewise wore fillets of the
female sea-ram. . . .

Arnobius Afer, *Seven Books Against the Gentiles* (Bryce-
Campbell translation).
      I, v: Did we [Christians] bring it about, that 10,000
years ago a vast number of men burst forth from the island
which is called the Atlantis of Neptune, as Plato tells us,
and utterly ruined and blotted out countless tribes?

Ammianus Marcellinus, *Roman History* (Yonge translation).
      XVII, vii, 13: Now earthquakes takes place in four
ways: either they are *brasmatidae,* which raise up the

ground in a terrible manner, and throw vast masses to the surface, as in Asia, Delos arose, and Hiera. . . . Or they are *climatiae*, which, with a slanting and oblique blow, level cities, edifices, and mountains. Or *chasmatiae*, which, suddenly, by a violent motion, open huge mouths, and so swallow up portions of the earth, as in the Atlantic Sea, on the coast of Europe, a large island was swallowed up. . . .

Lysios Proklos, *Commentaries on Plato's Timaios* (Taylor translation).

I: For the resumption of the discourse about a polity, and the narration respecting the Atlantic island, unfold through images the theory of the world. . . . There, therefore, the concise narrative of a polity, prior to physiology, ironically places us in the fabrication of the universe; but the history of the Atlanteans accomplishes this symbolically. . . . With respect to the whole of the narration about the Atlanteans, some say that it is a mere history, which was the opinion of Krantor, the first interpreter of Plato, who says that Plato was derided by those of his time as not being the inventor of the *Republic,* but transcribing what the Egyptians had written on this subject; and that he so far regards what is said by these deriders as to refer to the Egyptians this history about the Athenians and Atlanteans, and to believe that the Athenians lived conformably to this polity. Krantor adds that this is testified to by the prophets of the Egyptians, who assert that the particulars (which are narrated by Plato) are written on pillars which are still preserved. Others again say that the narration is a fable, and a fictitious account of this which by no means had an existence. . . . And of this, some refer to the analysis to the fixed stars and planets, assuming the Athenians analogous to the fixed stars but the Atlanteans to the planets. . . . Of this opinion therefore is the illustrious Amelius, who vehemently contends that this must be the case because it is clearly said in the *Kritias* that the Atlantic island was divided into seven circles. . . . Others again, like Origen, refer the analysis to the opposition of certain daemons, some of them being more but others less excellent. . . . But others refer it to the discord of souls . . . and this is the interpretation of Numenius. Before, however, souls descend into solid bodies, those theologists and Plato deliver the war of some with material daemons who are adapted to the west, since the west, as the Egyptians say, is the place of noxious daemons. Of this opinion is the philosopher Porphyrios. . . . These however are in my opinion very excellently corrected by the most divine Iamblichos.

According to him, therefore, and also to our preceptor Syrianos, this contrariety and opposition are not introduced for the purpose of rejecting the narrative, since on the contrary this is to be admitted as an account of the transactions that actually happened. . . .

Longinus doubts what was the intention of Plato in the insertion of this narration. For he does not introduce it either for the purpose of giving respite to the auditors, or as being in want of it. And he dissolved the doubt, he thinks, by saying that it is assumed by Plato prior to physiology, in order to allure the reader, and soften the severity of that kind of writing. But Origen says that the narration is indeed a fiction, and so far he agrees with Numantius and his followers, but he does not admit with Longinus that it was devised for the sake of pleasure.

That such and so great an island once existed, is evident from what is said by certain historians respecting what pertains to the external sea. For according to them, there were seven islands in that sea, in their times, sacred to Persephonë, and also three others of immense extent, one of which was sacred to Pluto, another to Ammon, and the middle (or second) of these to Poseidon, the magnitude of which was a thousand stadia. They also add that the inhabitants of it preserved the remembrance of their ancestors, or the Atlantic island that existed there, and was truly prodigiously great; which for many periods had domination over all the islands in the Atlantic Sea, and was itself likewise sacred to Poseidon. These things, therefore, Marcellus writes in his *Ethiopic History.* If however this be the case, and such an island once existed, it is possible to receive what is said about it as a history, and also as an image of a certain nature among wholes. . . .

And thus much concerning what is related of the magnitude of the Atlantic island, to show that it is not proper to disbelieve what is said by Plato, though it should be received as a mere history.

Kosmas Indikopleustes, *Christian Topography* (McCrindle translation).

XII: In like manner the philosopher Timaios also describes this earth as surrounded by the Ocean, and the Ocean as surrounded by the more remote land. For he supposed that there is to westward an island, Atlantis, lying out in the Ocean, in the direction of Gadeira, of an enormous magnitude, and relates that the ten kings having procured mercenaries from the nations in this island came from the coast far away, and conquered Europe and Asia, but were afterwards conquered by the Athenians, while that island itself was submerged by God under the sea. Both Plato and Aristotle praise this philosopher, and Proklos has written a commentary on him. He himself expresses views similar to ours with modifications. Moreover he mentions the ten generations as well as that land which lies below the Ocean. And in a word it is evident that all of them borrow from Moses, and publish his statements as their own. . . .

Timaios alone, who has already been mentioned, drawing from what source I know not, but perhaps from the Chaldeans, recast the story of the ten kings [of Chaldea, as given by Berossos] feigning that they came from the land beyond the Ocean into the island of Atlantis, which he says was sumberged below the sea, and that taking its inhabitants as mercenaries, they conquered Europe and Asia—all of which is a most manifest invention, for as he could not point out the island, he gave it out that God had consigned it to a watery grave.

APPENDIX B

Plato's family tree, according to Diogenes Laërtius:

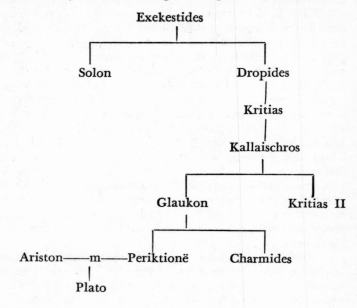

According to Iamblichos (*apud* Lysios Proklos):

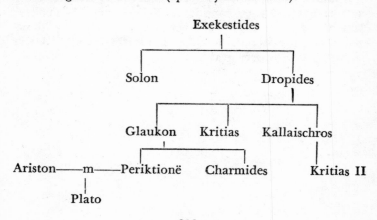

According to Lysios Proklos:

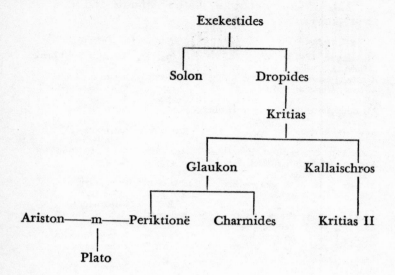

Plato also had other relatives of doubtful position such as a brother or cousin named Glaukon, according to Iamblichos (*apud* Proklos) one of the three sons of Kallaischros (the other two being Theon and the younger Kritias). Plato was also alleged to be descended from Poseidon via the Homeric hero Nestor and the semilegendary King Kodros of Athens.

APPENDIX C

List of interpreters of Plato's Atlantis and their interpretations.

Atlantist	*Date*	*Interpretation*
Abartiague, William d', Basque scholar	1937	Atlantic land-bridge
Acosta, José de, missionary	16th cent.	Atlantic island
Adcock, Norman, cultist	1937	Atlantic island
Albinus, historian	1598	America
Ali-Bey (Domingo Badia y Leblich, traveller)	1814	North Africa
Amelius, philosopher	3rd cent.	Astronomical symbolism
Ammianus Marcellinus, historian	4th cent.	Atlantic island
Anonymous writer quoted by Delisle de Sales	1779	Ceylon
Anville, J. B. B. d', geographer	18th cent.	Imaginary
Aristotle, philosopher	4th cent. B.C.	Imaginary
Arnobius Afer, rhetorician	3rd cent.	Atlantic island
Babcock, W. J., geographer	1922	Sargasso Sea
Bacon, Francis, philosopher	c. 1600	North America
Baër (or Bähr), C. F., writer	1762	Palestine
Baikie, James, Egyptologist	1900	Crete
Bailly, J. S., astronomer	18th cent.	Mongolia and Spitsbergen
Balch, E. S., explorer	1900	Crete
Ballard, G. W., cultist	1939	Atlantic island
Banning, P. W., cultist	1924	Atlantic island
Bartoli,	1829	Political allegory
Baudelot de Dairval, C. C.	1729	Atlantic island
Becman, J. C.	1680	Atlantic island
Bellamy, H. S., writer	1948	Atlantic island
Bérard, Victor, Classical scholar	1929	Carthage and Gades
Berlioux, E. F., geographer	1874	Atlas Mts.
Besant, Annie W., cultist	1897	Atlantic island
Bircherod, Jens	1683	America
Björkman, Edwin, scholar	1927	Tartessos
Blake, William, poet	18th-19th cent.	Atlantic island
Blavatsky, H. P., cultist	1888	Atlantic island
Bock, J. C.	1685	South Africa
Borchardt, Paul	1925	Tunisia
Bory de Saint-Vincent, J. B. G. M., naturalist	1803	Atlantic island
Bosco, Joseph, archeologist	1922	Malta
Braghine, A. P., writer	1940	Atlantic island
Bramwell, J. G., scholar	1938	Allegory
Brosse, R. P. E.	1892	Atlantic island
Brosses, Charles De, antiquary	18th cent.	Atlantic island
Buache, Philippe	1752	Atlantic island
Buffault, Pierre	1936	Atlantic island
Buffon, G. L. de, naturalist	1749	Doubtful

Atlantist	Date	Interpretation
Bunbury, E. H., geographer	1883	Imaginary
Butavand, F.	1925	Mediterranean, off the coast of Tunisia
Cadet de Gisancourt, L. C., chemist	1787	Atlantic island
Calahan, H. C., journalist	1946	Atlantic island
Carli, G. R. de, scholar	1785	Atlantic island
Cellarius, Christophe, geographer	1787	Atlantic island
Chatwin, C. P., naturalist	1940	North and South Atlantic islands
Churchward, James, writer	1926	Atlantic island
Clymer, R. S., cultist	1920	Atlantic island
Cotte, Charles	1919	Spain
Courcelle-Seneuil, J. L.	1907	Central France
Coussin, Paul, scholar	1928	Imaginary
Dawes, Charles G., banker and politician	1919	Atlantic island
Delisle de Sales (J.-B. C. Izouard), historian	1799	Caucasus & Western Mediterranean
Dévigne, Roger, writer	1924	Atlantic island
Diogenes Laërtius, biographer	3rd cent.	"Ethical dialogue"
Donnelly, Ignatius, politician	1882	Atlantic island
Duvillé, Daniel, draftsman	1936	Atlantic island and Ethiopia
Elgee, Capt. Frank	1908	Nigeria
Engel, Samuel	1767	Atlantic island
Erain, Jean d'	1914	The Arctic
Eurenius, Johannes, scholar	18th cent.	Palestine
Fabre d'Olivet, Antoine, writer	1801	Western Mediterranean
Fabricius, J. A., Classical scholar	17th- 18th cent.	Palestine
Fawcett, P. H., explorer	1925	Brazil
Fessenden, R. A., engineer	1923	Caucasus
Forrest, H. E., naturalist	1933	Atlantic land-bridge
Fortier d'Urban, Marquis de	1838	Atlantic island
Frobenius, Leo, explorer	1926	Nigeria
Frost, K. T., Classical scholar	1909	Crete
Gattefossé, R. M., perfume mfr.	1919	Atlantic island
Gentil, E. L., explorer	1921	Morocco & Eastern Mediterranean
Georg, Eugen, writer	1931	Atlantic island
Germain, M. L.	1911	Atlantic island
Gidon, F., professor	1935	North Sea
Ginguené, P. L., historian	18th- 19th cent.	Doubtful
Godron, D. A., botanist	1868	Sahara Desert
Gómara, F. L. de, historian	1553	North America
Gonzales y Saavedra, F. J.		Morocco
Gorsleben, Rudolf	1929	The Arctic
Gosselin	19th cent.	Imaginary
Grave, C. J. de	1806	Netherlands
Hafer	c. 1745	East Prussia and the Baltic

Atlantist	Date	Interpretation
Harris, T. L., occultist	1858	Atlantic island
Heer, Oscar	1856	Atlantic island
Heer, Oswald, paleobotanist	1835	Atlantic island
Hennig, Richard, historian	1925	Tartessos
Hermann, Albert	1925	Tunisia
Heumann, Fritz, Classical scholar	18th cent.	Imaginary
Hissman	18th cent.	Imaginary
Hosea, L. M.	1875	Atlantic island
Hörbiger, Hanns, engineer	1913	Atlantic island
Houvellique, M. L., geologist	20th cent.	Atlantic island
Humboldt, Alexander von, scientist	1807	Possibly America
Humboldt, K. W. von, philosopher	18th- 19th cent.	Imaginary, with historical basis
Iamblichos, philosopher	3rd- 4th cent.	Atlantic island and allegory
Jacolliot, Louis, mythologist	1879	Pacific continent
Jessen, Otto, geologist	1925	Tartessos
Jolibois, J. F.	1843	Atlantic island
Jowett, Benjamin, Classical scholar	1871	Political allegory
Kadner, Siegrief	1931	The Arctic
Karst, Joseph, orientalist	1930	North Africa and Indian Ocean
Khun de Prorok, Byron, explorer	1925	Ahaggar Mts.
Kircher, Athanasius, scientist	1665	Atlantic island
Kirchmaier, G. K., scholar	1685	South Africa
Knötel, A. F. R.	1893	Morocco
Kosmas, geographer	6th cent.	Biblical Flood
Krantor, philosopher	4th-3rd cent. B.C.	Atlantic island
Kruger, Jakob	19th cent.	America
la Borde, J. B. De	1791	South Pacific continent, incl. South America and Australia
Lagneau, Gustave	1876	Morocco
la Motte le Vayer, Francois de, writer	17th cent.	Greenland
Latreille, P. A., naturalist	1819	Political allegory and Iran
Launay, Louis de	1936	Mexico
Leaf, Walter, Classical scholar	1915	Crete
le Cour, Paul, professional Atlantist	1926	Atlantic island
Ledru, A. P., traveller	1810	Atlantic island
Leet, L. Don, seismologist	1948	Imaginary
Le Plongeon, Augustus, physician	1896	Atlantic island
Leslie, J. B., cultist	1911	Atlantic island
Letronne, A. J., Egyptologist	19th cent.	Imaginary
Lynch, George	1925	Brazil
MacCulloch, J. R., geographer	1841	Central America

Atlantist	*Date*	*Interpretation*
Magoffin, R. V. D., archeologist	1929	Crete
Mahoudeau, P. G., anthropologist	1913	Eastern Mediterranean
Mallinkrot, Bernard De	17th cent.	Imaginary
Manzi, Michel, cultist	1922	Atlantic island
Marque, Bernard	1933	North Africa
Martin, T. H., Classical scholar	1841	Imaginary
Martins, Charles	1864	Atlantic island
Mather, Cotton, theologian	1721	Atlantic island
Matthew, W. D., geologist	1920	Imaginary
Mentelle, Edmé	1787	Atlantic island
Mercator, Gerardus, cartographer	16th cent.	America
Merezhkovski, D. S., novelist	1931	Atlantic island
Miller, R. De W., writer	1949	Atlantic island
Montaigne, M. E. de, philosopher	16th cent.	Doubtful
Mortillet, Gabriel de, anthropologist	1897	Imaginary
Navarro, L. F.	1916	Atlantic island
Négris, Phocion, geologist	1924	Atlantic island
Numenius, philosopher	2nd cent.	Allegory
Olivier, C. M.	1726	Palestine
Origenes Adamantius, theologian	3rd cent.	Allegory
Ortelius, Abraham, cartographer	16th cent.	North America
Oviedo y Valdez, G. F., historian	1525	Iraq and Central America
Paiagua, André de	1900	Crimea
Pauw, C. de	1768	Central America
Phelon, W. P., physician	1903	West Indes
Philo Judaeus, theologian	1st cent.	Atlantic island
Pliny the Elder, encyclopedist	1st cent.	Doubtful
Plutarch, biographer	1st-2nd cent.	Doubtful
Pollet, Marcel	1923	Belgium and Netherlands
Porphyrios, philosopher	3rd cent.	Allegory
Poseidonios, philosopher	2nd-1st cent. B.C.	Atlantic island
Postel, Guillaume de	16th cent.	North America
Proklos, Lysios, philosopher	5th cent.	Atlantic island and allegory
Prutz, R. E., poet	1855	North America
Rafinesque, C. S., naturalist	1836	America
Raynal, G. T. F., historian	1775	Atlantic island
Reclus, Onésime	1918	North Africa
Rienzi, G. L. D. de	1832	Atlantic island
Rosny, Léon de	1875	Central America
Roux, Cloudius, librarian	1926	Sahara Desert
Rudbeck, Olof, scholar	1675	Sweden
Rutot, A. L., geologist	1920	Morocco
Sanson, Guillaume, cartographer	1689	North America
Schliemann, Heinrich, archeologist	19th cent.	Doubtful
Schliemann, Paul, writer	1912	Atlantic island
Schulten, Adolf, archeologist	1922	Tartessos
Schuchert, Charles, geologist	1917	Imaginary

Atlantist	*Date*	*Interpretation*
Scott-Elliot, W., cultist	1896	Atlantic island
Serranus (Jean de Serres), scholar	1570	Biblical Flood, Palestine
Silbermann, Otto	1930	North Africa
Sinnett, A. P., cultist	1884	Atlantic island
Smith, William, Classical scholar	1857	Imaginary
Snider-Pellegrine, Antoine	1859	Brazil
Spanuth, Juergen	1953	North Sea
Spence, Lewis, mythologist	1924	Atlantic island
Stacey-Judd, R. B., architect	1939	Atlantic island
Steiner, Rudolf, cultist	1923	Atlantic island
Stewart, J. A., Classical scholar	1905	Political allegory
Strabo, geographer	1st cent. B.C.-1st cent. A.D.	Doubtful
Susemihl, Franz, Classical scholar	c. 1850	Imaginary
Swan, John, scholar	1644	Atlantic island or extension of America
Sykes, Egerton, scholar	1949	Atlantic island
Syrianos, philosopher	5th cent.	Atlantic island and allegory
Taylor, A. E., Classical scholar	1929	Imaginary
Termier, Pierre, geologist	1912	Atlantic island
Tertullian, theologian	2nd-3rd cent.	Atlantic island
Tiedemann, Dietrich, historian	18th cent.	Imaginary
Tournefort, J. P. de botanist	1717	Atlantic island
Ukert, F. A., geographer	19th cent.	Imaginary
Unger, Franz, botanist	1860	Atlantic island
Vaillant, G. C., Americanist	1944	Imaginary
Van-Eys	18th cent.	Palestine
Vangoudy, Robert de, cartographer	1769	North America
Vasse, G. G. de	19th cent.	Malta
Verdaguer,* Jacinto, poet	19th cent.	Catalonia
Verneau, René, anthropologist	1888	Imaginary
Vibert, Théodor	1883	America
Vigers, Daphne, occultist	1944	Atlantic island
Vivarez, Mario	1925	Morocco
Voltaire, F. M. Arouet de, philosopher	18th cent.	Doubtful
von Ravn, C. I., occultist	1937	Atlantic island
Wallace, W. B.	1899	Atlantic land-bridge
Warren, W. F., scholar	1885	North Polar continent
Wegener, A. L., meteorologist	1924	North America
Wencker-Wildberg, Friedrich, historian	1925	Atlantic island
Whishaw, E. M., archeologist	1929	Atlantic island
Wilford, Francis, officer	1805	British Isles
Wilkins, H. T., writer	1946	Atlantic island
Wilson, Daniel, ethnologist	1886	Imaginary
Wirth, H. F., writer	20th cent.	The Arctic
Zschaetzsch, K. C., writer	1922	Atlantic island

APPENDIX D

Table of Geological Time

Era	Period	Years Ago	Mountain-making	Dominant life
Cenozoic	Pleistocene		Cascadian Revolution	Man
		1,000,000		
	Pliocene			
	Miocene		Minor disturbance	Mammals
	Oligocene			
	Eocene			
	Paleocene			
Mesozoic		60,000,000	Laramide Revolution	
	Cretaceous			
	Comanchean		Minor disturbance	
		135,000,000		
	Jurassic		Palisade disturbance	Reptiles
	Triassic			
		180,000,000	Appalachian Revolution	
	Permian		Minor disturbance	
	Pennsylvanian		Minor disturbance	Amphibians
	Mississippian		Arcadian disturbance	

Era	Period	Years Ago	Mountain-making	Dominant life
Paleozoic	Devonian			Fishes
	Silurian		Taconian disturbance	
	Ordovician			
	Cambrian	450,000,000	Killarney-Grand Canyon Revolution	Invertebrates
Proterozoic		550,000,000		
			Laurentian-Algoman Revolution	
Archeozoic		1,000,000,000		Single-celled organisms

Different authorities give different schemes of geological periods; for instance the older tables lump the Mississippian and Pennsylvanian periods together as the "Carboniferous," and consider the Comanchean part of the Cretaceous and the Paleocene part of the Eocene.

NOTES

CHAPTER I

Quotations: Edgar Allan Poe, *The City in the Sea*; James Elroy Flecker, *Gates of Damascus*; Plato, *Timaeus*, 20, 26; *Critias*, 116, 121; Homer, *Odyssey*, I, l. 52; Euripides, *Hippolytus*, l. 748ff; Diodorus, V, 55; Pliny, II, xcii; Plutarch, *Solon*, xxxvi; Bramwell, p. 235 (paraphrasing Beazley, I, p. 303).

Other main sources: For underground rivers see Lowes, p. 393; Pausanias, VIII, liv, 2; Seneca, III, xxvi, 4; VI, vii, 5; viii, 1-3; Strabo, VI, ii, 4, 9.

For the Atlas myths see Apollodorus, II, v, 11; III, x, 1; L. Cooper, pp. 12, 14 (Aeschylus, *Prometheus Bound*, 11. 348ff, 425ff); Diodorus, III, 60; IV, 27; Euripides, *Ion*, ll. 1-6; Hesiod, *Theogony*, ll. 506-519, 744-750; Ovid, LV, l. 627; Virgil, IV, l. 480ff.

For the Classical ideas of Atlantic islands, transatlantic continents, and the peoples of North Africa, see Aelianus, III, xviii; Apollodorus, II, v, 11; Cicero, *The Vision of Scipio*, xiii; de Camp & Ley; Diodorus, III, 53-60; Herodotus, IV, 184; Mela, III, vi, x; Pausanias, I, xxiii, 5; Pliny, IV, xxvii; V, viii; VI, xxxvii; Plutarch, *On the Cessation of Oracles*, xxvi; Strabo, VIII, iii, 19; XVII, iii, 2; Thévenin, pp. 9-16, 41, 49-54, 112-125; Tylor, II, p. 66.

The statement about the transoceanic origin of the ancestors of Zoroaster (Avestan *Zarathustra*, modern Persian *Zardusht*) is based on E. M. Butler's *The Myth of the Magus* (p. 22) but Professor Butler does not remember her source for the statement.

For Classical ideas about floods and the rising and sinking of lands see Apollodorus, I, vii, 2; Aristotle, *Meteorologica*, I, xiv; Diodorus, IV, 85; Herodotus, II, 12; King, p. 161ff; Ovid, I, ll. 125-415; Pindar, *Seventh Olympian Ode*, ll. 54-76; Pliny, II, xc; Seneca, II, xxvi; VI, xxix; Strabo, I, ii, 31; iii, 4, 10, 18; Tertullian, *De Pallio*, ii; Virgil, III, l. 419.

For the Achaean wall at Troy, see Homer, *Iliad*, VII, ll. 433, 441; XII, ll. 1-33; Strabo, II, iii, 6; XIII, i, 36. The first passage in Strabo, quoted from Poseidonios, does not name the original source, but this is supposed to have been Aristotle because the same sentence, quoted by Strabo in the second passage listed above and given in Appendix A, *is* attributed to Aristotle.

For other subjects discussed in this chapter see Apollodorus, III, xiv, 1; Bramwell, p. 64; Cicero, *De Natura Deorum*, III, xxiii; Diodorus, I, 96, 98; Diogenes, III; Disraeli, I, p. 320; *Hebrews* ix, 23-24; Kosmas, II & XII (pp. 33, 43, 376, 380); MacDougall, p. 137; Martin, pp. 246-261, 306; Plutarch, *Solon*, xxvi; *On Isis and Osiris*, ix, x; Proklos, I, pp. 4, 25, 64f, 70f, 113, 145, 148, 152, 165; Spence (1926) p. 32f; Stewart, p. 467ff; Strabo, XIII, i, 56; XVII, 29; Thucydides, I, xiii; *Time*, 12 Apr. 1948, p. 61.

Words: "Utopia" = Greek *ou topos*, "no place." Guillaume
Budé's name was commonly Latinized to *Budaeus*. "Plato" = Greek
Platōn, from *platys*, "broad." *Oreichalkōn* (or *oreichalkos*) may have
been a copper-silver alloy.

Poseidon was later merged by the Romans with their own river-
god Neptunus (Etruscan, *Nethuns*) so that "Poseidon" and "Neptune"
are used more or less interchangeably. The Hesperides were the nymphs
who guarded the golden apples that Ge and given Hera on her
marriage to Zeus. Apollonios Rhodios gives their names as Aiglë,
Hespera, and Erytheïs; in other versions their number is four or seven
and the names are different. Accounts of their parentage also differ, in
one version being the daughters of Atlas and Hesperis, and hence
called the Atlantides and confused or identified with the seven
daughters of Atlas who became the Pleiades. Their paradisiacal garden,
originally located in Arkadia, was moved west along with Atlas as the
Greeks became geographically sophisticated, and was finally placed
in the Western Ocean. For complete references on the Hesperides, see
the note on p. 220f of Vol. I of the LCL edition of Apollodoros. The
happy Hyperboreans were generally assumed to live in the Far North,
though Apollodoros (II, v, 11) confused this issue by saying that the
Golden Apples "were not, as some have said, in Libya, but on Atlas
among the Hyperboreans." Diodoros got his tale of Queen Myrina
and the Atlantioi, along with his version of the Argosy, from a
mythological poem that one Dionysios Skytobrachion had written in
Alexandria in the 2nd century B.C.

Some Atlantists have spoken of a "lost" part of *Kritias*, as if the
Kritias were but a fragment of what was once a complete work.
However, as Plutarch definitely states that Plato left the work unfinished (cf. Appendix A) and as Greek scholars assert that *Kritias* (in
contrast to the carefully-polished *Timaios*) has all the earmarks of a
rough draft, it is probable that Plato abandoned the *Kritias* in the
form in which we now have it, and that the manuscript was found
among his papers after his death and published without editing.

CHAPTER II

Quotations Shakespeare, *Othello*, Act. 1, Sc. iii, ll. 143-145;
Romans x, 18; Haugen, p. 66; Swan, p. 226f; Landa, pp. 16f, 169;
2 *Kings* xvii, 6; Brasseur (1869) I, p. 151; Pratt, p. 85; Donnelly, pp.
1f, 45, 133; Le Plongeon (1896) pp. xxif, 16, 95, 105; Scott-Elliot, I, p. 16
(p. 20 in the 1909 ed. of *The Story of Atlantis*); Bramwell, p. 23f;
Churchward (1926-1933) pp. 7, 46, 102; (1931-1935) p. 198f.

Other main sources: For early explorations and geographical
speculations see Aristotle, *De Mundo*, ii; Babcock, (1917), and (1922)
p. 7, chs. iv. and x; Beazley, I, pp. 230-240, 377; III, p. 532; Bramwell,
p. 237; Brasseur (1861) p. iv; Bunbury, II, p. 352; de Camp & Ley;
Diodorus, I, 40; Herodotus, IV, 36, 42, 43; Holand; Jastrow, p. 105ff;
Lowes, p. 116-123; MacDougall, p. 293; Macrobius, II, v-ix; Mela, III,
vii; Morgan; Morison; F. Osborn, p. 31; Plutarch, *Face in the Moon*, iv;
Ptolemy, IV, viii; VII, iii, v; Raglan (1936) ch. v; Spence (1925) p. 18;
(1933) p. 22; Spilhaus, p. 99; Strabo, I, iv, 6; II, iii, 4-5; v, 43; Tertullian,
De Pallio, ii; *Against Hermogenes*, xxv; Thévenin, pp. 56-80; Tozer;
Wallis, pp. 125-154; White, I, pp. 92-108; Wilson, ch. i.

For early Atlantist theories, the Lost Ten Tribes, &c., see Andrews, pp. 235-272; Bessmertny, pp. 20-34; Björkman, p. 31ff; Blom, p. 112; Bramwell, pp. 107-109; Brasseur (1869) I, pp. 189, 198; (1861) p. cxxix, note; Carli; Casson (1939) p. 112; Davidson; de Camp & Ley; De Voto, *Across the Wide Missouri*, pp. 5, 44f, 113; Erith; Gartenhaus; Georg, p. 222; Godbey; *Grande Encyclopédie, s. v. Brasseur;* Hewett, pp. 239-242, 345; Hospel; Hungerford; King, p. 47; Kingsborough; Kohut; Landa, pp. 16, 76, 169f; Launay; Magoffin, p. 205f; Martin, pp. 262-306; *The Maya and Their Neighbors,* pp. 139-148; Morley (1915) pp. i-32; (1946) pp. 259-265; C. F. Smith, p. 56; Spence (1913) pp. 1-5, 161; von Hagen, *Maya Explorer,* pp. 76-78.

For Donnelly and other recent Atlantists see Bessmertny, pp. 254-261; Bramwell, pp. 22-25; Churchward, *passim;* Conway, p. 53; Donnelly; Effler; *Encyclopaedia Britannica s. v. Shakespeare;* Evans; Holbrook; Hosea; Le Plongeon; Pratt, pp. 84-88; Thompson, p. 16f.

Words: *Avalon* = "Apple Island." *Antillia* (however spelled) is pronounced "ahn-*teel*-yah." Morrison derives it from *Gaetulia* in ancient North Africa. St. Brendan is also called *Brandan* or *Brennan Mac-Finnlonga. Edrisi* = *Abu Abdallah Muhammad ibn Muhammad ibn Abdallah ibn Idrisi* or *Edrisi;* c. 1099-1154. *Brazil* is also spelled *Brasil, Breasil, Hi-Brazil, Hy-Breasail, O'Brasil,* &c.; the name is probably of Keltic origin, meaning "excellent." *Magellan* = Portuguese *de Magalhães,* pronounced approximately "dee-mah-gahl-*yahnh*-ish." "Amerind" is an anthropological term for "American Indian." The Troano, Cortesianus, and Perezianus Codices are named after people who once owned or published them, *e.g.* the Troano Codex after Juan Troy Ortelano, a professor of paleography at the University of Madrid. "Fomorian" is pronounced "foworian," rhyming with "Gregorian."

The latest edition of Donnelly's *Atlantis* was published in 1949 by Harper, the original publisher, with notes and revisions by Egerton Sykes, F.R.G.S. Sykes (who inclines towards the Hörbiger theory) has corrected a few of Donnelly's many errors (like his parallel between the Latin alphabet and Landa's "Mayan alphabet") but has added some of his own.

CHAPTER III

Quotations: Blavatsky (1888-1938) I, p. 91f; III, pp. 28-32; Pindar, *Tenth Pythian Ode;* Scott-Elliot, I. pp. 28, 40; Cervé, p. 47ff; Blavatsky (1888-1938) I, pp. 150-191; Steiner, p. 7f; *Lemuria the Incomparable,* p. 9.

Other main sources: For the geologists' Lemuria see Fermor; Georg, p. 159; Schuchert (1908) p. 432; Suess, I, pp. 387, 405, 417f, 252ff.

For Mme. Blavatsky and Theosophy see Bechofer-Roberts; Besant, p. 301ff; Blavatsky (1923-1933) p. 149; (1888-1938) esp. I, p. 91f; III, pp. 19ff, 28f, 156-351, 392ff, 427-437; IV, pp. 311-347; Bramwell, pp. 192-213; Farquhar, p. 263; *Grande Encyclopédie, s. v. Louis Jacolliot;* Harris; Hort, Ince, and Swainson, p. 155ff; Jacolliot, p. 13ff; Lillie, p. 9ff; Manzi; Scott-Elliot; Sinnett (1884-1896) p. 106; (1896) p. 258ff; Solovyoff, pp. 352-362; Steiner; Williams (1931) pp. 270-276; (1946) *passim.*

For Hyperborea and Abaris see Apollodorus, II, v, 11; Diodorus, II, 47; Godwin, p. 47; Herodotus, IV, 36; Pausanias, I, iv; III, xiii; Pindar, *Tenth Pythian Ode*; Plato, *Charmides*; Pliny, IV, xxvi; Strabo, I, iv, 3-4; II, iv, 1; IV, v, 5; XV, i, 57.

For other occult lost-continent doctrines see Anderson, pp. 124-131, 143-172; Ballard; Banning; Bryan, esp. p. 107f; Cervé; Clymer; *Lemuria the Incomparable;* Leslie; Newbrough; Oliver; Phelon; Spence (1933) p. 104ff; Sugrue; Thévenin, pp. 106-125.

Words: "Seychelles" is pronounced *"say*-shell"; *Dzyan,* "John"; and "Cayce," "Casey." "Besant" rhymes with "pleasant," "Crowley" with "holy." "Helena Petrovna Blavatsky" = Russian *Yelena Petrovna Blavatskaya.* "Hyperborea" is usually derived from the Greek words for "beyond the North Wind"; see however *EB, s. v. Hyperboreans.* There is some question as to whether the larger of the two recently found giant ape-men from Asia should be called *Gigantanthropus* or *Gigantopithecus.* Judging from the size of the three teeth found so far, the creature must have weighed about 1,000 pounds! Furthermore, even more recently, remains of a similar beast have been found in South Africa.

CHAPTER IV

Quotations: Martin, p. 281; Fessenden, p. 1; Spence (1925) p. 17; (1924) pp. 112, 2, 25; (1933) p. 182; (1926) pp. 50, 208; (1933) p. 235.

Other main sources: For modern Atlantist societies and doctrines see Abartiague; Andersen, p. 22ff; Bellamy; Bessmertny, pp. 55f, 105-161, 181-185, 223-227; Braghine, pp. 6of, 82ff, 143, 188, 214, 218, 236 *et passim;* Bramwell, pp. 21-26, 109-119, 132-186, 252-260; Buffault; Calahan; Duvillé; *EB, s. v. Gibraltar, Tunisia;* Fessenden; R. M. Gattefossé; Gattefossé & Roux; Georg; Germain; Leet, p. 131; Ley (1947); MacDougall, pp. 121-130; Mahoudeau; Martin, pp. 266-277; Miller, xii; Mortillet; H. F. Osborn (1915-1923) pp. 451ff, 506-510; Spence (1924, 1925, 1926, 1933, 1942, 1943); Strabo, I, iii, 4-5; Vaillant (1940); Velikovsky; Verneau; Wells, pp. 113-116, 183f; Wilkins.

For new-World inscriptions, linguistic relationships, &c., see Apollodorus, III, xiv; Bessmertny, pp. 150-161; Braghine, pp. 160-181; de Camp & Ley; Delabarre; *EB, s. v. Poseidon;* Fawcett; Gattefossé, p. 66; Georg, p. 204; Gould, p. 145; Herodotus, VIII, 55; Holand, p. 106; MacDougall, p. 125; Magoffin, p. 208; Mason, p. 51; *The Maya and Their Neighbors,* pp. 52-70; Morison, p. 72; Rouse; Whorf (1940); Wilkins. For Fawcett's disappearance see also George M. Dyott's *Man Hunting in the Jungle,* Peter Fleming's *Brazilian Adventure,* Robert Churchward's *Wilderness of Fools,* and Charles E. Key's *Story of Twentieth-Century Exploration.*

Words: *Xingú* is pronounced "shing-*goo*"; *Zschaetzsch,* "chech"; *Cró-Magnon,* "kro-man-*yawnh*"; *Quetzalcoatl,* "kayt-sahl-*kwatl*" (or "ket-"). In the comparative table of numerals in various languages, the Hebrew numerals could have been transliterated *'eḥad, šenayim šelošah,* &c; systems of transliterating Hebrew vary widely. The (') in Hebrew words stands for *'aleph,* the glottal plosive (a cough) while (') stands for *'ayin,* the pharyngeal fricative (a kind of wheeze). The Chinese numerals in the Wade-Giles transcription are spelled *i, erh, san, szŭ, wu, liu, tsʻi, pa, kiu,* and *shih. I, tsi, giu, shï* are pronounced "yee," "chee," "jyoo," "shurr." The Japanese numerals in the phonemic

Rómazi system would be *hitotu, hutatu, mitu,* &c. The Japanese *u* is like the English *u* in *put*, and is sometimes dropped from pronunciation altogether. Japanese has another set of numerals borrowed from Chinese: *ichi, ni, san, shi,* &c. Mayan *x* = English *sh*; hence *uaxac* is pronounced "wah-shock." The Otomi numerals are in the Tepehua dialect.

The account of late Paleolithic culture (Crô-Magnon &c.) given in this chapter is that which was current at the time Spence wrote, in the 1920's. Since then our knowledge of these matters has been considerably expanded and reorganized; for an up-to-date summary see Ralph Turner's *The Great Cultural Traditions,* Vol. I, pp. 22-36.

Besides the Scandinavian lemmings alluded to in Chap. iv, the Alaskan lemmings commit a similar mass suicide by swimming into the Arctic Ocean (*N. Y. Times,* 18 June 1953, p. 31). Fessenden was one of the inventors of radio, and the inventor of the fathometer and other electrical and acoustic devices; see *Fessenden, Builder of Tomorrows* by Helen M. Fessenden. There are a number of ruined cities in Paraguay, about 800 miles south of where Fawcett was searching, but their origin is not mysterious. They were built in the 17th and 18th centuries by the Indians under the Jesuits' direction, and fell to pieces when the Spanish government expelled the Jesuits.

Chapter V

Quotations: *The Maya and Their Neighbors,* p. 185; Haugen, p. 76; Brinton, *Maya Chronicles,* p. 108; Barnes, p. 65; Brinton, *Chilan Balam,* p. 9.

Other main sources: For the re-discovery of the Mayas see von Hagen; for Mayan anthropology, language, and culture see Brinton, *Maya Chronicles,* pp. 13-90; Diringer, pp. 120-135; Gann, p. 206; Hewett, pp. 239-241; Jakeman, p. 184; Landa, p. 169f; Mason, pp. 51, 60f, 317f; *The Maya and Their Neighbors,* chs. v, ix; pp. 420, 457; Mitchell, p. 100; Morley (1915) pp. 26-32; (1946) chs. ii-iv, viii-xvii; Nordenskiöld, p. 48; Prescott, p. 34; Smith, Malinowski, *et al.,* pp. 70-95; Spence (1913) p. 16off; (1924) p. 129ff; Spinden, p. 114; E. H. Thompson, p. 77; Vaillant (1941) p. 170ff; (1944).

For Mayan literature, chronology, and history, see Blom, p. 199ff; Brasseur (1861); Brinton, *Maya Chronicles;* Gann, pp. 17, 81ff, 209; Hewett, pp. 209-302; Jakeman, p. 39ff; *The Maya and Their Neighbors,* pp. 128, 373; Mitchell, pp. 105ff, 140, 224ff; Morley (1915) pp. 44-195; (1946) p. 40ff; chs. v-vii, xii; Spence (1908); (1913) pp. 208-239; Spinden, pp. 108-133; Stirling; Vaillant (1941) pp. 26-50.

Words: *Ixtlilxochitl* is pronounced "eesh-tleel-*shaw*-cheetl"; *Uaxactún,* "wah-shahk-*toon*"; *Tuxtla,* "*toosh*-tlah"; *Tutulxiu,* "too-tool-shyoo"; *Hunac Ceel,* "hoo-nahk kayl"; *Ah Xupan Xiu,* "ahkh shoopahn shyoo"; *Chichén Itzá,* "chee-*chayn* eet-*sah*." The Aztec *tl* is a single phoneme, a laterally exploded *t*, pronounced about as in "butler"; there is no connection between Nahuatl words like *atl* and *Aztlan* and Greek words like *Atlas* and *Atlantis,* since in Greek the *t* and *l* are always separate phonemes. The term "Mayan language" is usually taken to mean Yucatecan Mayan or Mayathan; it is a matter of opinion which of the other speeches of the Mayan group should be deemed dialects and which separate languages. Homophones are words having the same sound but different meanings, like English *no* and *know*. The full name of Montezuma was *Montecuzomai-thuicamina*. *Quetzalcoatl* and *Gucumatz* both mean "feathered snake," the symbol of the god in question. "Hurricane" is derived from the name of the wind-god Hurakan. Many modern Americanists prefer to use the one form "Maya" for the plural "Mayas" and the adjective "Mayan," thus: "Maya culture" or "the culture of the Maya." Nordenskiöld (cited as an authority *supra*) is pronounced "*noor*-d'n-shöl." *Kukulcan* is also spelled *Kulkulcan*.

CHAPTER VI

Quotations from P. B. Shelley, *Ozymandias;* Raglan (1939) p. 40. Other main sources: For theories of the origin of the Amerinds, see Andersen, p. 22ff; Casson (1939) p. 112ff; De Voto, *Across the Wide Missouri*, pp. 2, 5, 44f, 133, 287, and *The Welsh Indians;* Diringer, p. 136; Heyderdahl; Holand, pp. 230-237; Hungerford, p. 73ff; *The Maya and Their Neighbors,* pp. 418, 434; Means, p. 41ff; Nordenskiöld, pp. 20-36; G. E. Smith (1924) p. 103f; Joseph Smith; Spence (1913) p. 5; (1914) p. 4; (1925) p. 172; *Time,* 21 Apr. 1947, p. 85; Verrill & Verrill; Wissler (1940) pp. 154-163, 167, 176f, 264, 270f. See also the review of Heyderdahl's book by Ekholm and Bird in *Natural History* for Dec. 1950, and the further remarks by Riesenfeld on p. 50 of the Feb. 1951 issue of the same magazine; Gordon F. Ekholm's article *Is the American Indian Culture Asiatic?* in *Natural History* for Oct. 1950, p. 344; and George F. Carter's *Plant Evidence for Early Contacts with America* in *Southwestern Journal of Anthropology,* Vol. VI, No. 2 (Summer, 1950) p. 161. The Verrill book, one of most eccentric yet, propounds a theory of Sumerian Indians.

For the diffusionist controversy see Blom, pp. 191, 204-209; Daunt; Delafosse, p. 267; Diringer, pp. xi, 54, 58, 95, 148-157, 175-185; Dixon, pp. 47, 86, 122, 131f, 141f, 157-162, 185, 244-251; *EB s. v. Blowgun, Printing;* Gladwin, *passim;* Goldenweiser, pp. 35-152; Linton, p. 313; Lowie, pp. 156-195; Mitchell, p. 87f; *N. Y. Times,* 15 Mar. 1947, Sec. IV, p. 9; Nordenskiöld, p. 2f; Perry, p. 18f; Raglan (1939); Rivers, pp. 109-149; G. E. Smith; Smith, Malinowski, *et al.;* von Hagen (1946); Wallis, p. 111. See also Daniel S. Davidson's *The Antiquity of Man in the Pacific and the Question of Trans-Pacific Migrations,* and E. D. Merrill's *Domesticated Plants in Relation to the Diffusion of Culture,* in *Early Man* (a symposium, Lippincott, 1937).

For the history of the Amerinds see Gladwin; Hibben; *The Maya and Their Neighbors,* pp. 7f, 35f, 419; Spinden; Vaillant (1941); and Wissler. See also several of the contributions to *Early Man,* cited *supra.*

Words: *Madoc* is also spelled *Madog.* The Welsh word for "cow" is *buwch,* pronounced "*bee*-ookh" or "bewkh." The spear-thrower was called an *atlatl* in Aztec and a *hulche* in Maya.

CHAPTER VII

In the quotation from Seneca's *Medea* at the head of this chapter, some manuscripts name Tiphys (the helmsman of the Argonauts) instead of Tethys.

Main sources: For the history of geology see Adams, p. 1of *et passim; EB s. v. Geology; also Giants of Geology* by C. L. & M. A. Fenton. For the structure of the earth and former continents and land-bridges see Bonnier; Dacqué; Daly (1926) pp. 101-124; (1938), (1942); Dévigne; Fermor; Forrest; Georg; Gladwin, p. 49f; Gregory; Joleaud; Joly; Leet, p. 149; Ley, *The Lungfish and the Unicorn*, ch. ix; *New York Times*, 18 Nov. 1948, p. 23; F. Osborn, p. 22ff; H. F. Osborn; Schuchert (1908), (1917), (1931), and *Gondwana Land Bridges*; Scott, pp. 241-253; Simpson; Spence (1924) pp. 31-39; Suess, I, pp. 419, 596; II, pp. 254-258; III, p. 19f; Termier; *Theory of Continental Drift*, p. 72; Willis.

For the continental-drift hypothesis see Chaney; Chatwin; Dacqué (1915); Daly (1926); Fermor; Schuchert (1931) pp. 4, 102; Spence (1924) p. 43f; *Theory of Continental Drift; Time*, 19 Jul. 1948, p. 68 (which tries to connect the flight of migratory birds with continental drift); Wegener.

For the causes of inundations see Ammianus, XV, ix, 3-4; Bessmertny, p. 215ff; Bramwell, pp. 125-131, 183f; Daly (1926); Diodorus, V, 22; Ewing; Florus, I, xxxviii; Gould, pp. 124-130; Leet; Ley (1939); *The Maya and Their Neighbors*, p. 422; *New York Times*, 18 Jan. 1948, Sec. IV, p. 11; H. F. Osborn (1910-1921) pp. 19-22; Schuchert, *The Periodicity of Ocean Spreading*; Spence (1924) p. 221f; Strabo, VII, ii, 1-2; Suess, I, p. 43ff; Thévenin, p. 99f; *Time*, 16 Dec. 1946, p. 34f; 2 Jan. 1948, p. 64; Wells, pp. 113-116; Woolley, pp. 2-26.

Words: "Sima" rhymes with "I'm a"; "sial" with "Hi, All!"; "nife" (the name given the nickel-iron core of the earth) with "wifey"; "orogeny" with "progeny." The original term of Suess (whose name is also spelled *Süss*) for sial was *sal*, but for various reasons modern geologists prefer the other. *Angara* is pronounced "ahng-gah-rah"; *Tethys*, "teeth-iss"; *tsunami*, "tsuh-*nah*-mee"; *loess*, "*lo*-iss" or "löss"; *Dahut*, "da-üt" or "da-ü." In some versions Dahut is called "Ahes." "Rann of Kachh" is pronounced and sometimes spelled "Runn of Cutch." Andesite is a kind of rock mixed sima-sial composition. The legendary Ys is also called "Ker-is," *ker (caer, car)* meaning "city" or "fort" in the Keltic languages.

If you wonder why pressure solidifies lava, though it melts ice, remember that nearly all substances shrink as they freeze, while only a few like water and bismuth swell as they solidify. The tendency of continental blocks to float in the crust at heights determined by their natural buoyancy compared with sima is called *isostasy*. Schuchert (*Theory of Continental Drift*, p. 143f) also classifies the three main Paleozoic land-masses as "Holarctis," "Equatoris" (= Gondwanaland) and "Antarctis," while Du Toit, a Wegenerite, assumes two original continents, a northern "Laurasia" and a southern "Gondwanaland." Suess also believed that the earth was once covered by a single universal ocean, which he called *Panthalassa*, and that land appeared above the water only when the originally smooth crust became wrinkled (a theory lately revived by Urey — see *Time*, 7 Nov. 1949, p. 59f.) The report read by Termier on Mt. Pelée was by Lacroix, and the land-bridge across the South Atlantic, the theory adopted by Neumayr, originated with Marcou.

On the subject of submergences and inundations, the Chinese Emperor Yao was probably legendary. A subsidence of as much as two inches a month has occurred in Long Beach, Calif., but this is due to human action — the pumping of oil out from the beds underlying the region. (*N. Y. Times*, 21 May, 1949, p. 15.) Dr. Ewing, in the course of his expeditions in the *Atlantis,* has brought up from the Atlantic, at depths of two to three and a half miles, unexplained deposits of what looks like beach sand in his coring device. Atlantists may get what comfort they can from the fact; Ewing himself does not believe there is any connection with Plato's Atlantis.

Estimates of the raising and lowering of sea-level as a result of changes in climate, and the consequent enlargement and shrinkage of glaciers and ice caps, vary widely. For one thing the thickness of the Antarctic ice-cap has been measured in a few places only, so that the total volume of Antarctic ice cannot yet be accurately estimated. R. L. Daly estimated the rise of sea-level resulting from the melting of all the world's ice at 164 feet. (See W. F. Heald, *We Live in an Ice Age,* in *Natural History,* XLVIII, No. 5 (Dec. 1941) pp. 296-302.)

Chapter VIII

Quotations: Bramwell, p. 218; Spence (1924) p. 17; *Isaiah* xxiii, 1; 1 *Kings* x, 22; *Ezekiel* xxvii, 12; Pliny; II, lxvii; Bunbury, II, p. 703 (my translation); Bramwell, p. 155 (summarizing Whishaw, pp. 176-184); *Odyssey,* I, l. 54; V-IX, XIII; Leaf, p. 183; Polybius, XXXIV, ix.

Other main sources: For various identifications of Atlantis see Baer; Bessmertny, pp. 27-32; Björkman, p. 32f; Bramwell, pp. 214-229; Gattefossé & Roux; Hospel; Hungerford; Martin, pp. 271-300.

For Atlantis in North Africa see Benoît; Berlioux; Bessmertny, pp. 20-63; Bramwell, pp. 109-122; Bunbury, I, p. 323; Burton, IV, p. 176f; VI, pp. 83-122; Butavand; Cory, p. 37ff; Georg, p. 221ff; Herodotus, IV, 196; Khun de Prorok, pp. 60-66, 70, 104, 114, 170f, 238, and other works by the same author; Marque; Plato, *Critias,* 115D; Pliny, VI, xxxvi; Rawlinson, pp. 258-304, 389; Spilhaus, p. 36; Strabo, I, iii, 2.

For Atlantis in Crete, Carthage, and Tartessos see *American Journal of Archaeology,* Vol. LII, No. 3 (July-Sep. 1948) pp. 474-480; Apollodorus, I, ix, 26; II, v, 7, 10; III, i, 1-4; xv, 708; *Epitome,* I, 7-15; and refs. given *loc. cit.* in LCL ed.; Appian, VI (*The Wars in Spain*) I (2); Aristotle, *Marvelous Things,* 135, 136; Arrian, II, 16; Babcock (1922) pp. 27-31; Balch; Bérard; Bessmertny, pp. 32, 62-78; Björkman, pp. 73-102; Bonsor; Bramwell, pp. 71-90, 122-125, 154-158; Bunbury, I, p. 104; II, p. 72f; 2 *Chronicles* ix, 21; xx, 36; Cory, p. 281f; Dawson, pp. 217, 228f; Diringer, pp. 249-252; Diodorus, IV, 18, 59-61, 75-79; *EB s. v. Carthage, Crete, Spain;* *Ezekiel* xxxviii, 13; Georg, p. 222; Herodotus, I, 163-165; IV, 43, 152, 192; Isaiah ii, 16; xxxiii, 6, 10, 14; lx, 9; lxvi, 19; Janvier; Jeremiah x, 9; Jonah i, 3; Knapp; Leaf, p. 119; Ley (1943); Magoffin, p. 102; Paterculus, I, ii, 3; Pausanias, VI, xix; Pliny, III, iii; IV, xxxvi; *Psalms* xlviii, 7; lxxii, 10; Rawlinson, pp. 98-125 and refs. given *loc. cit.;* Spence (1924) p. 207; Strabo, I, ii, 26; III, i, 6; ii, 11-15; XVII, i, 19; Tacitus, *Agricola,* x; *Germania,* xlv; Whishaw.

For Scheria see Apollodorus, I, ix, 25; Björkman, pp. 37-63; Bramwell, pp. 68ff, 76, 79ff, 123; Butler, p. 225f; *EB s. v. Ithaca;* Leaf, pp. 140-183; *Odyssey, passim;* Shewan, pp. 1-102; Strabo, VI, ii, 4; VII, iii, 6; VIII, ii, 2; iii, 26; X, ii, 7-20; XVII, i, 16.

For possible lost civilizations see *American Journal of Archaeology,* Vol. LI, No. 3 (Jul.-Sep. 1947) p. 322f; Casson, (1934) p. 92ff;

G. Cooper; Delafosse, pp. 42-96; *New York Times*, 4 Jan. 1947, p. 7; 28 Mar. 1947; 8 Apr. 1947, p. 26; *Time*, 12 Nov. 1945, p. 50; 27 Jan. 1947, p. 74ff. For European prehistory see Dawson and Woolley; also other works by these authors and by V. Gordon Childe, C. F. Hawkes, H. J. E. Peake, &c.

Words: Olof Rudbeck's given name is also spelled *Olf* and *Olaüs*; the title of his book was *Atlantis sive Manheim, vera Japheti posteriorum sedes ac patria*, or (in Swedish) *Atland eller Manheim dedan japhetz afkomme*. Bailly's *Caf* is also spelled *Kaf* or *Qaf*. The *Ahaggar* is also spelled *Hoggar*. In French, *Atlantis* appears variously as *Atlantis*, *Atlantide* (from the Greek genitive *Atlantidos*) and *Atlantie*. *Tuareg* (or *Touareg*) is the plural of *Tarqi* (or *Targui*). The *Shott el Jerid* (or *Shatt*, or *Chott*, or *Djerid*) is also called the *Shott el Kebir*, *el Farun*, or *el Hammeina*. *Nebuchadrezzar* is also spelled *Nebuchadnezzar* or *Nabochodonosor* (Akkadian, *Nabu-kudurri-ushur*). The Carthaginian *suffete* (Heb. *shophet*) was one of a pair of elected magistrates with powers like those of a Roman consul. The *Tarshish* of Genesis x, 4 may be Tarsus in Anatolia instead of Tartessos. *Guadalquivir* is pronounced "gwa-*dhal*-kee-*veer*"; *Guadalete*, "gwa-dha-*lay*-tay" (or "-*leh*-tay"); Ceuta, "*thay*-oo-tah." *Scheria* is also spelled *Scherië*, and pronounced "skerry-uh" or "skerry-ee." (Could there be a connection between *Scheria* and *Euskera* or *Eskuara*, the Basques' word for their own unique language? Probably not, but it's an intriguing idea.) *Rhadamanthos* and *Samë* also appear as *Rhadamanthys* and *Samos*. Greek *Odysseus* = Latin *Ulyxes*, *Ulysses*; Greek *Alkinoös* = Latin *Alcinoüs*, &c. Diodorus (IV, 72) says the father of Alkinoös was, not Nausithoös, but Phaiax, for whom the Phaiakes were named.

Andersen (*op. cit.*) gives the length of a large Polynesian catamaran as 150 ft. with a complement of 400 persons, compared to 75 ft. and 120 men for a trireme of the time of Salamis according to W. L. Rogers, *Greek and Roman Naval Warfare*, p. 46. Both would be considered fair-weather craft by modern standards because of their low freeboard.

CHAPTER IX

Quotations: Diogenes, III; Plato, *Timaeus*, 26E; Babcock (1922) p. 11; Chambers, p. 11; *Odyssey*, XII, ll. 12ff; Hesiod, *Works and Days*, ll. 170-169d; Pindar, *Olympian Odes*, II, ll. 68-80; Plutarch, *Theseus*, i; Virgil, *Aeneid*, IV, ll. 48off; Thucydides, III, lxxxix; Apollodorus, I, i, 1-2; Herodotus, II, 4, 99; Manetho, fragm. 6 from Syncellus, p. 27.

Other main sources: For the nature of *Timaeus* and *Critias* see Bramwell, pp. 102-105; Lowes, *passim;* Plato, *Timaeus*, vii (20-21), xxvi (26); *Critias*, vii (113), and other works mentioned.

For the Homeric question see *American Journal of Archaeology*, LII, No. 1 (Jan.-Mar., 1948) and LIV, No. 3 (Jul., 1950) *passim;* Björkman, pp. 12, 50-69; Bunbury, I, pp. 1-79; Butler; Chadwick, pp. 159-307; de Camp & Ley, ch. ii; Diringer, p. 89; *EB s. v. Achaeans*, *Hittites, Homer;* Fiske, pp. 174-208; Geddes, pp. 5-69; Homer, *passim;* Leaf, pp. 60-183; Ley, *Scylla Was a Squid*, and (1945); Murray; Shewan, pp. 1-111, 237; Strabo, I, ii, 3; J. A. K. Thompson; Tozer, ch. ii; Whibley, pp. 117-127.

For the growth of Classical geography see Apollodorus, I, vi, 1; Aristotle, *Marvelous Things*, 84, 136; *Meteorologica*, II, i, 354a (22); *De Mundo*, iii, 392b; Björkman, pp. 16f, 54-61; Bunbury, I, pp. 33-404; ii, pp. 173-185, 523; Casson (1939) pp. 27, 55, 83; Cicero, *De Natura Deorum*, III, xxiii; Diodorus, III, 11, 72; IV, 4, 18, 27; Florus, II, x; Georg. p. 136; Herodotus, II, 4-15; Hesiod, fragm. 5; Mela, III, x; Paterculus, II, 51; Pausanias, I, xxiii, 5; VIII, xxxiii; Pindar, *Olympian Odes*, III, ll. 42-45; VII, ll. 54-76; *Nemean Odes*, III, ll. 19-26; IV, ll. 69-72; *On Delos;* Pliny, V, v, viii; VI, xxxvi, xxxvii; VII, ii; Plutarch, *Sertorius*, viii-ix; *Face in the Moon*, xviii, xxvi; Ptolemy, I, viii, 5; W. Smith, *s. v. Atlanticum Mare;* Spilhaus, pp. 26-44; Strabo, I, iii, 20; III, v, 3; VIII, iii, 19; Tozer; Tylor, II, p. 66.

For the general discussion of Plato's sources see Adams, p. 10f; Aristotle, *Constitution of Athens* (editor's introd., *GH*, II, p. 676); Baikie, II, pp. 381-387; Bramwell, pp. 90-105; Bury, chs. i, ii, iv, xv; Cory, p. 72; Martin, p. 252.

Words: *Atarissiyas* is also spelled *Attarishshijash*, &c. *Akhiyawas* is variously spelled *Akhkhiyavâ, Ahhiyawas, Ahijjawas, Ahhijawas*, &c.; they are probably the same as the Biblical Hivites (*Joshua* ix, &c.), the Egyptian Akkaiwasha or Ekwesh, and the Homeric Achaeans, who called themselves *Achaiwoi* before the *w* (*F*, digamma) dropped out of Ionic Greek. All existing mss. of Herodotus call both Atlantes and Atarantes *Atlantes,* but two later Greek scholars corrected the name of the second of these to *Atarantes,* saying that Herodotos wrote it thus. (See W. Smith *s. v. Atarantes;* Bunbury, I, p. 279.) He also described the two tribes distinguished by Pliny as the Garamantes and Gamphasantes, but called both *Garamantes.* Diodoros (I, xliii, xlv, xciii) calls Aha-Mena *Mēnas* and *Mneuēs. Silenos* = *Greek Silēnos, Seilēnos;* Latin *Silenus.*

Anaximandros (who thought we lived on one of the flat surfaces of a drum-shaped earth) was not the first map-maker. The Babylonians had drawn crude maps on slabs of clay, and the art is known to some primitive peoples. His, however, was the first-known *world* map (Strabo, I, i, 11).

CHAPTER X

Quotations: Raglan (1936) p. 11; Spence (1926) p. 111; Donnelly, p. 2; Raglan, *op cit., pp.* 59-127; Fiske, p. 178; Plato, *Republic*, VIII (Jowett, II, p. 308); *Isiah* xi, 7; Andrews, p. 275f.

Other main sources: For mythology see Aelianus, VII, i; Andersen, p. 349ff; Apollodorus, I, vii, 2-3; III, viii, 2; xiv, 5, and notes, *loc. cit.,* in LCL ed.; and *Epitome,* vi, 2-4, 19; Boas, pp. 397-524; Cicero, *De Natura Deorum*, I, xlii; Cory, pp. 53-62; Diodorus, II, iv-xx; *EB s. v. Deluge;* Fiske; *Genesis* vi-ix; Guerber; Herodotus, I, 184; King, pp. 81, 161-173; MacBain, pp. 16-19; Magoffin, p. 75; Plato, *Republic*, II (Jowett, II, p. 47f); Plutarch, *Isis* and *Osiris*, xxiii; Schnabel; Seneca, III, xxvii-xxx; Smith, Malinowski, *et al.*; Spinden, p. 210; Stewart, p. 8; Stocking; Strabo, XIV, i, 27; iv, 3; v, 16; Tylor, I, p. 279; Vaillant (1941) p. 214; Woolley, p. 21ff.

For Plato's writing see Bramwell, pp. 52-59, 230-234; *EB s. v.*

Lost Continents

Plato; *Hertzler*, pp. 99-120; Plato, *passim* (esp. Jowett, I; II, pp. 53ff, 73ff; III, pp. 26off, 442); Plutarch, *Lycurgus*; Stewart.

For utopianism see Andrews; Arnobius, I, v; Bloomfield; Bramwell, pp. 230-248; *Cambridge Ancient History*, VII, p. 265; VIII, p. 810; Diodorus, II, lv-lx; V, xlii-xlvi; Florus, I, xxxv; Paterculus, II, iv, 1; Plato, *The Republic; Timaeus; Critias; The Laws*, I and V; Pliny, V, xxiv; Plutarch, *Tiberius Gracchus;* Ptolemy, VII, iv; Spence (1942); Strabo, II, i, 14; XIV, i, 38; XV, i, 14-15; Tarn, p. 104; and the utopian novels by Wells, Wright, and other authors mentioned.

Words: The name of the Sumerian Flood-hero Ziusudra appears in the fragments of Berossos as *Xisouthros* and *Sisithros*, and possibly also as *Xouthos*, the grandson of Deukalion, in the Greek Deluge-legend. Xouthos is described (Apollodorus, I, vii) as the son of Hellen and the father of Achaios and Ion, the eponymous ancestors of the Hellenes, Achaeans, and Ionians respectively. The Babylonians changed Ziusudra's name to *Utnapishtim* — the first syllable of which, however, can also be read *Shamash-, Pir-, Tsit-*, and several other ways (Cory, pp. 53, 6off; King, pp. 161-173).

In ancient times Ceylon was also known as Kabodanea and Salika. The Babylonians named the five visible planets after their gods Nabu, Ishtar, Nerigal, Marduk, and Ninurta. Under Babylonian influence the Greeks in turn named these planets after the corresponding gods of their own pantheon: respectively Hermes (or Apollo), Aphrodite, Ares (or Herakles), Dios (Zeus), and Kronos. In the same way the Romans later adapted the system to their own language and pantheon: Mercurius, Venus, &c. Plato's reference to the "Star of Hermes" is in *Timaeus* 38D. *Berossos* = Latin *Berossus*, Greek *Bērōssos, Bērōsos, Bērosos*, &c. (See Schnabel.) *Euemeros* = Greek *Euēmeros*, Latin *Euhemerus, Evemerus*. *Dietrich (Theodoric)* is spelled 85 known ways: *Tidrick, Thiŏrek, Thiŏŏrekr, Thierry*, &c.; the original Gothic was *Thiudareiks*. The German *Wittich* (also spelled *Witege)* = Gothic *Witigis*, Latin *Vitiges, Cúchulainn* is pronounced with a German-type *ch*, to rhyme with "Sue Bullen."

CHAPTER XI

Quotations: Verne, p. 176f; Wheatley (1936) p. 12.

For the lost-continent theme in fiction see also Ashton; Baring-Gould; Benoît; Bond; Coblentz; Cox; Cradock, de Camp; Doyle; Foster; Hatfield; Howard; Hyne; Livingston; Mitchell (1932); Mullen; Oliver; Parry; Roy; Rutter; Shaver; Sibson; C. A. Smith; Vigers; von Ravn; Wheatley; and Wright.

Additional titles can be found in *The Checklist of Fantastic Literature* by Bleiler & Korshak (Shasta Publishers, 1948). Besides the stories mentioned in Chapter XI, lost-continent stories appear in the Checklist under *Anderson, Olaf W.;* Aubrey (1899); *Bachelder; Birkmaier; Brown, Joseph M.; Bruce, Muriel; Burroughs, Edgar Rice (Tarzan & the Jewels of Opar); Campbell, John W. Jr.; Chester, George R.* (1912); *Crawford, Isabell C.; Dail* (1890); *Dunn, Waldo H.; Gale; Holtby; Horniman* (1900); *Kline, Otis Adalbert* (1937); *Lang* (1886); *Laurie* (1896); *Legge, Margaret; Leiber; Le Plongeon, Alice; Lewis, L. A.*

(pp. 133-152); *Merritt* (1919); *Metchim; Morrow, Lowell H.; Muddock; Mundy, Talbot (The Nine Unknown; Jimgrim); Odell* (1889); *Scoggins, C. E.* (1931); *Serviss* (1912); *Simms; Stutfield; Vivian, E. Charles (City of Wonder); Whitehead* (1946); & *Williams, M. Sheldon.* The novels by Rousseau and Stilson mentioned in Chapter XI have not appeared as books, but were recently reprinted in *Fantastic Novels* (May, 1949 & Sep. 1950 respectively).

Words: Clark Ashton Smith informs me that he means "Zothique" to rhyme with "antique." "Ptath" is pronounced "tahth," rhyming with "Goth."

BIBLIOGRAPHY

(This is not a complete bibliography on the subject of lost continents. For additional titles see the bibliographies in Abartiague, Bessmertny, R.-M. Gattefossé, and Gattefossé & Roux—especially the last with its 1700 titles. For further study of the subject, the best books I know of are Babcock (1922), Bessmertny, Björkman, Bramwell, Bunbury, Hibben, Leet, Ley *(The Lungfish & the Unicorn)*, Lowie, Martin, Morley (1946), Murray (1924), Schuchert (1931), Silbermann, Stewart, *Theory of Continental Drift*, and, among the pro-Atlantis books, Spence (1924). Abbreviations: ANCL—*Ante-Nicene Christian Library*, Edinburgh: Clark & Clark, 1867 *sqq.*, 24 vols. GH—*The Greek Historians*, N. Y.: Random House, 1942, 2 vols. LCL—*Loeb Classical Library*, London: Heinemann. ML—*Modern Library*, N. Y.: Random House.)

ABARTIAGUE, William d': *L'Atlantide et les Basques: Essai de Bibliographie*, Bayonne: Courrier, 1937.

ADAMS, Frank Dawson: *The Birth & Development of the Geological Sciences*, Baltimore: Williams & Wilkins, 1938.

AELIANUS, Claudius ("Aelian"): *His Various History*, London: Dring, 1665.

AMMIANUS Marcellinus ("Ammian"): *Roman History*, London: Bohn, 1862.

ANDERSEN, Johannes C.: *Myths & Legends of the Polynesians*, N. Y.: Farrar & Rinehart, 1928.

ANDERSON, Wing: *Seven Years that Change the World*, L. A.: Kosmon, 1940-1944. A book of prophecies.

ANDREWS, Charles M.: *Ideal Empires & Republics (Rousseau's Social Contract, More's Utopia, Bacon's New Atlantis, Campanella's City of the Sun)*, Washington: Dunne, 1901.

APOLLODORUS of Athens: *The Library*, LCL. A mythological treatise.

APPIAN (Appianos of Alexandria): *Roman History*, LCL.

ARISTOTLE (Aristoteles of Stagyra): *Meteorologica & De Mundo* in *Works*, Oxford: Clarendon, 1931; *The Constitution of Athens*, GH; *On Marvelous Things Heard*, LCL.

ARNOBIUS Afer: *The Seven Books of Arnobius Adversus Gentes*, ANCL.

ARRIAN (Flavius Arrianus): *Anabasis of Alexander*, GH.

ASHTON, Francis: *Alas, That Great City*, London: Dakers, A novel.

——: *The Breaking of the Seals*, London: Dakers, 1946. A novel.

BABCOCK, William H.: *Atlantis & Antillia*, in *The Geographical Review*, Vol. III, No. 5 (May, 1917) pp. 392-395.

——: *Legendary Islands of the Atlantic (A Study in Medieval Geography*, N. Y.: Amer. Geogr. Soc., 1922.

BAER, Frédéric-Charles: *Essai historique et critique sur l'Atlantique des anciens*, Avignon: Seguin, 1835.

BAIKIE, James: *A History of Egypt (From the Earliest Times to the End of the XVIIIth Dynasty)*, London: Black, 1929.

BALCH, Edwin Swift: *Atlantis or Minoan Crete*, in *The Geographical Review*, Vol. III, No. 5 (May, 1947) pp. 388-392.

BALLARD, Guy Warren ("Godfré Ray King"): *Unveiled Mysteries,* Chicago: Saint Germain, 1939. The bible of the I AM cult.

BANNING, Pierson Worrall: *Maker, Man & Matter,* L. A.: Internat. Book Concern, 1924. Occult Atlantism.

BARING-GOULD, William S.: *Little Superman, What Now?* in *Harper's Magazine,* Sep. 1946, p. 283ff.

BARNES, Harry Elmer: *A History of Historical Writing,* Norman: Univ. of Okla. Pr., 1937.

BEAZLEY, C. Raymond: *The Dawn of Modern Geography (A History of Exploration & Geographical Science),* London: Murray, 1897-1903, 3 vols.

BECHOFER-ROBERTS, Carl Eric: *The Mysterious Madame: Helena Petrovna Blavatsky,* N. Y.: Brewer & Warren, 1931.

BELLAMY, Hans Schindler: *The Atlantis Myth,* London: Faber & Faber, 1948. Hörbiger's Cosmic Ice Theory.

——: *Built Before the Flood (The Problem of the Tiahuanaco Ruins),* London: Faber & Faber, 1942.

——: *Moons, Myths, & Man (A Reinterpretation),* London: Faber & Faber, 1936.

BENOîT, Pierre: *L'Atlantide,* Paris: Michel, 1920. A novel.

——: *The Queen of Atlantis,* London: Hutchinson. A translation of *L'Atlantide.*

BÉRARD, Victor: *L'Atlantide de Platon,* in *Annales de Géographie,* Vol. XXXVIII, No. 213 (15 May 1929) pp. 193-205.

BERLIOUX, Étienne-Félix: *Les Atlantes: Histoire de l'Atlantis et de l'Atlas Primitif,* Paris: Leroux, 1883.

BESANT, Annie Wood: *The Ancient Wisdom (An Outline of Theosophical Teachings),* Adyar: Theosoph. Pub. House, 1897-1939.

BESSMERTNY, Alexandre: *L'Atlantide (Exposé des hypothèses relatives à l'énigme de l'Atlantide),* Paris: Payot, 1935. A translation from German to French of Bessmertny's *Das Atlantis-Rätsel* by F. Gidon, with appendix giving Gidon's own theory.

BJöRKMAN, Edwin: *The Search for Atlantis,* N. Y.: Knopf, 1927.

BLAVATSKY, Helena Petrovna (Yelena Petrovna Blavatskaya): *The Mahatma Letters to A. P. Sinnett (from the Mahatmas M. & K. H., Transcribed, & with an Introduction by A. T. Barker),* London: Rider, 1923-1933.

——: *The Secret Doctrine (The Synthesis of Science, Religion, & Philosophy),* Adyar: Theosoph. Pub. House, 1888-1938, 6 vols.

BLOM, Frans: *The Conquest of Yucatan,* Boston: Houghton Mifflin, 1936. The Mayas.

BLOOMFIELD, Paul: *Imaginary Worlds (or the Evolution of Utopia),* London: Hamilton, 1932.

BOAS, Franz: *Race, Language, & Culture,* N. Y.: Macmillan, 1940.

BOND, Nelson S.: *Exiles of Time,* Phila.: Prime Pr., 1949. A novel.

BONNIER, Gaston: *L'Atlantide et les Continents Disparus,* in *La Nouvelle Revue,* Ser. 3, Vol. I (15 Jan. 1908) pp. 145-157.

BONSOR, George: *Tartesse,* N. Y.: Hispanic Soc. of Amer., 1922.

BRAGHINE, Col. Alexander Pavlovitch: *The Shadow of Atlantis,* N. Y.: Dutton, 1940.

BRAMWELL, James Guy: *Lost Atlantis,* N. Y.: Harper, 1938.

BRASSEUR de Bourbourg, Abbé Charles Étienne: *Manuscrit Troano (Études sur la Système et Langue des Mayas),* Paris: Impériale, 1869, 2 vols.

——: *Popol Vuh (le Livre Sacré et les Mythes de l'Antiquité Américaine, avec les Livres Héroïques des Quichés)*, Paris: Durand, 1861.
BRINTON, Daniel G.: *The Books of Chilan Balam (The Prophetic & Historical Records of the Mayas of Yucatan)*, Phila.: Stern, 1882. Reprint of a lecture.
——: *The Maya Chronicles*, Phila.: Brinton, 1882. Text & translation of the *Books of Chilan Balam*.
BRYAN, Gerald B.: *Psychic Dictatorship in America*, L. A.: Truth Research, 1940. An exposé of the I AM cult.
BUFFAULT, Pierre: *L'Atlantide de Platon*, in *Revue Philomathique*, Vol. XXXIX, No. 2 (Apr.-Jun. 1936) pp. 49-63.
BUNBURY, Edward Herbert: *A History of Ancient Geography (Among the Greeks & Romans from the Earliest Ages till the Fall of the Roman Empire)*, London: Murray, 1883, 2 vols.
BURTON, Richard F. (transl.): *The Book of the Thousand Nights & a Night*, London: Burton Club, 1886, 17 vols.
BURY, John B.: *A History of Greece*, ML.
BUTAVAND, F.: *La Véritable Histoire de l'Atlantide*, Paris: Chiron, 1925.
BUTLER, Samuel: *The Authoress of the Odyssey*, London: Jonathan Cape, 1897-1922.
CALAHAN, Harold Augustin: *Geography for Grown-ups*, N. Y.: Harper, 1946.
Cambridge Ancient History, Cambridge: Cambr. Univ. Pr., Vols. VII & VIII.
CARLI, Comte J. R. (de) (Giovanni Rinaldo Carli): *Lettres Américaines*, Paris: Buisson, 1788.
CASSON, Stanley: *The Discovery of Man (The Story of the Inquiry into Human Origins)*, N. Y.: Harper, 1939.
——: *Progress of Archaeology*, London: Whittlesey, 1934.
CERVÉ, Wishar S.: *Lemuria, the Lost Continent of the Pacific*, San Jose: Rosicrucian Pr., 1931.
CHADWICK, Hector Munro: *The Heroic Age*, Cambridge: Cambr. Univ. Pr., 1912. A study of epic poetry.
CHAMBERS, Edmund Kerchever: *The History & Motives of Literary Forgery*, Oxford: Blackwell, 1891.
CHANEY, Ralph W.: *Tertiary Forests & Continental History*, in *Bull. of the Nat. Geol. Soc. of Amer.*, Vol. LI (1 Mar. 1940) pp. 469-488.
CHATWIN, C. P.: *The Lost Atlantis*, in *The South-Eastern Naturalist & Antiquary*, Vol. XLIV (1940) pp. 41-51.
CHURCHWARD, James: *The Children of Mu*, N. Y.: Washburn, 1931-1945.
——: *The Lost Continent of Mu*, N. Y.: Washburn, 1926-1933.
——: *The Sacred Symbols of Mu.* N. Y.: Washburn, 1933-1945.
CICERO, Marcus Tullius: *De Natura Deorum*, LCL; *The Vision of Scipio*, in *Cicero's Offices*, London: Bohn, 1850.
CLYMER, R. Swinburne: *The Philosophy of Fire*, Quakertown: Philos. Pub. Co., 1920.
COBLENTZ, Stanton A.: *The Sunken World*, L. A.: Fantasy Pub. Co., 1948. A novel.
CONWAY, Moncure Daniel: *My Pilgrimage to the Wise Men of the East*, Boston: Houghton Mifflin, 1906. Baconianism & Theosophy.
COOPER, Gordon: *Dead Cities and Forgotten Tribes*, N. Y.: Philosoph. Lib., 1952.

COOPER, Lane (transl.): *Fifteen Greek Plays,* N. Y.: Oxford Univ. Pr., 1943.

CORY, Isaac Preston: *Cory's Ancient Fragments (of the Phoenician, Carthaginian, Babylonian, Egyptian and Other Authors),* London: Reeves & Turner, 1826-1876.

COX, Erle: *Out of the Silence,* Melbourne: Robertson & Mullins, 1947.

——: *The Eternal Echo,* London: Dakers, n.d.

CRADOCK, Phyllis: *Gateway to Remembrance,* London: Dakers, 1949. A novel.

DACQUÉ, Edgar: *Grundlagen und Methoden der Paläogeographie,* Jena: Fischer, 1915.

——: *Paläogeographie,* in *Enzyklopädie der Erdkunde,* Leipzig: 1926.

DALY, Reginald Aldworth: *Architecture of the Earth,* N. Y.: Appleton-Century, 1938.

——: *The Floor of the Ocean (New Light on Old Mysteries),* Chapel Hill: Univ. of N. Car. Pr., 1942.

——: *Our Mobile Earth,* N. Y.: Scribner's, 1926.

DAUNT, Hew Dalrymple: *The Centre of Ancient Civilization (Discoveries in Ancient Geography & Mythologies),* London: Bodley Head, 1926. A theory of the East Indian origin of civilization.

DAVIDSON, David, & Herbert Aldersmith: *The Great Pyramid: Its Divine Message,* London: Williams & Norgate, 1924.

DAWSON, Christopher: *The Age of the Gods (A Study in the Origins of Culture in Prehistoric Europe & the Ancient East),* London: Murray, 1928.

DE CAMP, L. Sprague: *Science-Fiction Handbook (The Writing of Imaginative Fiction),* N. Y.: Hermitage House, 1953.

——: *The Tritonian Ring (and Other Pusadian Tales),* N. Y.: Twayne Publ., 1953.

——& Willy Ley: *Lands Beyond,* N. Y.: Rinehart, 1952.

DELABARRE, Edmund Burke: *Dighton Rock (A Study of the Written Rocks of New England),* N. Y.: Neale, 1928.

DELAFOSSE, Maurice: *The Negroes of Africa (History & Culture),* Washington: Associated, 1931.

DÉVIGNE, Roger: *Un Continent Disparu (L'Atlantide, Sixième Partie du Monde),* Paris: Crès, 1924.

DE VOTO, Bernard: *Across the Wide Missouri,* Boston: Houghton Mifflin, 1947. The Mandans & the Welsh Indians.

——: *The Welsh Indians.* Lecture at Cooper Union, N. Y., 12 Dec. 1947.

DIODORUS Siculus (Diodoros of Sicily): *Library of History,* LCL.

DIOGENES Laërtius: *Lives of the Eminent Philosophers,* LCL, 2 vols.

DIRINGER, David: *The Alphabet (A Key to the History of Mankind),* N. Y.: Philos. Lib., 1948.

DISRAELI, Isaac: *Curiosities of Literature,* London: Routledge, Warnes, & Routledge, 1859, 3 vols.

DIXON, Roland B.: *The Building of Cultures,* N. Y.: Scribner's, 1928.

DONNELLY, Ignatius T. T.: *Atlantis: The Antediluvian World,* N. Y.: Harper, 1882. (Also 1949; but page-numbers in notes refer to the 1882 edition.)

DOYLE, Sir Arthur Conan: *The Maracot Deep (& Other Stories),* Garden City: Doubleday, Doran, 1928. A novel.

DUVILLÉ, D.: *L'Aethiopia Orientale ou Atlantide, Initiatrice des Peuples Anciens,* Paris: Soc. Française d'Ed. Littéraires & Techniques, 1936.

EFFLER, Louis R.: *The Spirals of Mu-Land*, Toledo: 1943. Pamphlet.

Encyclopaedia Britannica, 14th ed., esp. *s. v. Achaeans, Atlantis, Atlas, Bible, Blowgun, Carthage, Chronology, Crete, Deluge, Egypt, Geography, Geology, Gibraltar, Gondwanaland, Hittites, Homer, Hyperboreans, Ithaca, North America, Plato, Poseidon, Printing, Pyramid, Sahara, Shakespeare, Spain, Tunisia, Utopia.*

ERITH, Lionel Edward Patrick: *The British-Israel Fallacy*, London: Mowbray, 1921.

EURIPIDES: *The Bacchanals & Other Plays*, London: Routledge, 1888.

EVANS, Bergen: *Was Shakespeare Really Shakespeare?* in *The Saturday Review of Literature*, Vol. XXXII, No. 19 (7 May 1949) p. 7ff.

EWING, Maurice: *Exploring the Mid-Atlantic Ridge*, in *The National Geographic Magazine*, Vol. XCIV, No. 3 (Sep. 1948) p. 275ff.

——: *New Discoveries on the Mid-Atlantic Ridge*, in *The National Geographic Magazine*, Vol. XCVI, No. 5 (Nos. 1949) pp. 611-640.

FARQUHAR, John Nicol: *Modern Religious Movements in India*, N. Y.: Macmillan, 1915. Theosophy.

FAWCETT, Col. Percy Harrison: *Lost Trails, Lost Cities*, N. Y.: Funk & Wagnalls, 1953.

FERMOR, Sir Lewis L.: *Gondwanaland, a Former Southern Continent*, in *Proc. of the Bristol Naturalists' Soc.*, Vol. IX (1945) pp. 483-493.

FESSENDEN, Reginald Aubrey: *The Deluged Civilization of the Caucasus Isthmus*, Boston: Russell, 1923.

FISKE, John: *Myths & Myth-Makers (Old Tales & Superstitions Interpreted by Comparative Mythology)*, Boston: Houghton Mifflin, 1872-1901.

FLORUS, Lucius Annaeus: *Epitome of Roman History*, LCL.

FORREST, H. Edward: *The Atlantean Continent (Its Bearing upon the Great Ice Age & the Distribution of Species)*, London: Witherby, 1933.

FOSTER, George C.: *The Lost Garden*, London: Chapman & Hall, 1930. A novel.

GANN, Thomas, & J. Eric Thompson: *The History of the Maya (From the Earliest Times to the Present Day)*, N. Y.: Scribner's, 1931.

GARTENHAUS, Jacob: *The Ten Lost Tribes (A Discussion of British Israelism)*, Atlanta: Home Mission Board, 1938.

GATTEFOSSÉ, René-Maurice: *La Vérité sur l'Atlantide*, Lyon: Legendre, 1923.

GATTEFOSSÉ, Jean, & Claudius Roux: *Bibliographie de l'Atlantide et des questions connexes*, Lyon: Bosc & Riou, 1926.

GEDDES, William D.: *The Problem of the Homeric Poems*, London: Macmillan, 1878.

GEORG, Eugen: *The Adventure of Mankind*, N. Y.: Dutton, 1931.

GERMAIN, Louis: *La Problème de l'Atlantide et la Zoologie*, in *Annales de Géographie*, Vol. XXII, No. 123 (15 May 1913) pp. 209-226.

GLADWIN, Harold Sterling: *Men Out of Asia*, N. Y.: Whittlesey, 1947.

GODBEY, Allen H.: *The Lost Tribes a Myth (Suggestions Towards Rewriting Hebrew History)*, Durham: Duke Univ. Pr., 1930.

GODWIN, William: *Lives of the Necromancers*, London: Chatto & Windus, 1876.

GOLDENWEISER, Alexander: *History, Psychology, & Culture*, N. Y.: Knopf, 1933.

GOULD, Charles: *Mythical Monsters*, London: Allen, 1886.

La Grande Encyclopédie, Paris: Lamirault, 1890; *s. v. Brasseur, Jacolliot.*

GREGORY, John Walter: *Presidential Address*, in *Proc. of the Geol. Soc. of Lon.*, 1929-1930, pp. lxii-cxxxvi. Pacific geology.

GUERBER, Hélène Adeline: *Myths & Legends of the Middle Ages (Their Origin & Influence on Literature & Art)*, London: Harrap, 1931.

HARRINGTON, James: *The Oceana of James Harrington & His Other Works (Som wherof are now first publish'd from his own Manuscripts . . .)*, London: 1700.

HARRIS, Thomas Lake: *Arcana of Christianity (An Unfolding of the Celestial Sense of the Divine World through T. L. Harris)*, N. Y.: New Church Pub. Co. & Brotherhood of the New Life, 1858-1867.

HATFIELD, Richard: *Geyserland (Empiricism in Social Reform)*, Washington: Privately printed, 1908. A novel.

HAUGEN, Einar (transl.): *Voyages to Vinland: The First American Saga*, N. Y.: Knopf, 1942.

HERODOTUS of Halicarnassus: *The History*, GH & LCL.

HESIOD (Hesiodos): *Works & Days, Theogony, & Fragments*, LCL.

HETZLER, Joyce Oramel: *The History of Utopian Thought*, London: Allen & Unwin, 1923.

HEWETT, Edgar L.: *Ancient Life in Mexico & Central America*, N. Y.: Bobbs-Merrill, 1936.

HEYERDAHL, Thor: *Kon-Tiki*, N. Y.: Rand McNally, 1950.

HIBBEN, Frank C.: *The Lost Americans*, N. Y.: Crowell, 1946. The origin of the Amerinds.

HOLAND, Hjalmar R.: *America, 1355-1364 (A New Chapter in Pre-Columbian History)*, N. Y.: Duell, Sloane, & Pearce, 1946. Pre-Columbian Norse voyages.

HOLBROOK, Stewart: *A Congressman Rediscovers Atlantis*, in *The New York Times Book Review*, 30 Jul. 1944, p. 2. Donnelly's life.

HOMER (Homeros): *The Iliad & The Odyssey*, LCL & several other editions.

HORT, Gertrude M.; Richard Basil Ince; & William Perkes Swainson: *Three Famous Occultists (Dr. John Dee, Franz Anton Mesmer, & Thomas Lake Harris)*, Phila.: McKay.

HOSEA, L. M.: *Atlantis: A Statement of the "Atlantic" Theory Respecting Aboriginal Civilization*, in *The Cincinnati Quarterly Journ. of Sci.*, Vol. II, No. 3 (Jul. 1875) pp. 193-211.

HOSPEL, Paul: *Histoire des idées géographiques relatives à l'Atlantide*, in *Bull. de la Société Royale Belge de Géographie*, 30 Sep. 1939, pp. 185-215.

HOWARD, Robert Ervin: *The Coming of Conan*, N. Y.: Gnome Pr., 1953.

——: *Conan the Conqueror*, N. Y.: Gnome Pr., 1950.

——: *King Conan*, N. Y.: Gnome Pr., 1953.

——: *Skull-Face & Others*, Sauk City, Wis.: Arkham House, 1946. Collected stories.

——: *The Sword of Conan*, N. Y.: Gnome Pr., 1952. Collected stories.

HUNGERFORD, Edward B.: *Shores of Darkness*, N. Y.: Columbia Univ. Pr., 1941. Mythological speculations of the 18th & early 19th centuries.

HYNE, Charles J. Cutcliffe: *The Lost Continent*, N. Y.: Harper, 1900.

Lost Continents

339

JACOLLIOT, Louis: *Histoire des Vierges,* Paris: Lacroix, 1879. A study of mythology.

JAKEMAN, M. Wells: *The Origin & History of the Mayas (A General Reconstruction, in the Light of Basic Documentary Sources & Latest Archaeological Discoveries),* L. A.: Research, 1945.

JANVIER, Thomas A.: *In the Sargasso Sea,* N. Y.: Harper, 1898. A novel.

JASTROW, Joseph (ed.): *The Story of Human Error,* N. Y.: Appleton-Century, 1936. Terra Australis.

JOLEAUD, Léonce: *Atlas de Paléobiogéographie,* Paris: Lechevalier, 1939.

JOLY, John: *The Surface-History of the Earth,* Oxford: Clarendon, 1925.

KHUN de PROROK, Byron: *Mysterious Sahara (The Land of Gold, of Sand, & of Ruin),* Chicago: Reilly & Lee, 1929.

KING, Leonard William: *First Steps in Assyrian (A Book for Beginners . . .),* London: Kegan, Paul, Trench, Trübner, 1898.

KINGSBOROUGH, Lord (Edward King, Third Earl of Kingston): *Antiquities of Mexico (comprising Fac-similes of Ancient Mexican Paintings & Hieroglyphics . . .),* London: Bohn, 1848, 9 vols.

KNAPP, Philip Coombs: *Crete & Atlantis,* in *The Geographical Review,* Vol. VIII, No. 2 (Aug. 1919) pp. 126-129.

KOHUT, George Alexander: *The Lost Ten Tribes in America,* Portland, Ore.: The Jewish Tribune, 1909.

KOSMAS Indikopleustes: *The Christian Topography of Cosmas, an Egyptian Monk,* London: Hakluyt Soc., 1897.

LANDA, Diego de: *Landa's Relación de las Cosas de Yucatan,* Cambridge, Mass.: Peabody Mus., 1941. A translation.

LAUNAY, Louis de: *Où était l'Atlantide?* in *La Revue de France,* Vol. XVI, No. 9 (1 May 1936) pp. 91-113.

LEAF, Walter: *Homer & History,* London: Macmillan, 1915.

LEET, L. Don: *Causes of Catastrophe (Earthquakes, Volcanoes, Tidal Waves, & Hurricanes),* N. Y.: McGraw-Hill, 1948.

Lemuria the Incomparable—the Answer, Milwaukee: Lemurian Pr., 1939. Pamphlet.

Le PLONGEON, Augustus: *Queen Móo & the Egyptian Sphinx,* London: Kegan, Paul, Trench, Trübner, 1896.

——: *Sacred Mysteries among the Mayas & Quiches 11,500 Years Ago (Their Relation to the Sacred Mysteries of Egypt, Greece, Chaldea, & India . . .),* N. Y.: Macoy, 1886.

LESLIE, J. Ben: *Submerged Atlantis Restored,* Rochester: Austin, 1911. Mediumistic revelations.

LEY, Willy: *. . . Giants in Those Days,* in *Astounding Science Fiction,* Vol. XXXVI, No. 3 (Nov. 1945) p. 111ff. Polyphemos & Gigantanthropus.

——: *Ice Age Ahead?* in *Astounding Science Fiction,* Vol. XXII, No. 6 (Feb. 1939) p. 86ff.

——: *The Lungfish & the Unicorn (An Excursion into Romantic Zoology),* N. Y.: Modern Age, 1941. Gondwanaland.

——: *Pseudoscience in Naziland,* in *Astounding Science Fiction,* Vol. XXXIX, No. 3 (May, 1947) p. 90ff. Hörbigerism.

——: *Scylla Was a Squid,* in *Natural History,* Vol. XLVII, No. 1 (Jun. 1941) p. 11ff. The *Odyssey.*

——: *Sea of Mystery,* in *Astounding Science Fiction,* Vol. XXXI, No. 4 Jun. 1943) p. 97ff. The Sargasso Sea.

——: *The Secret Giant,* in *Astounding Science Fiction,* Vol. XXXV. No. 5 (Jul. 1945) p. 101ff. The giant squid.

LILLIE, Arthur: *Madame Blavatsky & Her "Theosophy" (A Study),* London: Sonnenschein, 1895.

LINTON, Ralph: *The Study of Man (An Introduction),* N. Y.: Appleton-Century, 1936.

LIVINGSTON, Marjorie: *Island Sonata,* London: Dakers, 1944. A novel.

LOWES, John Livingston: *The Road to Xanadu (A Study in the Ways of the Imagination),* Boston: Houghton Mifflin, 1927. Coleridge's sources.

LOWIE, Robert H.: *The History of Ethnological Theory,* N. Y.: Farrar & Rinehart, 1937.

MacBAIN, Alexander: *Celtic Mythology & Religion,* N. Y.: Dutton, 1917.

MacDOUGALL, Curtis D.: *Hoaxes,* N. Y.: Macmillan, 1940.

MACROBIUS, Ambrosius Theodosius: *Commentary on "Scipio's Dream" from Cicero's "Republic,"* in *Macrobe, Varron, & Pomponius Méla,* Paris: Dubochet, 1845. In French.

MAGOFFIN, Ralph V. D., & Emily C. Davis: *The Romance of Archaeology,* Garden City: Garden City Pub. Co., 1929. Previously published as *Magic Spades.*

MAHOUDEAU, P. G.: *Les Traditions Relatives à l'Atlantide et la Grèce Préhistorique transmises par Platon,* in *Revue Anthropologique,* 1913, pp. 102-108.

MANETHO, *apud* Malalas, Syncellus, Africanus, & Eusebius, LCL.

MANZI, Michel: *Le Livre de l'Atlantide,* Paris: Glomeau, 1922. A French version of Scott-Elliot's Theosophical Atlantist doctrine.

MARQUE, Bernard: *L'Atlantide,* in *Bull. de la Société des Lettres, Sciences, & Arts de la Corrèze,* Vol. L (Jan.-Jun. 1933) pp. 5-26.

MARTIN, T. Henri: *Études sur le Timée de Platon,* Paris: Ladrange, 1841, 2 vols.

MASON, Gregory: *Columbus Came Late,* N. Y.: Appleton-Century, 1931. Amerind civilizations.

MATTHEW, William Diller: *Plato's Atlantis in Palaeogeography,* in *Proc. of the Natl. Acad. of Sciences,* Vol. VI (1920) p. 17f.

The Maya & Their Neighbors, by W. W. Howells, O. G. Ricketson, Jr., Ralph Linton, *et al.,* N. Y.: Appleton-Century, 1940. A symposium.

MEANS, Philip Ainsworth: *Ancient Civilization of the Andes,* N. Y.: Scribner's, 1931-1936.

MELA, Pomponius: *Géographie de Pomponius Mela,* Paris: Panckoucke, 1843. Latin & French.

MEREJKOWSKY, Dmitri (Dmitri Sergyeëvitch Merezhkovski): *The Secret of the West,* N. Y.: Brewer, Warren, & Putnam, 1931.

MILLER, R. De Witt: *Forgotten Mysteries,* Los Angeles: R. D. Miller, 1947.

MITCHELL, James Leslie: *The Conquest of the Maya,* N. Y.: Dutton, 1935. A diffusionist view.

——: *Three Go Back,* Indianopolis: Bobbs-Merrill, 1932. A novel.

MORGAN, Arthur E.: *Nowhere Was Somewhere (How History Makes Utopias & How Utopias Make History),* Chapel Hill: Univ. of N. Car. Pr., 1946. An interpretation of More's *Utopia.*

MORISON, Samuel Eliot: *Portuguese Voyages to America in the Fifteenth Century,* Cambridge, Mass.: Harvard Univ. Pr., 1940.

MORLEY, Sylvanus Griswold: *The Ancient Maya*, Stanford: Stanford Univ. Pr., 1946.
——: *An Introduction to the Study of Maya Hieroglyphics*, Washington: Smithsonian Inst., 1915.
——: Rev. of De Gruyter's *A New Approach to Maya Hieroglyphics*, in *Amer. Journ. of Archaeol.*, Vol. L, No. 3 (Jul.-Sep. 1946) p. 444f.
MORTILLET, Gabriel de: *L'Atlantide*, in *Bull. de la Société d'Anthropologie de Paris*, Ser. IV, No. 8 (1897) pp. 447-451.
MULLEN, Stanley: *Kinsmen of the Dragon*, Chicago: Shasta Pub., 1951.
MUMFORD, Lewis: *The Story of Utopias*, N. Y.: Boni & Liveright, 1922.
MURRAY, Gilbert: *The Rise of the Greek Epic*, Oxford: Clarendon, 1924.
——: *The Wanderings of Odysseus*, in *Quart. Rev.* (London) No. 403 (Apr. 1905) pp. 344-370. Rev. of Bérard's *Les Phéniciens et l'Odyssée*.
NÉGRIS, Phocion: *Glaciers et Atlantes*, Athens: Sakellarios, 1924. Pamphlet; follows Dévigne.
NEWBROUGH, John Ballou: *Oahspe: A Kosmon Revelation in the Words of Jehovih & His Angel Ambassadors*, L. A.: Kosmon, 1932.
NORDENSKIÖLD, Erland: *Origin of the Indian Civilizations in South America*, in *Comparative Ethnographical Studies*, No. 9, pp. 1-155; Göteborg: Elanders, 1931.
OLIVER, Frederick Spencer: *A Dweller on Two Planets, by "Phylos,"* L. A.: Borden, 1894-1940. A novel.
OSBORN, Fairfield (ed.): *The Pacific World (Its Vast Distances, Its Lands & the Life upon Them, & Its Peoples)*, N. Y.: Norton, 1944.
OSBORN, Henry Fairfield: *The Age of Mammals (in Europe, Asia, & North America)*, N. Y.: Macmillan, 1910-1921.
——: *Men of the Old Stone Age*, N. Y.: Scribner's, 1915-1923.
——: *The Origin & Evolution of Life*, N. Y.: Scribner's, 1916-1921.
OVID (Publius Ovidius Naso): *Metamorphoses*, LCL.
PARRY, David M.: *The Scarlet Empire*, Indianapolis: Bobbs-Merrill, 1906. A novel.
PATERCULUS, Gaius Velleius: *Compendium of Roman History*, LCL.
PAUSANIAS: *Description of Greece*, London: Bell, 1886, 2 vols., & LCL.
PERRY, William James: *The Children of the Sun (A Study in the Early History of Civilization)*, N. Y.: Dutton, 1923. Diffusionism.
PHELON, William P.: *Our Story of Atlantis*, S. F.: Hermetic Book Concern, 1903; & Quakertown, Penn.: Philosoph. Pub. Co., 1937. Occult revelations.
PINDAR (Pindaros of Kynoskephalai): *The Odes of Pindar*, LCL.
PLATO: *The Republic, Phaedo, Gorgias, The Statesman, Phaedrus, Timaeus, Critias,* & *The Laws*, in the *Works of Plato*, N. Y.: Tudor (Jowett transl.) & LCL.
PLINY the Elder (Gaius Plinius Secundus): *Natural History*, London: Bohn, 1855, 6 vols., & LCL.
PLUTARCH (Ploutarchos of Chaironeia): *The Lives of the Noble Grecians & Romans*, or *Parallel Lives*, ML & LCL.
——: *On Isis & Osiris; On the Cessation of Oracles; On the Apparent Face in the Moon's Orb;* in *Plutarch's Morals (Theosophical Essays)*, London: Bell, 1889, & LCL.
POLYBIUS (Polybios of Megalopolis): *The Histories of Polybius*, London: Macmillan, 1889, 2 vols.

PRATT, Fletcher: *Secret & Urgent (The Story of Codes & Ciphers)*, Garden City: Blue Ribbon, 1939-1942. Baconianism.

PRESCOTT, William H.: *The Conquest of Mexico*, Garden City: Blue Ribbon, 1943.

PROKLOS Diadochos ("Lysius Proclus"): *The Commentaries of Proclus on the Timæus of Plato*, London: Valpy, 1820.

PTOLEMY, Claudius (Klaudios Ptolemaios): *The Geography of Claudius Ptolemy*, N. Y.: Pub. Lib., 1932.

RAGLAN, Lord (Fitz Roy Richard Somerset, Baron Raglan): *The Hero (A Study in Tradition, Myth, & Drama)*, London: Methuen, 1936.

——: *How Came Civilization?*: London: Methuen, 1939.

RAWLINSON, George: *History of Phoenicia*, London: Longmans, Green, 1889.

RIVERS, William H. R.: *Psychology & Politics (& Other Essays)*, London: Kegan, Paul, Trench, Trübner, 1923. Diffusionism.

ROUSE, Irving: *Petroglyphs*, in *Bulletin 143*, vol. V, Bureau of American Ethnology, Smithsonian Inst., pp. 493-502.

ROY, Lillian Elizabeth: *The Prince of Atlantis*, N. Y.: Educational Pr., 1929. A novel.

RUTTER, Owen: *The Monster of Mu*, London: Benn, 1932. A novel.

SCHNABEL, Paul: *Berossos und die Babylonisch-Hellenistische Literatur*, Leipzig: Teubner, 1923.

SCHUCHERT, Charles: *Atlantis & the Permanency of the North Atlantic Ocean Bottom*, in *Proc. of the Natl. Acad. of Sciences*, Vol. III (Feb. 1917) pp. 65-72.

——: *Gondwana Land Bridges*, in *Bull. of the Geol. Soc. of Amer.*, Vol. XLIII, No. 4 (Dec. 1932) pp. 876-915.

——. *Outlines of Historical Geology*, N. Y.: Wiley, 1931.

——: *Paleogeography of North America*, in *Bull. of the Geol. Soc. of Amer.*, Vol. XX (1908) pp. 427-606.

——: *The Periodicity of Ocean Spreading, Mountain-Making & Paleogeography*, in *Bull. of the Natl. Research Council*, No. 85 (1932) pp. 537-557.

SCOTT, William Berryman: *A History of Land Mammals in the Western Hemisphere*, N. Y.: Macmillan, 1913-1937.

SCOTT-ELLIOT, W.: *The Story of Atlantis & The Lost Lemuria*, London: Theosoph. Pub. House, 1896-1925.

SENECA, Lucius Annæus: *Physical Science in the Time of Nero (Being a Translation of the* Quaestiones Naturales *of Seneca)*, London: Macmillan, 1910.

SHAVER, Richard Sharpe: *I Remember Lemuria & The Return of Sathanas*, Evanston: Venture, 1948. Two short novels.

SHEWAN, Alexander: *Homeric Essays*, Oxford: Blackwell, 1935.

SIBSON, Francis H.: *The Stolen Continent*, London: Melrose, 1934.

——: *The Survivors*, Garden City: Doubleday, Doran, 1932. A novel.

SILBERMANN, Otto: *Un Continent Perdu: L'Atlantide*, Paris: Genet, 1930.

SIMPSON, George Gaylord: *Holarctic Mammalian Faunas & Continental Relationships During the Cenozoic*, in *Bull. of the Geol. Soc. of Amer.*, Vol. LVIII (Jul. 1947) pp. 613-688.

——: *Mammals & the Nature of Continents*, in *The Amer. Journ. of Science*, Vol. CCXLI, No. 1 (Jan. 1943) pp. 1-31.

SINNETT, Arnold Percy: *Esoteric Buddhism*, Boston: Houghton Mifflin, 1884-1896. Theosophy.

——: *The Growth of the Soul (A Sequel to "Esoteric Buddhism")*, N. Y.: Theosoph. Pub. Soc., 1896.

SMITH, Charlotte Fell: *John Dee (1527-1608)*, London: Constable, 1909.

SMITH, Clark Ashton: *The Double Shadow & Other Fantasies*, Auburn, Calif.: Privately printed, 1933.

——: *Genius Loci & Other Tales*, Sauk City, Wis.: Arkham House, 1948.

——: *Lost Worlds*, Sauk City, Wis.: Arkham House, 1944.

——: *Out of Space & Time*, Sauk City, Wis.: Arkham House, 1942.

SMITH, Edward E.: *Triplanetary*, Reading: Fantasy Pr., 1948.

SMITH, Sir Grafton Elliot: *The Ancient Egyptians (& the Origin of Civilization)*, London: Harper, 1923.

——: *Elephants & Ethnologists*, London: Kegan Paul, Trench, Trubner, 1924.

——: *In the Beginning (The Origin of Civilization)*, N. Y.: Morrow, 1928.

——: *The Migrations of Early Culture (A Study of the Significance of the Distribution of the Practice of Mummification . . .)*, Manchester Univ. Pr., 1929.

——, Bronislaw Malinowsky, Herbert J. Spinden, & Alexander Goldenweiser: *Culture (The Diffusionist Controversy)*, N. Y.: Norton, 1927. A symposium.

SMITH, Joseph, Jr.: *The Book of Mormon (An Account Written by the Hand of Mormon Upon Plates Taken from the Plates of Nephi)*, Salt Lake City: Church of Jesus Christ of Latter-day Saints, 1830-1921.

SMITH, William (ed.): *Dictionary of Greek & Roman Geography*, Boston: Little, Brown, 1857, 2 vols., *s. v. Atalanta, Atarantes, Atlantes, Atlanticum Mare, Atlantis*, & *Atlas.*

SOLOVYOFF, Vsevolod Sergyeevich: *A Modern Priestess of Isis*, London: Longmans, Green, 1895. Mme. Blavatsky.

SPENCE, J. Lewis T. C.: *Atlantis in America*, N. Y.: Brentano's, 1925.

——: *The History of Atlantis*, London: Rider, 1926.

——: *The Myths of Mexico & Peru*, N. Y.: Crowell, 1913.

——: *Myths of the North American Indians*, N. Y.: Farrar & Rinehart, 1914.

——: *The Occult Sciences in Atlantis*, London: Rider, 1943.

——: *The Popol Vuh (The Mythic & Heroic Sagas of the Kichés of Central America)*, London: Nutt, 1908.

——: *The Problem of Atlantis*, N. Y.: Brentano's, 1924.

——: *The Problem of Lemuria (The Sunken Continent of the Pacific)*, Phila.: McKay, 1933.

——: *Will Europe Follow Atlantis?* London: Rider, 1942.

SPILHAUS, Margaret Whiting: *The Background of Geography*, Phila.: Lippincott, 1935.

SPINDEN, Herbert J.: *Ancient Civilizations of Mexico & Central America*, N. Y.: Amer. Mus. of Nat. Hist., 1922.

STEINER, Rudolf: *Atlantis & Lemuria*, London: Anthroposoph. Pub. Co., 1923.

STEWART, John Alexander: *The Myths of Plato (Translated with Introductory & Other Observations)*, London: Macmillan, 1905.

STIRLING, Matthew W.: *Indian Tribes of Pueblo Land*, in *Natl. Geog. Mag.*, Vol. LXXVIII, No. 5 (Nov. 1940) p. 549ff.

STOCKING, Hobart E.: *Bombs from Interstellar Space*, in *Natural History*, Vol. LIII, No. 1 (Jan. 1944) p. 31ff.

STRABO (Strabon of Amasia): *Geography,* London: Bell, 1881, & LCL.

SUESS, Eduard: *The Face of the Earth (Das Antlitz der Erde),* Oxford: Clarendon, 1906, 5 vols.

SUGRUE, Thomas: *There Is a River (The Story of Edgar Cayce),* N. Y.: Holt, 1942. A credulous & fictionized biography.

SWAN, John: *Speculum Mundi, or, A Glasse representing the Face of the World . . .,* Cambridge: John Williams, 1644.

TACITUS, Publius Cornelius: *Complete Works,* ML.

TARN, William Woodthorpe: *Hellenistic Civilization,* London: 1927. Early Utopias.

TERMIER, Pierre: *Atlantis,* in the *Annual Rep.* of the *Board of Regents of the Smithsonian Inst.,* 1915, pp. 219-234.

TERTULLIAN (Quintus Septimus Florens Tertullianus): *De Pallio,* or *On the Ascetics' Mantle,* & *Against Hermogenes,* ANCL.

Theory of Continental Drift (A Symposium on the Origin & Movement of Land Masses . . .) by W. A. J. M. van Waterschoot van der Gracht, Bailey Willis, R. T. Chamberlin, *et al.,* Tulsa: Amer. Assn. of Petrol. Geologists, 1928.

THÉVENIN, René: *Les Pays Légendaires (Devant la Science),* Paris: Presses Universitaires, 1946.

THOMPSON, Edward Herbert: *People of the Serpent (Life & Adventure Among the Mayas),* Boston: Houghton Mifflin, 1932.

THOMPSON, J. A. K.: *Studies in the Odyssey,* Oxford: Clarendon Pr., 1914.

THUCYDIDES (Thoukydides of Athens): *The Peloponnesian War,* GH & ML.

TOZER, Henry Fanshawe: *A History of Ancient Geography,* Cambridge: Cambr. Univ. Pr., 1897-1935.

TYLOR, Edward Burnett: *Primitive Culture (Researches into the Development of Mythology, Philosophy, Religion, Language, Art, & Custom),* N. Y.: Holt, 1889, 2 vols.

VAILLANT, George C.: *Aztecs of Mexico (Origin, Rise & Fall of the Aztec Nation),* Garden City: Doubleday, Doran, 1941.

——: Broadcast over Station WGI, 4 Oct. 1944.

——: Rev. of Braghine's *Shadow of Atlantis,* in *Natural History,* Vol. XLV, No. 5 (May, 1940) p. 313.

VELIKOVSKY, Immanuel: *Worlds in Collision,* N. Y.: Macmillan, 1950.

VERNE, Jules: *Twenty Thousand Leagues Under the Sea* & *The Blockade Runners,* N. Y.: Book League, 1940. A novel.

VERNEAU, René: *A propos de l'Atlantide,* in *Bull. de la Société d'Anthropologie de Paris,* Ser. IV, No. 8 (1897) pp. 447-451.

VERRILL, A. Hyatt & Ruth Verrill: *America's Ancient Civilizations,* N. Y.: Putnam, 1953.

VIGERS, Daphne: *Atlantis Rising,* London: Dakers, 1944. An "astral" account of Atlantis.

VIRGIL (Publius Vergilius Maro): *Aeneid,* Boston: Houghton Mifflin, 1902.

von HAGEN, Victor Wolfgang: *Maya Explorer (John Lloyd Stephens & the Lost Cities of Central America & Yucatan),* Norman: Univ. of Okla. Pr., 1947.

——: *Mr. Catherwood Also Is Missing,* in *Natural History,* Vol. LVI, No. 3 (Mar. 1947) p. 104ff.

——: *Waldeck,* in *Natural History,* Vol. LVI, No. 10 (Dec. 1946) p. 450ff.

VON RAVN, Clara Iza: *Selestor's Men of Atlantis,* Boston: Christopher, 1937. A mediumistic account of Atlantis.

WALLIS, Wilson D.: *Messiahs: Their Role in Civilization,* Washington: Amer. Council of Pub. Affairs, 1943.

WEGENER, Alfred Lothar: *The Origins of Continents & Oceans,* N. Y.: Dutton, 1924.

WELLS, Herbert George: *The Outline of History,* Garden City: Garden City Pub. Co., 1920-1931 (Revised ed.).

WHEATLEY, Dennis: *The Man Who Missed the War,* London: Hutchinson, 1945. A novel.

——: *They Found Atlantis,* Phila.: Lippincott, 1936. A novel.

WHIBLEY, Leonard: *A Companion to Greek Studies,* Cambridge: Univ. of Cambr. Pr., 1916.

WHISHAW, Ellen Mary Addy-Williams: *Atlantis in Andalucia (A Study of Folk Memory),* London: Rider, 1929.

WHITE, Andrew Dickson: *A History of the Warfare of Science with Theology in Christendom,* N. Y.: Appleton, 1908, 2 vols.

WHORF, Benjamin Lee: *Decipherment of the Linguistic Portion of The Maya Hieroglyphics,* in *Smithsonian Rep.* for 1941, pp. 479-502.

——: *Science & Linguistics,* in *Technology Rev.,* Vol. XLII, No. 6 (Apr. 1940) p. 229ff.

WILKINS, Harold T.: *Mysteries of Ancient South America,* London: Rider, 1946. A writer influenced by Spence, Braghine, & Bellamy.

WILLIAMS, Gertrude Marvin: *The Passionate Pilgrim,* N. Y.: Coward-McCann, 1931. A biography of Annie Besant.

——: *Priestess of the Occult: Madame Blavatsky,* N. Y.: Knopf, 1946.

WILLIS, Bailey: *Isthmian Links,* in *Bull. of the Geol. Soc. of Amer.,* Vol. XLIII (Dec. 1932) pp. 917-952.

WILSON, Sir Daniel: *The Lost Atlantis & Other Ethnographic Studies,* Edinburgh: Douglas, 1892.

WISSLER, Clark: *Indians of the United States (Four Centuries of Their History & Culture),* N. Y.: Doubleday, Doran, 1940.

——: *Wheat & Civilization,* in *Natural History,* Vol. LII, No. 4 (Nov. 1943) p. 172ff.

WOOLLEY, C. Leonard: *Ur of the Chaldees (A Record of Seven Years of Excavation),* N. Y.: Scribner's, 1930.

WRIGHT, S. Fowler: *Dawn,* N. Y.: Cosmopolitan, 1929. A novel.

——: *Deluge: A Romance,* N. Y.: Cosmopolitan, 1928. A novel.

INDEX

Aac & Coh, 44
Aahmes or Amasis, 55
Abaris, 57
Abilene man, 142f
Abraham, 229, 238
Achaeans or Akhiyawas, 187, 214, 238; wall of, 17, 245
Achilles, 21, 215, 221, 238
Adepts, 58, 63, 69
Adhémar, A.-J., 260
Adonis, 243
Aegean Sea, 212, 216f
Aegipani or Goat-Pans, 226
Aelfwine or Alboin, 239
Africa, Atlanteans in, 40, 275; Atlantis in, 80-83, 178-86; Lost Tribes in, 32; Atlas in, 225; size of, 26
Agadir, 180, 193
Age, Bronze, 40, 85, 168, 229, 231; Copper, 58; Dark, 19, 21, 120; Golden, 1, 58, 229, 231, 242, 253, 275; Heroic, 227ff; Hyborian, 261; Ice, 37, 159f, 174; Neolithic, 83, 108, 193; of Mammals, 59, 62, 147ff, 159; of Reptiles, 52, 156, 159; Silver, 58; Stone, 62; Viking, 178
Aguilar, Jeronimo de, 114
Ahaggar Mts., 181-84, 223
Aiaia, 14, 199, 217, 221
Aias or Ajax, 242
Aiken, Conrad, 1, 233, 256
Aiolos, 199, 216
Aircraft, 9, 67, 71f, 266
Akkadians, 63f
Alas, That Great City, 268
Alaska, 31, 125, 142, 171-74
Albion, 180
Aldabra Islands, 52, 155
Alexander the Great, 132, 240
Alexandria, 17, 197; library of, 210f
Algeria, 83, 181ff
Alice in Wonderland, 232
Alkinoös or Alcinous, 40, 184, 198-204, 231
Alph or Alpheios River, 8, 200, 211, 231
Alphabets, Atlantean, 40; Etruscan, 190; European, 40; Greek, 45,

190; Iberian, 190; Latin, 31, 35, 42, 100; Mayan, 32, 33, 36f, 40, 42; Muvian, 47; Panic, 70; Phoenician, 40, 42; Spanish, 31f, 119; Tartessian, 190, 192
Amazing Stories, 274f; Quarterly, 263
Amazon River, 39, 49, 73
Amazons, 14, 82, 85, 184, 226
Ameinokles, 9
America, Atlanteans in, 39, 86, 143f; Atlantis in, 30, 80, 161f, 178, 211, 250; discovery of, 22, 30, 123f; knowledge of, 61, 220, 224; Muvians in, 143; Norsemen in, 22; plants of, 41, 111f, 143
American Indians (see Amerinds)
American Indians, 125
Amerinds, 23, 30, 40-43, 76, 78, 94, 98, 105-11, 131f, 136-39, 225; cities of, 104; culture of, 30, 40f, 105, 123, 139, 143; existence of, 114; history of, 142; migration of, 133; origin of, 30f, 72, 101f, 123-27, 130, 142; originality of, 144; societies of, 121
Amis d'Atlantis, 77
Ammon or Siwah, Oasis of, 225
Amometos, 247
Anaximandros, 216
Andeans, 111, 126f, 204; civilization of, 109, 127; metallurgy of, 143
Andersen, Johannes C., 98
Anderson, Wing, 70
Andesite line, 48, 153
Andreä, Johann V., 251
Angaraland, Frontispiece, 152, 154
Anglo-Israelites or British-Israelites, 34
Animals, distribution of, 41, 147-50, 162; domestic, 111, 113
Anostos, 15
Antarctica, 25, 27, 150, 156
Anthropology & anthropologists, 206, 240, 255, 275f; evolutionary, 129f
Anthroposophy, 67
Antichthon, 26
Antillia or Antilles, 21f, 92, 93, 95